IEEE Recommended Practice for Electric Power Distribution for Industrial Plants

Published by
The Institute of Electrical and Electronics Engineers, Inc

Distributed in cooperation with
Wiley-Interscience, a division of John Wiley & Sons, Inc

IEEE
Std 141-1976

IEEE Recommended Practice for Electric Power Distribution for Industrial Plants

Sponsor

Industrial and Commercial Power Systems Committee
of the
IEEE Industry Applications Society

Library of Congress Catalog Number 76-42860

Foreword

This publication is prepared by the Industrial Plants Power Systems Subcommittee of the IEEE Industrial and Commercial Power Systems Committee, which is a technical committee of the IEEE Industry Applications Society. It has been approved by the IEEE Standards Board as an IEEE standards document to provide current information and recommended practices for the design, construction, operation, and maintenance of electric power systems in industrial plants.

It was thirty-one years ago that "Electric Power Distribution for Industrial Plants" was first published by the AIEE. It was given the nickname of the "Red Book" because of the color of its cover ... and it became the first of the present IEEE Color Book series.

The second edition of the Red Book was produced in 1956, and identified as AIEE No. 952. This was followed by the third edition in 1964; it was also identified as IEEE No. 141. The fourth edition was produced in 1969; it was approved as a Recommended Practice of the Institute and identified as IEEE Std 141-1969. This, the fifth edition, is now identified as IEEE Std 141-1976. It was initiated in 1970 with the participation of more than fifty electrical engineers from industrial plants, consulting firms, and equipment manufacturers.

This IEEE Recommended Practice continues to serve as a companion publication to the following other Recommended Practices prepared by the IEEE Industrial and Commercial Power Systems Committee:

Recommended Practice for Grounding of Industrial and Commercial Power Systems (IEEE Green Book), IEEE Std 142-1972

Recommended Practice for Electric Power Systems in Commercial Buildings (IEEE Gray Book), IEEE Std 241-1974

Recommended Practice for Protection and Coordination of Industrial and Commercial Power Systems (IEEE Buff Book), IEEE Std 242-1975

Recommended Practice for Emergency and Standby Power Systems (IEEE Orange Book), IEEE Std 446-1974

Comments, corrections, and suggestions for the next revision of this publication are welcome and should be submitted to the IEEE Standards Board, 345 East 47 Street, New York, NY 10017.

Electric Power Distribution for Industrial Plants

5th Edition

Working Group Members and Contributors

Donald T. Michael, *Working Group Chairman*

Chapter 1 — Introduction: Donald T. Michael, *Chairman*; Daniel L. Goldberg, R. Gerald Irving

Chapter 2 — System Planning: John F. Baum, *Chairman*; Richard H. Kaufmann, Carl N. Noel, Ned V. Poer

Chapter 3 — Voltage Considerations: Donald T. Michael, *Chairman*; Walter C. Bloomquist, Willard H. Dickinson, Ralph H. Lee

Chapter 4 — System and Equipment Protection: M. Shan Griffith, *Chairman*; Kao Chen, Frank L. Colvin, John W. Courter, Hugo W. Fahrenthold, Richard H. Kaufmann, D. P. LaMorte, David H. Smith

Chapter 5 — Fault Calculations: Calvin W. Boice, Jr, *Chairman*; John W. Courter, Walter C. Huening, Richard H. Kaufmann, George L. Nuss

Chapter 6 — Grounding: Richard H. Kaufmann, *Chairman*; A. C. Crymble, John A. Hart, Terrell W. Haymes, Donald W. Zipse

Chapter 7 — Power Factor: Walter C. Bloomquist and Myron Zucker, *Cochairmen*; Graydon Bauer, Frank L. Colvin, James A. Ferner, M. Shan Griffith, John Harder, Herman Moderau, Lester Sever, George W. Walsh

Chapter 8 — Power Switching, Transformation, and Motor-Control Apparatus: Harold C. Miles, *Chairman*; Lucas G. Ananian, William P. Burt, Kao Chen, John C. Conte, John W. Courter

Chapter 9 — Instruments and Meters: John A. Hart, *Chairman*; Donald N. Bigbie

Chapter 10 — Cable Systems: Andrew Hvizd, Jr, *Chairman*; Graydon Bauer, Thomas W. Brook, Wesley C. Dorothy, Hugo W. Fahrenthold, Roger S. Keith, Ralph H. Lee, Charles Maguire, Norman Shackman, Eugene R. Smith, Dale E. Wagner, Bruce H. Zacherle

Chapter 11 — Busways: Frank C. Johnston, *Chairman*; Lucas G. Ananian, Neil R. Bjornson, Russell S. Davis, Lawrence E. Fisher, Russell O. Ohlson

Editorial Consultant: Ingeborg M. Stochmal

Editorial Review Board: Donald S. Brereton, *Chairman*; Lucas G. Ananian, Warren H. Cook, Derio Dalasta, Daniel L. Goldberg, Donald T. Michael, William J. Neiswender

Production: Mary B. Goulding; Artwork: Dorothy Brosterman; Jacket Design: T. Dale Goodwin

low-voltage system (electric power). An electric system having a maximum rms ac voltage of 1000 volts or less.

medium-voltage system (electric power for industrial and commercial systems only). An electric system having a maximum rms ac voltage above 1000 volts to 72 500 volts.

high-voltage system (electric power for industrial and commercial systems only). An electric system having a maximum rms ac voltage above 72 500 volts to 242 000 volts.

extra-high-voltage system (electric power). An electric system having a maximum rms ac voltage above 242 000 volts to 800 000 volts.

Voltage Class	NOMINAL SYSTEM VOLTAGE (volts)			Maximum System Voltage	Reference
	Two-Wire	Three-Wire	Four-Wire		
Low Voltage		Single-Phase Systems			IEEE Std 100-1972 (ANSI C42.100-1972) defines low-voltage system as "less than 750 V." It is now being proposed as "maximum rms ac voltage of 1000 V."
	(120)	120/240		127 or 127/254	
		Three-Phase Systems			
			208Y/120	220	
	(240)		240/120	254	
		480	480Y/277	508	
	(600)			635	
Medium Voltage	(2400)			2 540	These voltages are listed in ANSI C84.1-1970 but are not identified by voltage class.
	4160			4 400	
	(4800)			5 080	
	(6900)			7 260	
			12 470Y/7200	13 200	
			13 200Y/7620	13 970	Note that additional voltages in this class are listed in ANSI C84.1-1970 but are not included because they are primarily oriented to electric-utility system practice.
	13 800		(13 800Y/7970)	14 520	
	(23 000)			24 340	
			24 940Y/14 400	26 400	
	(34 500)		34 500Y/19 920	36 510	
	(46 000)			48 300	
	69 000			72 500	
High Voltage	115 000			121 000	ANSI C84.1-1970 identifies these as "higher voltage three-phase systems" (as well as 46 kV and 69 kV).
	138 000			145 000	
	(161 000)			169 000	
	230 000			242 000	
Extra High Voltage	345 000			362 000	ANSI C92.2-1974
	500 000			550 000	
	735 000—765 000			800 000	

System Voltages shown without brackets are preferred.
See paragraph 2.2 on page 8 of ANSI C84.1-1970.

NOTES:

(1) Voltage class designations were approved for use in documents published by the IEEE Industrial and Commercial Power Systems Committee at its Committee meeting on May 16, 1973 (and on June 6, 1974 and January 28, 1975).

(2) Limits for the 600-volt Nominal System Voltage established by ANSI C84.1a-1973.

(3) The maximum and minimum voltage ranges for system voltages from 120 V to 230 kV are given in ANSI C84.1-1970, "Voltage Ratings for Electric Power Systems and Equipment (60 Hz)."

(4) It is appropriate to observe that the voltage class of Medium Voltage has had long acceptance in industrial practice. In contrast, the term Medium Voltage does not as easily apply to electric-utility practice because of the traditional use of overlapping voltage areas of transmission and distribution (as defined in the IEEE Std. 100-1972, "Standard Dictionary of Electrical and Electronics Terms").

(5) The preferred maximum voltage for three-phase system voltages from 345 to 735-765 kV are given in ANSI C92.2-1974, "Preferred Voltage Ratings for Alternating-Current Electrical Systems and Equipment Operating at Voltages Above 230 Kilovolts Nominal." When ANSI C92.2-1974 superseded ANSI C92.2-1967 the specific identification of the Nominal System Voltages for the EHV systems were withdrawn. They are now only identified as "Typical 'Nominal' Voltages." (The limits for the EHV systems which appear in ANSI C84.1-1970, page 9, are superseded by ANSI C92.2-1974.)

Contents

7

11

18

1. Introduction

1.1 Institute of Electrical and Electronics Engineers (IEEE). IEEE Std 141-1976, Electric Power Systems for Industrial Plants, commonly called the IEEE Red Book, is published as a recommended practice by the Institute of Electrical and Electronics Engineers,[1] a nonprofit, transnational professional society divided into 31 groups and societies covering the various sectors of electrical engineering. One of these is the Industry Applications Society, comprising 25 technical committees involved in the application of electric systems to industry. Of these, the Industrial and Commercial Power Systems Committee is concerned with electric systems for industrial plants and commercial buildings. As sponsor for IEEE Std 141, it has assigned technical responsibility to its Industrial Plants Power Systems Subcommittee. Approval as an IEEE standard is conferred by the IEEE Standards Board.

The IEEE Standards Board requires that all IEEE standards be updated within five years of the date of publication. Notification of errors, suggestions for improvement, and offers of assistance in making this review of IEEE Std 141 are welcome and should be submitted to the IEEE Standards Office, 345 East 47 Street, New York, NY 10017.

1.2 IEEE Meetings and Publications. The IEEE and its constituent groups and societies hold more than 75 technical conferences each year

[1] 345 East 47 Street, New York, NY 10017.

for the presentation of technical papers. Most of these are published in the *IEEE Transactions* or in Conference Records. Local sections of the IEEE and local chapters of the component groups and societies hold local meetings on various subjects of interest to members. Electrical engineers find these meetings and publications invaluable in keeping up to date on the latest technical developments in their particular areas of specialization. Many of these papers are referred to in this document. Copies may be obtained from the IEEE.[1]

1.3 Standards, Recommended Practices, and Guides. Electrical engineers use standards, guides, and recommended practices extensively in designing, installing, operating, and maintaining electric systems for industrial plants. Standards establish specific requirements such as definitions of electrical terms, methods of measurement and test procedures, and dimensions and ratings of equipment. Recommended practices suggest the best method of accomplishing an objective for specified conditions. Guides specify the factors which must be considered in accomplishing a specific objective. All are grouped together as standards documents.

All standards carry the date of publication as a part of their title. All IEEE and ANSI standards are required to be reviewed and updated within five years of this date. If a copy of a standard referred to in this document is obtained, the date should always be checked. If it is older, the standard is obsolete and should be discarded or conspicuously marked "obsolete" and a copy of the current standard obtained if needed. If the date on the standard is later, the material referred to in the reference has been revised and should be checked. If the date on a standard is followed by a later date preceded by the letter R, the standard has been reaffirmed as of the R date without change so that the old standard is still in effect.

1.4 IEEE Standards Documents. The IEEE publishes several hundred standards documents covering various fields of electrical engineering. Basic standards of general interest include the following:

IEEE Std 91-1973, Graphic Symbols for Logic Diagrams (Two-State Devices) (ANSI Y32.14-1973)

IEEE Std 100-1972, IEEE Standard Dictionary of Electrical and Electronics Terms (ANSI C42.100-1972)

IEEE Std 260-1967, Letter Symbols for Units Used in Science and Technology (ANSI Y10.19-1969)

IEEE Std 268-1973, Units in Published Scientific and Technical Work

IEEE Std 280-1968, Letter Symbols for Quantities Used in Electrical Science and Electrical Engineering (ANSI Y10.5-1968)

IEEE Std 315-1975, Graphic Symbols for Electrical and Electronics Diagrams (ANSI Y32.2-1975) (CSA Z99-1975)

The IEEE publishes several standards documents of special interest to electrical engineers involved with industrial plant electric systems which are sponsored by the Industrial and Commercial Power Systems Committee of the IEEE Industry Applications Society:

IEEE Std 142-1972, Grounding of Industrial and Commercial Power Systems (ANSI C114.1-1973) (IEEE Green Book)

IEEE Std 241-1974, Electric Power Systems in Commercial Buildings (IEEE Gray Book)

IEEE Std 242-1975, Protection and Coordination of Industrial and Commercial Power Systems (IEEE Buff Book)

IEEE Std 446-1974, Emergency and Standby Power Systems (IEEE Orange Book)

IEEE standards catalogs may be obtained and IEEE standards documents purchased from the IEEE Standards Office.[1]

1.5 National Electrical Manufacturers Association (NEMA) Standards. The National Electrical Manufacturers Association[2] prepares standards which establish dimensions, ratings, and performance requirements for electric equipment between manufacturers. Their standards are widely used in the preparation of purchase specifications.

1.6 National Fire Protection Association (NFPA) Standards Documents. The National Fire Protection Association[3] publishes standards documents specifying requirements for fire protection and safety. Of interest to industrial plant electrical engineers are the following:

NFPA No 70, National Electrical Code (1975), (ANSI C1-1975), (NEC)

The NFPA *Handbook of the National Electrical Code*, sponsored by the NFPA and published by McGraw-Hill, contains the complete NEC text plus explanations edited to correspond with each edition of the NEC.

NFPA No 70B, Electrical Equipment Maintenance (1975), (ANSI C132.1-1975)

1.7 Underwriters Laboratories, Inc, (UL) Standards. Underwriters Laboratories, Inc,[4] prepares safety standards for electric equipment including appliances and tests equipment for compliance with these standards. Manufacturers who have their products approved by UL as meeting the standards are authorized to use the UL label on the equipment. UL publishes lists periodically of approved equipment.

1.8 American National Standards Institute (ANSI). The American National Standards Institute[5] does not write standards. It promotes and coordinates the development of American National Standards and approves as American National Standards those documents which have been prepared in accordance with ANSI regulations.

Standards which have been approved by other organizations and then approved as Amer-

---

[2]155 East 44 Street, New York, NY 10017.
[3]470 Atlantic Avenue, Boston, MA 02110.
[4]207 E. Ohio Street, Chicago, IL 60611.
[5]1430 Broadway, New York, NY 10018.

ican National Standards will carry the identification numbers of both organizations and may be purchased from either. The sponsoring organization retains the responsibility for keeping the standard up to date. Standards carrying only an ANSI number were prepared by American National Standards committees organized and administered by other organizations in accordance with ANSI regulations. These committees are generally used to coordinate participation by a large number of organizations.

ANSI standards of interest to industrial plant electrical engineers include the following:

ANSI Y1.1-1972, Abbreviations for Use on Drawings and in Text

ANSI Y32.9-1972, Graphic Symbols for Electrical Wiring and Layout Diagrams Used in Architecture and Building Construction

1.9 OSHA. Legislation by the U.S. Federal Government has had the effect of giving standards, such as those of the American National Standards Institute (ANSI), the impact of law. The Occupational Safety and Health Act, administered by the U.S. Department of Labor, permits federal enforcement of codes and standards. The Occupational Safety and Health Administration (OSHA) has adopted the 1971 (or later) NEC for new electrical installations and equipment within the scope of subpart S — Electrical — of OSHA regulations and also for major replacements, modification, or repair installed after March 5, 1972. Some articles and sections of the NEC apply to all electrical installations and utilization equipment, and thus are retroactive.

1.10 Environmental Considerations. In all branches of engineering, an increasing emphasis is being placed on social, ecological, and environmental concerns. Today's engineer must consider air, water, noise, and all other items which have an environmental impact. The limited availability of energy sources and the steadily increasing cost of electric energy require a concern with energy conservation on the part of the engineer.

The electrical engineer may participate in studies such as total energy compared to utility power, electric heating versus fossil fuel, a comparison of boilers, purchased steam, and the heat pump. The use of steam turbines compared to absorption units and electric drives for air-conditioning may be evaluated. In these studies the effects of noise, vibration, exhaust gases, cooling methods, and energy requirements must be considered in relationship to the immediate and sometimes general environment.

1.11 Edison Electric Institute (EEI). The Edison Electric Institute,[6] the trade association of the privately owned electric utilities, publishes the following handbooks:

Electric Heating and Cooling Handbook

A Planning Guide for Architects and Engineers

Electric Space-Conditioning

Industrial and Commercial Power Distribution

Industrial and Commercial Lighting

1.12 Handbooks. The following handbooks have, over the years, established reputations in the electrical field. This list is not intended to be all inclusive, and other excellent references are available, but are not listed here because of space limitations.

CROFT, T., CARR, C., and WATT, J. *American Electricians' Handbook* (New York: McGraw-Hill, 1970). The practical aspects of equipment, construction, and installation are covered.

ASHRAE Handbook [American Society of Heating, Refrigerating and Air-Conditioning Engineers[7] (ASHRAE)]. This series of reference books in four volumes, which are periodically updated, detail the electrical and mechanical aspects of space-conditioning and refrigeration.

Electrical Transmission and Distribution Reference Book (Westinghouse Electric Corporation,[8] 1964). The design and application of electric systems are outlined.

Electric Utility Engineering Reference Book, vol 3, *Distribution Systems* (Westinghouse Electric Corporation,[8] 1965). The application of high-voltage equipment and the design of high-voltage and network systems are covered in detail.

---

[6] 90 Park Avenue, New York, NY 10017.
[7] 345 East 47 Street, New York, NY 10017.
[8] Printers Division, Forbes Road, Trafford, PA 15085.

McPARTLAND, J., and the editors of *Electrical Construction and Maintenance* magazine. *How to Design Electrical Systems* (McGraw-Hill, 1968); McPARTLAND, J., and NOVAK, W. *Electrical Design Details* (McGraw-Hill, 1960). These references offer an insight for the younger engineer into the systems approach to electrical design.

BEEMAN, D.L., Ed. *Industrial Power Systems Handbook* (McGraw-Hill, 1955). A text on electrical design with emphasis on equipment, including that applicable to commercial buildings.

Lighting Handbook [Illuminating Engineering Society[9] (IES), 1972]. All aspects of lighting, including seeing, recommended lighting levels, lighting calculations, and design, are included in extensive detail in this comprehensive text.

FINK, D.G., and CARROLL, J.M., Eds. *Standard Handbook for Electrical Engineers* (McGraw-Hill, 1968). Virtually the entire field of electrical engineering is treated, including equipment and systems design.

Transformer Connections (General Electric Company, Publication GET-2).

Underground Systems Reference Book (Edison Electric Institute, 1957). The principles of underground construction and the detailed design of vault installations, cable systems, and related power systems are fully illustrated; cable splicing design parameters are thoroughly covered. Unfortunately, it has not been updated for some time.

Electrical Maintenance Hints (Westinghouse Electric Corporation, 1975)

SHAW, E.T. *Inspection and Test of Electrical Equipment* (Westinghouse Electric Corporation, 1967)

1.13 Periodicals. *Spectrum*, the basic monthly publication of the IEEE, covers all aspects of electrical and electronic engineering. It contains references to IEEE books and other publications, technical meetings and conferences, IEEE group, society, and committee activities, abstracts of papers and publications of the IEEE and other organizations, and other material essential to the professional advancement of the electrical engineer.

Following are some other well-known periodicals:

Actual Specifying Engineer, 205 East 42 Street, New York, NY 10017

Electrical Construction and Maintenance, 1221 Avenue of the Americas, New York, NY 10020

Electrical Consultant, 1760 Peachtree Road NW, Atlanta, GA 30309

Lighting Design and Application, Illuminating Engineering Society, 345 East 47 Street, New York, NY 10017

Plant Engineering, 1301 South Grove Avenue, Barrington, IL 60010

Power, 1221 Avenue of the Americas, New York, NY 10020

Power Engineering, 1301 South Grove Avenue, Barrington, IL 60010

1.14 Manufacturers' Data. The electrical industry through its associations and individual manufacturers of electrical equipment issues many technical bulletins and data books. While some of this information is difficult for the individual to obtain, copies should be available to each major design unit. The advertising sections of electrical magazines contain excellent material, usually well illustrated and presented in a clear and readable form, concerning the construction and application of equipment. Such literature may be promotional; it may present the advertiser's equipment or methods in a "best light" and should be carefully evaluated. Manufacturers' catalogs are a valuable source of equipment information. Some of the larger manufacturers' complete catalogs are very extensive, covering dozens of volumes. However, these companies may issue abbreviated or condensed catalogs which are adequate for most applications. Data sheets referring to specific items are almost always available from the sales offices. Some technical files may be kept on microfilm for use either by projection or by printing at larger design offices. Manufacturers' representatives, both sales and technical, can do much to provide complete information on a product.

[9] 345 East 47 Street, New York, NY 10017.

2. System Planning

2.1 Introduction. The continuity of production in an industrial plant is only as reliable as its electric power distribution system. This chapter outlines the procedures for system planning and presents a guide to the use of the succeeding chapters.

No standard electric distribution system is adaptable to all industrial plants because two plants rarely have the same requirements. The specific requirements must be analyzed qualitatively for each plant and the system designed to meet its electrical requirements. Due consideration must be given to both the present and future operating and load conditions.

2.2 Basic Design Considerations. Any approach to the problem must include several basic considerations which will affect the overall design.

2.2.1 *Safety.* Safety of life and preservation of property are two of the most important factors in the design of the electric system. Safety to personnel involves no compromise; only the safest system can be considered. Following established codes in the selection of material and equipment is imperative.

2.2.2 *Reliability.* The continuity of service required is dependent on the type of manufacturing or process operation of the plant. Some plants can tolerate interruptions while others require the highest degree of service continuity. The system should be designed to isolate faults with a minimum disturbance to the system and should have features to give the maximum dependability consistent with the plant requirements and justifiable cost.

2.2.3 *Simplicity of Operation.* Simplicity of operation is very important in the safe and reliable operation and maintenance of the industrial power system. The operation should be as simple as possible to meet system requirements.

2.2.4 *Voltage Regulation.* Poor voltage regulation is detrimental to the life and operation of electric equipment. Voltage at the utilization equipment must be maintained within equipment tolerance limits under all load conditions.

2.2.5 *Maintenance.* The distribution system must include preventive maintenance requirements in the design. Accessibility and availability for inspection and repair with safety are important considerations in selecting equipment. Space must be provided for inspection, adjustment, and repair in clean, well-lighted, and temperature-controlled areas.

2.2.6 *Flexibility.* Flexibility of the electric system means the adaptability to development and expansion as well as to changes to meet varied requirements during the life of the plant. Consideration of the plant voltages, equipment ratings, space for additional equipment, and capacity for increased load must be given serious study.

2.2.7 *First Cost.* While first costs are important, safety, reliability, voltage regulation, maintenance, and the potential for expansion must also be considered in selecting the best from alternate plans.

2.3 Planning Guide for Distribution Design. The following procedure will guide the engineer in the design of an electric distribution system for any industrial plant. The system designer should have or acquire knowledge of the plant's processes in order to select the proper system and its components.

2.3.1 *Load Survey.* Obtain a general layout, mark it with the major loads at various locations, and determine the approximate total plant load in kilowatts or kilovolt-amperes. Initially the amount of accurate load data may be limited. Some loads such as lighting and air-conditioning may be estimated from generalized data. The majority of industrial plant loads are a function of the process equipment, and such information will have to be obtained from process and equipment designers. Since their design is often concurrent with power system design, initial information will be subject to change. It is therefore important that there be continuing coordination with the other design disciplines. For example, a change from electric powered to absorption refrigeration or a change from electrostatic to high-energy scrubber air-pollution control can change the power requirements for these devices by several orders of magnitude. The power system load estimates will require continual refinement until job completion.

2.3.2 *Demand.* The sum of the electrical ratings of each piece of equipment will give a total connected load. Because some equipment operates at less than full load and some intermittently, the resultant demand upon the power source is less than the connected load.

Standard definitions for these load combinations and their ratios have been devised.

demand. The electric load at the receiving terminals averaged over a specified interval of time. *Note:* Demand is expressed in kilowatts, kilovolt-amperes, amperes, or other suitable units. The interval of time is generally 15 min, 30 min, or 1 h.

peak load. The maximum load consumed or produced by a unit or group of units in a stated period of time. It may be the maximum instantaneous load or the maximum average load over a designated period of time.

maximum demand. The greatest of all demands which have occurred during a specified period of time. *Note:* For utility billing purposes the period of time is generally 1 month.

demand factor. The ratio of the maximum demand of a system to the total connected load of the system.

diversity factor. The ratio of the sum of the individual maximum demands of various subdivisions of the system to the maximum demand of the complete system.

load factor. The ratio of the average load over a designated period of time to the peak load occurring in that period.

coincident demand. Any demand that occurs simultaneously with any other demand, also the sum of any set of coincident demands.

Information on these factors for the various loads and groups of loads is useful in designing the system. For example, the sum of the connected loads on a feeder, multiplied by the demand factor of these loads, will give the maximum demand which the feeder must carry. The sum of the individual maximum demands on the circuits associated with a load center or panelboard, divided by the diversity factor of those circuits, will give the maximum demand at the load center and on the circuit supplying it. The sum of the individual maximum demands on the circuits from a transformer, divided by the diversity factor of those circuits, will give the maximum demand on the distribution transformer. The sum of the maximum demand on all distribution transformers, divided by the diversity factor of the transformer loads, will give the maximum demand on their primary feeder. By the use of the proper factors, as outlined, the maximum demands on the various parts of the system from the load circuits to the power source can be estimated.

2.3.3 *Systems.* Investigate the various types of distribution systems and select the system or systems best suited to the requirements of the plant.

A variety of basic circuit arrangements is available for industrial plant power distribution. Selection of the best system or combination of systems will depend upon the needs of the manufacturing process. In general, system costs increase with system reliability if component quality is equal. Maximum reliability per unit

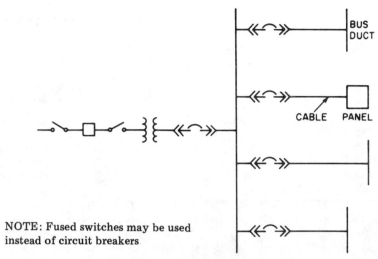

NOTE: Fused switches may be used
instead of circuit breakers

Fig 1
Simple Radial System

investment will be secured by using properly applied and well designed components.

The first step is the analysis of the manufacturing process to determine its reliability need and potential losses in the event of power interruption. Some processes are little affected by interruption. Here a simple radial system is satisfactory. Other processes may sustain long-term damage by even a brief interruption. Here a more complex system with an alternate power source for critical load may be justified.

A need for circuit redundancy may exist in continuous-process industry to allow equipment maintenance. Although the reliability of electric power distribution equipment is high, optimum reliability and safety of operation require routine maintenance. A system that cannot be maintained because of the needs of a continuous process is improperly designed.

Far more can be accomplished by the proper selection of the circuit arrangement than by economizing on equipment details. Cost reductions should never be made at the sacrifice of safety and performance by using inferior apparatus. Reductions should be obtained by using a less expensive distribution system with some sacrifice in reserve capacity and reliability.

(1) *Simple Radial System* (Fig 1). Distribution is at the utilization voltage. A single primary service and distribution transformer supply all the feeders. There is no duplication of equipment. System investment is the lowest of all circuit arrangements.

Operation and expansion are simple. If quality components are used, reliability is high. Loss of a cable, primary supply, or transformer will cut off service. Equipment must be shut down to perform routine maintenance and servicing.

This system is satisfactory for small industrial installations where process allows sufficient down time for adequate maintenance and the plant can be supplied by a single transformer.

(2) *Expanded Radial System* (Fig 2). The advantages of the radial system may be applied to larger loads by using a radial primary distribution system to supply a number of unit substations located near the centers of load supplying the load through radial secondary systems.

The advantages and disadvantages are the same as those described for the simple radial system.

(3) *Primary Selective System* (Fig 3). Protection against loss of a primary supply can be gained through use of a primary selective system. Each unit substation is connected to two separate primary feeders through switching equipment to provide a normal and an alternate source. Upon failure of the normal source the

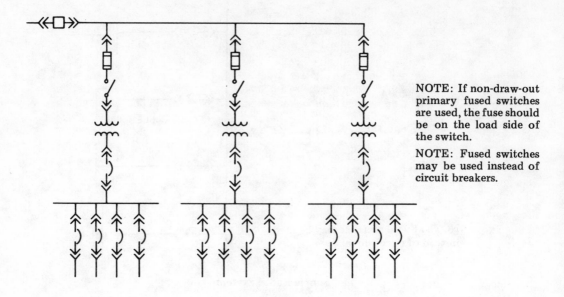

NOTE: If non-draw-out primary fused switches are used, the fuse should be on the load side of the switch.

NOTE: Fused switches may be used instead of circuit breakers.

**Fig 2
Expanded Radial System**

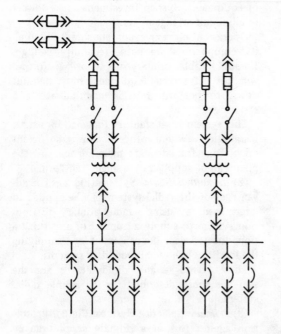

NOTE: If non-draw-out fused switches are used, the fuse should be on the load side of the switch.

NOTE: An alternate arrangement uses a primary selector switch with a single fused interrupter switch (which may not have certified current-switching ability).

**Fig 3
Primary Selective System**

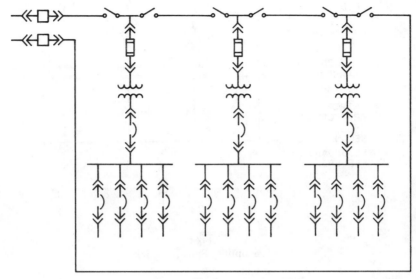

Fig 4
Primary Loop System

distribution transformer is switched to the alternate source. Switching can be either manual or automatic, but there will be an interruption until load is transferred to the alternate source.

If the two sources can be paralleled during switching, some maintenance of primary cable and, in certain configurations, switching equipment may be performed with little or no interruption of service. Cost is somewhat higher than for a radial system because of duplication of primary cable and switchgear.

(4) *Primary Loop System* (Fig 4). This system offers the same advantages and disadvantages as the primary selective system. The failure of the normal source of a primary cable fault can be isolated and service restored by sectionalizing. However, finding a cable fault in the loop may be difficult. The system may be dangerous because the quickest way to find a fault is to sectionalize the loop and reclose. This may involve several reclosings on the fault. Also potentially dangerous is the fact that a section may be energized from both ends.

Cost may be somewhat less than for the primary selective system. However, it is doubtful that the savings are justified in view of the disadvantages.

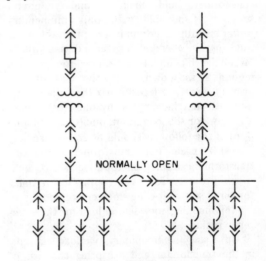

Fig 5
Secondary Selective System

(5) *Secondary Selective System* (Fig 5). If pairs of unit substations are connected through a normally open secondary tie circuit breaker, the result is a secondary selective system. If the primary feeder or a transformer fails, the main secondary circuit breaker on the affected transformer is opened and the tie circuit breaker closed. Operation may be manual or

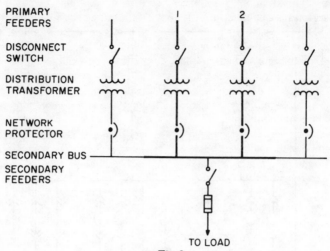

PRIMARY
FEEDERS

DISCONNECT
SWITCH

DISTRIBUTION
TRANSFORMER

NETWORK
PROTECTOR

SECONDARY BUS
SECONDARY
FEEDERS

TO LOAD

**Fig 6
Secondary Spot Network**

automatic. Normally the stations operate as radial systems. Maintenance of primary feeders, transformer, and main secondary circuit breakers is possible with only momentary power interruption or no interruption if the stations may be operated in parallel during switching, although complete station maintenance will require a shutdown. With the loss of one primary circuit or transformer the total substation load may be supplied by one transformer. To allow for this condition, one (or a combination) of the following should be considered.

(a) Oversizing both transformers so that one transformer can carry the total load

(b) Providing forced-air cooling to the transformer in service for the emergency period

(c) Shedding nonessential load for the emergency period

(d) Using the temporary overload capacity in the transformer and accepting the loss of transformer life

A distributed secondary selective system has pairs of unit substations in different locations connected by a tie cable and a normally open circuit breaker in each substation. The designer must balance the cost of the additional tie circuit breaker and the tie cable against the cost advantage of putting the unit stations nearer the load center.

The secondary selective system may be combined with the primary selective system to pro-

vide a high degree of reliability. This reliability is purchased with additional investment and addition of some operating complexity.

(6) *Secondary Spot Network* (Fig 6). In this system two or more distribution transformers are each supplied from a separate primary distribution feeder. The secondaries of the transformers are connected in parallel through a special type of circuit breaker, called network protector, to a secondary bus. Radial secondary feeders are tapped from the secondary bus to supply utilization equipment.

If a primary feeder fails, or a fault occurs on a primary feeder or distribution transformer, the other transformers start to feed back through the network protector on the faulted circuit. This reverse power causes the network protector to open and disconnect the supply circuit from the secondary bus. The network protector operates so fast that there is a minimal exposure of secondary equipment to the associated voltage drop.

The secondary spot network is the most reliable power supply for large loads. A power interruption can only occur when there is a simultaneous failure of all primary feeders or when a fault occurs on the secondary bus. There are no momentary interruptions caused by the operation of the transfer switches which occur on primary selective, secondary selective, or loop systems. Voltage dips caused by faults on the system or large transient loads are materially reduced.

Networks are expensive because of the extra cost of the network protector and duplication of transformer capacity. In addition, each transformer connected in parallel increases the short-circuit-current capacity and may increase the duty ratings of the secondary equipment.

2.3.4 *Equipment Locations.* In cooperation with the architect and process personnel select locations for distribution transformers and major utilization voltage switching centers. In general the closer the transformer to the load center of the area served, the lower the distribution system cost.

2.3.5 *Voltage.* Select the best voltages for the various system voltage levels. The most common utilization voltage in United States industrial facilities is 480 V. Other voltage levels depend upon motor size, utility voltage available, total load served, potential expansion requirements, voltage regulation, and cost. Chapter 3 is a guide for correct voltage selection. The system must be capable of providing power to all equipment within published voltage limits under all normal operating conditions.

2.3.6 *Utility Service.* As soon as practical in the design sequence, a conference should be arranged with the supplying utility to determine the requirements for electric service. The utility should be furnished with the following data:

(1) Plan of the plant property showing buildings and other structures

(2) Plant load, preferably maximum demand, in kilovolt-amperes

(3) Preferred point of delivery of electric service

(4) Preferred service voltage

(5) Preferred utility supply arrangement

(6) Construction and startup schedule

(7) Unusual requirements such as high-speed reclosing being unacceptable

(8) Any unusually large motors on the system

(9) Anticipated power factor

(10) Nature of connected load

The utility will be able to furnish the following data:

(1) Supply voltage or voltages available

(2) Point of delivery and line route

(3) Billing rate or rates available

(4) Options available on ownership of supply transformers

(5) Space requirements for the distribution substation if supplied by the utility

(6) Space required for the distribution transformer if plant is supplied at utilization voltage

(7) Short-circuit duty at the point of delivery and system characteristics

(8) Requirements for metering

(9) Type of grounding on the supply system

(10) Requirements for coordination with the utility protection system

(11) Performance data on reliability of supply as necessary

(12) Back-up supply circuitry as necessary

2.3.7 *Generation.* Determine whether parallel, standby, or emergency generation will be required. If so, the following will have to be considered:

(1) Generator load in kilovolt-amperes

(2) Generator voltage

(3) Relaying and generator protection

(4) Metering

(5) Voltage regulation

(6) Synchronizing

(7) Grounding

(8) Cost

(9) Maintenance requirements

(10) Largest motor to be started

The complete design must be coordinated with the utility if parallel operation with their system is anticipated.

2.3.8 *One-Line Diagram.* A complete one-line or single-line diagram, in conjunction with a physical plan of the installation, should present sufficient data to plan and evaluate the electric power system. Figs 37 and 76 in Chapters 4 and 5 represent one-line diagrams containing information required for system-protection design and fault-current analysis.

Symbols commonly used in one-line diagrams are defined in IEEE Std 315-1975, Graphic Symbols for Electrical and Electronics Diagrams (ANSI Y32.2-1975).

The following items should be shown on any one-line diagram:

(1) Power sources including voltages and available short-circuit currents

(2) Size, type, ampacities, and number of all conductors

(3) Capacities, voltages, impedance, connections, and grounding methods of transformers

(4) Identification and quantity of protective devices (relays, fuses, and circuit breakers)

(5) Instrument transformer ratios

(6) Type and location of surge arresters and capacitors

(7) Identification of all loads

(8) Identification of any other distribution system equipment

The one-line diagram should show proposed future additions, and the effect of such additions should be a part of the original system planning.

The actual drawing should be kept as simple as possible. It is a schematic diagram and need not show geographical relationship. Duplication should be avoided.

2.3.9 *Short-Circuit Analysis.* Calculate short-circuit currents available at all system components. Chapter 5 provides a detailed guide to making these calculations.

2.3.10 *Protection.* Using the data presented in Chapter 4, design the required protective systems. System protection design must be an integral part of the total system design and not superimposed on a system after the fact.

2.3.11 *Expansion.* If expanding an existing plant, determine if all the existing equipment is adequate for additional load and short-circuit requirements. Check such ratings as voltage, interrupting capacity, short-circuit withstand ratings, momentary ratings, switch close and latch requirements, and current-carrying capacity. Coordinate new and existing system protective devices. Carefully review the new design to ensure that maximum safety levels are maintained.

Determine the best method of connecting the new part of the power system to the existing system to minimize production loss and construction cost.

2.3.12 *Other Requirements.* Investigate unusual loads or conditions such as the following:

(1) Starting requirements of large motors

(2) Arc furnace operation

(3) Welder operation

(4) Loads that must be kept in operation under all conditions

(5) Sensitive loads such as computers and high-gain laboratory equipment which may be affected by voltage and frequency transients that do not affect other equipment

(6) Equipment producing high noise levels, (Federal law requires that personnel not be exposed continuously for an eight hour period to noise pressure levels of more than 90 dBA. Higher levels are allowed up to 115 dBA with a reduction in exposure time.)

(7) The possibility of demand limiting to reduce power cost

(8) Coordination of the electric power system with other energy systems

2.3.13 *Safety.* Make sure that adequate safety features are incorporated into all parts of the system. Safety of life and preservation of property are two of the most important factors in the design of the electric system.

Listed below are items which should be considered in order to provide safe working conditions for personnel.

(1) Interrupting devices must be able to function safely and properly under the most severe duty to which they may be exposed.

(2) Protection must be provided against accidental contact with energized conductors by enclosing the conductors, installing protective barriers, or installing the conductors at sufficient height to avoid accidental contact.

(3) Isolating switches must not be operated while they are carrying current, unless designed to do so. They should be equipped with interlocks or warning signs if load break or fault closing capability is not provided.

(4) In many instances it is desirable to isolate a power circuit breaker using disconnect switches. In such cases the circuit breaker must be opened first, then the disconnect switches. Interlocks to ensure this sequence are usually desirable.

(5) The system must be designed so that maintenance work on circuits and equipment can be accomplished with the particular circuits and equipment de-energized and grounded. System design must provide for locking out circuits or equipment for maintenance. A written procedure must be established to provide instructions on tagging or locking out circuits during maintenance and re-energizing after completion.

(6) Electric-equipment rooms, especially those

containing apparatus over 600 V such as transformers, motor controls, or motors, must be equipped and located to eliminate or minimize the need for access by nonelectrical maintenance or operating personnel. Convenient remotely located exits must be provided to allow quick exit during an emergency.

(7) Electric apparatus located outside special rooms must be provided with protection against mechanical damage. The area must be accessible to maintenance and operating personnel for emergency operation of protective devices.

(8) Warning signs have to be installed on electric equipment accessible to unqualified personnel, on fences surrounding electric equipment, on doors giving access to electrical rooms, and on conduits or cables above 600 V in areas which include other equipment or pipelines.

(9) An adequate grounding system must be installed.

(10) Emergency lights have to be provided where necessary to protect personnel against sudden lighting failure. In areas with high populations, exit routes, process control locations, and electric switching centers are particularly important.

(11) Operating and maintenance personnel must be provided with complete operating and maintenance instructions including wiring diagrams, equipment ratings, and protective device settings. Spare fuses of the correct ratings should be stocked.

2.3.14 *Communications.* Any plan for the protection of a plant must include a reliable communication system. This can be accomplished either by a self-contained and self-maintained interplant system of telephones, alarms, etc, and may include modern radio and television equipment, or by a joint system tied into the existing communication services.

Fire and smoke alarm circuits, whether self-contained or connected to city alarm systems, should be installed in a manner that will be least affected by faults and changes in buildings or plant operations. Circuits should be arranged to provide easy means of testing and of isolating portions of the system without interference with the balance of the system.

Watchman circuits, including television and radio equipment, are used in many plants for the purpose of providing a ready means for the individual watchman to report unusual circumstances to his supervisor without delay. Such systems are frequently combined with loudspeaker paging systems and other alarm methods.

Annunciator systems are available for alerting operations to abnormal situations in critical areas. The operator can dispatch others to investigate the malfunction or disorder or take corrective action with controls furnished to him.

2.3.15 *Maintenance.* Electric equipment must be selected and installed with attention to adequacy of performance, safety, and reliability. To preserve these features, a maintenance program must be established, tailored to the type of equipment and the details of the particular installation. Some items require daily attention, some weekly, and others can be tested or checked annually or less frequently.

Requirements of a maintenance program should be incorporated in the electrical design to provide working space, easy access for inspection, facilities for sampling and testing, disconnecting means for protection of the workmen, lighting, and standby power. The maintenance program should have the following objectives.

(1) *Cleanliness.* Dirt and dust accumulation affects the ventilation of equipment and causes excess heat which reduces the life of the insulation. Dirt and dust also build up on the surfaces of insulators to form paths for leakage which may result in arcing faults. Insulated surfaces must be regularly cleaned to minimize these hazards.

(2) *Moisture Control.* Moisture reduces the dielectric strength of many insulating materials. Unnecessary openings should be closed and necessary openings should be baffled or filtered to prevent the entrance of moisture, especially light snow. Also, even though equipment is adequately housed and indoors, condensation from weather changes must be minimized by supplying heat, usually electric, to the enclosure interiors. From 5 to 7.5 W per square foot of enclosure external surface is usually effective when placed at the bottom of each space affected. A small amount of ventilation (or breathing) outdoors is necessary even with

heating to avoid condensation damage and insulation failure.

(3) *Adequate Ventilation.* Much electric equipment is designed with paths for ventilating air to pass over insulating surfaces to dissipate heat. Filters must be changed, fans inspected, and equipment cleaned often enough to keep such ventilating system operating properly.

(4) *Reduced Corrosion.* Corrosion destroys the integrity of equipment and enclosures. As soon as evidence of corrosion is noted, action should be taken to clean the affected surfaces and inhibit future deterioration.

(5) *Maintenance of Conductors.* Conducting surfaces reveal problems caused by overheating, wear, or misalignment of contact surfaces. These conditions should be corrected by tightening bolts, correcting excessive operations, aligning contacts, or whatever action is necessary.

(6) *Regular Inspections.* Inspections should be scheduled on a regular basis depending on equipment needs and process requirements. External inspection can often be made and reveal significant information without process shutdown. However, a complete inspection will require a shutdown. Plans for repairs should be based on such inspections so that necessary manpower, tools, and replacement parts will be available as needed during the shutdown.

(7) *Regular Testing.* Performance of protective devices depends on the accuracy of the sensing devices and the integrity of the control circuits. Periodic tests of such devices as well as of the dielectric strength of insulating systems, and the color and acidity of the insulating oils, etc, will reveal deteriorating conditions which cannot be determined by visual inspection. Necessary adjustments or corrections can be made before failure occurs.

(8) *Adequate Records.* An organized system of records of inspection, maintenance, tests, and repairs provides a basis for trouble-shooting, predicting equipment failures, and selecting future equipment.

2.3.16 *Codes and Standards.* Throughout the design the engineer must comply with all national and local laws, codes, and standards.

2.4 Standards References. The following standards publications were used as references in preparing this chapter.

IEEE Std 277-1975, Cement Plant Power Distribution

IEEE Std 315-1975, Graphic Symbols for Electrical and Electronics Diagrams (ANSI Y32.2-1975)

NFPA No 70B, Electrical Equipment Maintenance (1975), (ANSI C132.1-1975)

2.5 Bibliography

[1] BJORNSON, N. R. How Much Redundancy and What It Will Cost. *IEEE Transactions on Industry and General Applications,* vol IGA-6, May/Jun 1970, pp 192-195.

[2] HEISING, C. R. Reliability and Availability Comparison of Common Low-Voltage Industrial Power Distribution Systems. *IEEE Transactions on Industry and General Applications,* vol IGA-6, Sept/Oct 1970, pp 416-424.

[3] IEEE COMMITTEE REPORT. Report on Reliability Survey of Industrial Plants. Parts I—III. *IEEE Transactions on Industry Applications,* vol IA-10, Mar/Apr 1974, pp 213-252.

[4] JOHNSON, G. T., and BRENIMAN, P. E. Electrical Distribution System for a Large Fertilizer Complex. *IEEE Transactions on Industry and General Applications,* vol IGA-5, Sept/Oct 1969, pp 566-577.

[5] McFADDEN, R. H. Power-System Analysis: What It Can Do for Industrial Plants. *IEEE Transactions on Industry and General Applications,* vol IGA-7, Mar/Apr 1971, pp 181-188.

[6] REGOTTI, A. R., and TRASKY, J. G. What to Look for in a Low-Voltage Unit Substation. *IEEE Transactions on Industry and General Applications,* vol IGA-5, Nov/Dec 1969, pp 710-719.

[7] SHAW, E. T. *Inspection and Test of Electrical Equipment.* Pittsburgh, PA: Westinghouse Electric Corporation, Electric Service Division, Publ MB3051, 1967.

[8] YUEN, M. H., and KNIGHT, R. L. On-Site Electrical Power Generation and Distribution for Large Oil and Gas Production Complex in Libya. *IEEE Transactions on Industry and General Applications,* vol IGA-7, Mar/Apr 1971, pp 273-289.

3. Voltage Considerations

3.1 Introduction. An understanding of system voltage nomenclature and the preferred voltage ratings of distribution apparatus and utilization equipment is essential to proper voltage identification throughout a power distribution system. The dynamic characteristics of the system must be recognized and the proper principles of voltage control applied so that satisfactory voltage will be supplied to all utilization equipment under all conditions of operation.

    3.1.1 *Voltage Standard for the United States.* The voltage standard for electric power systems in the United States is ANSI C84.1-1970, Voltage Ratings for Electric Power Systems and Equipment (60 Hz), including Supplement C84.1a-1973. This standard lists all the standard nominal system voltages in general use in the United States and specifies the tolerance limits for these voltages at the point of delivery by the supplying utility, as well as at the point of connection to utilization equipment. Two sets of tolerance limits are provided, range A which specifies the limits under most operating conditions and range B which allows minor excursions outside of the range A limits. The rated nameplate voltages of common items of utilization equipment are listed together with the nominal voltage of the system on which they should be used. Furthermore reference voltage standards are listed for common types of utilization equipment and distribution apparatus.

    3.1.2 *Voltage Standard for Canada.* The voltage standard for Canada, Preferred Voltage Levels for AC Systems, 0 to 50 000 V — Recommended Reference Standard for Canadian Practice, is published by the Canadian Standards Association.[10] This standard differs from the United States standard in both the list of standard nominal voltages and the tolerance limits.

3.2 Voltage Control in Electric Power Systems
    3.2.1 *Principles of Power Transmission and Distribution in Utility Systems.* To understand the principles of voltage control required to provide satisfactory voltage to utilization equipment, a general understanding of the principles of power transmission and distribution in utility systems is necessary since most industrial plants obtain most of their electric power from the local electric utility. Fig 7 shows a one-line diagram of a typical utility power generation, transmission, and distribution system.

Most utility generating stations are located near convenient sources of fuel and water. Generated power, except for station requirements, is transformed in a substation located at the generating station to a transmission voltage generally 69 000 V or more for transmission to

---

[10]178 Rexdale Boulevard, Rexdale, Ont, Canada.

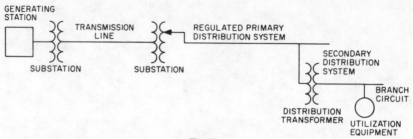

Fig 7
Typical Utility Generation, Transmission, and Distribution System

major load areas. Transmission lines are classed as unregulated because the voltage at the generating station is controlled only to keep the lines operating within normal voltage limits and to facilitate power flow. ANSI C84.1-1970 specifies only the nominal and maximum values for systems over 34 500 V.

Transmission lines supply distribution substations equipped with transformers which step the transmission voltage down to a primary distribution voltage generally in the range from 4160 through 34 500 V with 12 470 and 13 200 V in widest use.

NOTE: When an industrial plant transforms to a voltage in this medium-voltage range, 13 800 V is usually selected.

There is an increasing trend toward the use of 34 500 V for primary distribution in some locations.

It is at this point that voltage control is applied when necessary for the purpose of supplying satisfactory voltage to the terminals of utilization equipment. The transformers used to step the transmission voltage down to the primary distribution voltage are generally equipped with tap-changing-under-load equipment which changes the ratio of the transformer under load in order to maintain the primary distribution voltage constant at the substation regardless of fluctuations in the transmission voltage. Separate step or induction regulators may also be used.

Generally the regular controls are equipped with compensators which raise the voltage as the load increases and lower the voltage as the load decreases to compensate for the voltage drop in the primary distribution system which extends out radially from the substation to supply the surrounding area. Thus a fixed average voltage is maintained along the primary distribution system, and the voltage is prevented from rising to peak values during light-load conditions when the voltage drop along the primary distribution system is low. This is illustrated in Fig 8. Note that plants close to the substation will receive voltages which will on the average be higher than those received by plants at a distance from the substation. See Section 3.2.8 on the use of distribution transformer taps.

The primary distribution system supplies distribution transformers which step the primary distribution voltage down to utilization voltages generally in the range of 120 through 600 V to supply a secondary distribution system to which the utilization equipment is connected. Small transformers used to step a higher utilization voltage down to a lower utilization voltage such as 480 V to 208Y/120 V are considered part of the secondary distribution system.

The supply voltages available to an industrial plant depend upon whether the plant is connected to the distribution transformer, the primary distribution system, or the transmission system, which, in turn, depends on the size of the plant load.

Small plants with loads up to several hundred kilovolt-amperes and all plants supplied from low-voltage secondary networks are connected to the distribution transformer, and the secondary distribution system consists of the connections from the distribution transformer to the plant service and the plant wiring.

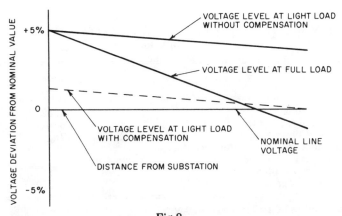

Fig 8
Effect of Regulator Compensation on Primary
Distribution System Voltage

Medium-size plants with loads of a few thousand kilovolt-amperes are connected to the primary distribution system, and the plant provides the portion of the primary distribution system within the plant, the distribution transformers, and the secondary distribution system.

Large plants with loads of more than a few thousand kilovolt-amperes are connected to the transmission system, and the plant provides the primary distribution system, the distribution transformers, the secondary distribution system, and it may provide the substation.

Details of the connection between the utility system and the plant system will depend on the policy of the supplying utility. If power is supplied by in-plant generation, the generators will replace wholly or in part the utility supply system up to the distribution transformer if generation is over 600 V, and it will also replace the distribution transformer if generation is at 600 V or below.

3.2.2 *Development of the Voltage Tolerance Limits for ANSI C84.1-1970.* The voltage tolerance limits in ANSI C84.1-1970 are based on NEMA MG1-1972, Motors and Generators, which established the voltage tolerance limits of the standard induction motor at ± 10 percent of nameplate ratings of 230 and 460 V. Since motors represent the major component of utilization equipment, they were given primary consideration in the establishment of the voltage standard.

The best way to show the voltages in a distribution system is in terms of a 120 V base. This cancels the transformation ratios between systems so that the actual voltages vary solely on the basis of the voltage drops in the system. Any voltage may be converted to a 120 V base by dividing the actual voltage by the ratio of transformation to 120 V. For example, the ratio of transformation for the 480 V system is 480/120 or 4, so 460 V in a 480 V system would be 460/4 or 115 V.

The tolerance limits of the 460 V motor in terms of the 120 V base become 115 V plus 10 percent, or 126.5 V, and 115 V minus 10 percent, or 103.5 V. The problem is to decide how this tolerance range of 23 V should be divided between the primary distribution system, the distribution transformer, and the secondary distribution system which make up the regulated distribution system. The solution adopted by American National Standards Committee C84 is shown in Table 1.

The tolerance limits of the standard motor on the 120 V base of 126.5 V maximum and 103.5 V minimum were raised ½ V to 127 V maximum and 104 V minimum to eliminate the fractional volt. These values became the tolerance limits for range B in the standard. An allowance of 13 V was allotted for the voltage drop in the primary distribution system. Deducting this voltage drop from 127 V estab-

Table 1
Standard Voltage Profile for Low-Voltage Regulated Power Distribution System,
120 V Base

	Range A (volts)	Range B (volts)
Maximum allowable voltage	126(125*)	127
Voltage drop allowance for primary distribution line	9	13
Minimum primary service voltage	117	114
Voltage drop allowance for distribution transformer	3	4
Minimum secondary service voltage	114	110
Voltage drop allowance for plant wiring	6(4†)	6(4†)
Minimum utilization voltage	108(110†)	104(106†)

*For utilization voltage of 120 through 600 V.
†For building wiring circuits supplying lighting equipment.

lishes a minimum of 114 V for utility services supplied from the primary distribution system. An allowance of 4 V was provided for the voltage drop in the distribution transformer and the connections to the plant wiring. Deducting this voltage drop from the minimum primary distribution voltage of 114 V provides a minimum of 110 V for utility secondary services from 120 through 600 V. An allowance of 6 V, or 5 percent, for the voltage drop in the plant wiring, as provided in NFPA No 70, National Electrical Code (1975), (ANSI C1-1975), (NEC), completes the distribution of the 23 V tolerance zone down to a minimum utilization voltage of 104 V.

The range A limits for the standard were established by reducing the maximum tolerance limits from 127 to 126 V and increasing the minimum tolerance limits from 104 to 108 V. The spread band of 18 V was then allotted as follows: 9 V for the voltage drop in the primary distribution system to provide a minimum primary service voltage of 117 V; 3 V for the voltage drop' in the distribution transformer and secondary connections to provide a minimum utility secondary service voltage of 114 V; and 6 V for the voltage drop in the plant wiring to provide a minimum utilization voltage of 108 V.

Four additional modifications were made in this basic plan to establish ANSI C84.1-1970. The maximum utilization voltage in range A was reduced from 126 to 125 V for low-voltage

systems in the range from 120 through 480 V because there should always be sufficient load on the distribution system to provide at least 1 V drop on the 120 V base under most operating conditions. This maximum voltage of 125 V is also a practical limit for lighting equipment because the life of the 120 V incandescent lamp is reduced by 42 percent when operated at 125 V (see Table 8). The voltage drop allowance of 6 V on the 120 V base for the drop in the plant wiring was reduced to 4 V for circuits supplying lighting equipment. This raised the minimum voltage limit for utilization equipment to 106 V in range B and 110 V in range A because the minimum limits for motors of 104 V in range B and 108 V in range A were considered too low for satisfactory operation of lighting equipment. The utilization voltages for the 6900 and 13 800 V systems in range B were adjusted to coincide with the tolerance limits of ± 10 percent of the nameplate rating of the 6600 and 13 200 V motors used on these respective systems.

The tolerance limits for the service voltage provide guidance to the supplying utility for the design and operation of its distribution system. The service voltage is the voltage at the point where the utility conductors connect to the user conductors, although in practice it is generally measured at the service switch for services of 600 V and below and at the billing meter potential transformers for services over 600 V. The tolerance limits for the voltage at the point of connection of utilization equipment provides guidance to the user for the design and operation of the user distribution system, and to utilization equipment manufacturers for the design of utilization equipment.

Electric supply systems are to be designed and operated so that most service voltages fall within the range A limits. User systems are to be designed and operated so that, when the service voltages are within range A, the utilization voltages are within range A. Utilization equipment is to be designed and rated to give fully satisfactory performance within the range A limits for utilization voltages.

Range B is provided to allow limited excursions of voltage outside the range A limits which necessarily result from practical design and operating conditions. The supplying utility is expected to take action within a reasonable time to restore service voltages to range A limits. The user is expected to take action within a reasonable time to restore utilization voltages to range A limits. Insofar as practicable, utilization equipment may be expected to give acceptable performance at voltages outside range A but within range B. When voltages occur outside the limits of range B, prompt corrective action should be taken.

To convert the 120 V base voltage to equivalent voltages in other systems, the voltage on the 120 V base is multiplied by the ratio of the transformer which would be used to connect the other system to a 120 V system. In general, oil-filled distribution transformers for systems below 15 000 V have nameplate ratings which are the same as the standard system nominal voltages; so the ratio of the standard nominal voltage may be used to make the conversion. However, for primary distribution voltages over 15 000 V the primary nameplate rating of oil-filled distribution transformers is not the same as the standard system nominal voltages. Also most distribution transformers are equipped with taps which can be used to change the ratio of transformation. So if the primary distribution voltage is over 15 000 V, or taps have been used to change the transformer ratio, then the actual transformer ratio must be used to convert the base voltage to another system.

For example, the maximum tolerance limit of 127 V on the 120 V base for the service voltage in range B is equivalent, on the 4160 V system, to $4160/120 \times 127 = 4400$ V to the nearest 10 V. However, if the 4160 to 120 V transformer is set on the plus 2½ percent tap, the voltage ratio would be $4160 + (4160 \times 0.025) = 4160 + 104 = 4264$ to 120. The voltage on the primary system equivalent to 127 V on the secondary system would be $4264/120 \times 127 = 35.53 \times 127 = 4510$ V to the nearest 10 V. If the maximum primary distribution voltage of 4400 V is applied to the 4264 to 120 V transformer, the secondary voltage would be $4400/4260 \times 120 = 124$ V. So the effect of using a plus 2½ percent tap is to lower the secondary voltage by 2½ percent.

3.2.3 *Definition of Nominal System Voltage.* The term "nominal system voltage" designates not a single voltage, but a range of voltages over which the actual voltages at any point on the system may vary and still provide satisfactory operation of equipment connected to the system. The voltage range for all standard nominal system voltages in the utilization and distribution range of 120 through 34 500 V is specified in ANSI C84.1-1970 for two critical points on the distribution system: the point of delivery by the supplying utility and the point of connection to utilization equipment. For transmission voltages over 34 500 V only the maximum voltage is specified, because the voltages are normally unregulated and only a maximum voltage is required to establish the design insulation level for the line and associated apparatus.

The actual voltage measured at any point on the system will vary depending on the location of the point of measurement and the time the measurement is made. Fixed voltage changes take place in transformers in accordance with the transformer ratio, while voltage variations occur from the operation of voltage control equipment and the changes in voltage drop between the supply source and the point of measurement as a result of changes in the current flowing in the circuit.

3.2.4 *Voltage Profile Limits for a Regulated Distribution System.* Fig 9 shows the voltage profile of a regulated power distribution system using the limits of range A in Table 1. Assuming

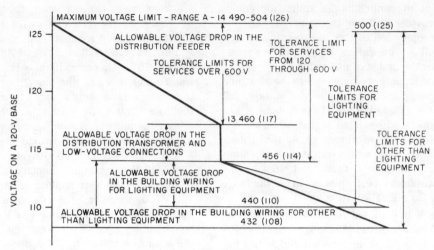

Fig 9
Voltage Profile of Limits of Range A, ANSI C84.1-1970

a nominal primary distribution voltage of 13 800 V, range A in Table 1 shows that this voltage should be maintained by the supplying utility between a maximum of 126 V and a minimum of 117 V on a 120 V base. Since the base multiplier for converting from the 120 V system to the 13 800 V system is 13 800/120 or 115, the actual voltage limits for the 13 800 V system are 115 × 126 or 14 490 V maximum and 115 × 117 or 13 460 V minimum.

If a distribution transformer with a ratio of 13 800 to 480 V is connected to the 13 800 V distribution feeder, range A of Table 1 requires that the nominal 480 V secondary service be maintained by the supplying utility between a maximum of 126 V and a minimum of 114 V on the 120 V base. Since the base multiplier for the 480 V system is 480/120 or 4, the actual values are 4 × 126 or 504 V maximum and 4 × 114 or 456 V minimum.

Range A of Table 1 as modified for utilization equipment of 120 through 600 V provides for a maximum utilization voltage of 125 V and a minimum of 110 V for lighting equipment and 108 V for other than lighting equipment on the 120 V base. Using the base multiplier of 4 for the 480 V system, the maximum utilization voltage would be 4 × 125 or 500 V and the minimum for other than lighting equipment would be 4 × 108 V or 432 V. For lighting

equipment connected phase to neutral, the maximum voltage would be 500 V divided by the square root of 3 or 288 V and the minimum voltage would be 4 × 110 V or 440 V divided by the square root of 3 or 254 V.

3.2.5 *System Voltage Nomenclature.* The nominal system voltages in Table 2 are designated in the same way as on the nameplate of the transformer for the winding or windings supplying the system.

(1) *Single-Phase Systems*

120   Indicates a single-phase two-wire system in which the nominal voltage between the two wires is 120 V.

120/240   Indicates a single-phase three-wire system in which the nominal voltage between the two phase conductors is 240 V, and from each phase conductor to the neutral it is 120 V.

(2) *Three-Phase Systems*

240/120   Indicates a three-phase four-wire system supplied from a delta-connected transformer. The midtap of one winding is connected to a neutral. The three-phase conductors provide a nominal 240 V three-phase system, and the neutral and the two adjacent phase

Single number Indicates a three-phase three-wire system in which the number designates the nominal voltage between phases.

Two numbers separated by a Y/ Indicates a three-phase four-wire system from a wye-connected transformer in which the first number indicates the nominal phase-to-phase voltage and the second number indicates the nominal phase-to-neutral voltage.

conductors provide a nominal 120/240 V single-phase system.

NOTES: (1) All single-phase systems and all three-phase four-wire systems are suitable for the connection of phase-to-neutral load.

(2) See Chapter 6 for methods of system grounding.

(3) The following chart gives an overview of voltage relationships for 480 V three-phase systems and 120/240 V single- and three-phase systems.

3.2.6 *Nonstandard Nominal System Voltages.* Since ANSI C84.1-1970 lists only the standard nominal system voltages in common use in the United States, system voltages will frequently be encountered which differ from the standard list. A few of these may be so widely different as to constitute separate systems in too limited use to be considered standard. However, in most cases the nominal system voltages will differ by only a few percent as shown in Table 2(b). A closer examination of the table shows that these differences are due mainly to the fact that some voltages are multiples of 110 V, others are multiples of 115 V, some are multiples of 120 V, and a few are multiples of 125 V.

The reasons for these differences go back to the original development of electric power distribution systems. The first utilization voltage was 100 V. However, the supply voltage had to be raised to 110 V in order to compensate for the voltage drop in the distribution system.

Voltage Relationships Based on Voltage Ranges in ANSI C84.1-1970

480 V Three-Phase System					120/240 V Single-Phase and Three-Phase Systems				
Voltage at Point of Service Entrance			Voltage at Point of Utilization Equipment		Voltage at Point of Service Entrance			Voltage at Point of Utilization Equipment	
Range B	Range A	Voltage	Range A	Range B	Range B	Range A	Voltage	Range A	Range B

———— Lighting or Combination Lighting and Power Circuits
– – – Power Circuits

Table 2
Voltage Classes and Nominal System Voltages

(a) Voltage Classes and Definitions as Applied to Industrial and Commercial Power Systems

Voltage Classes	Two-Wire	Nominal System Voltages (volts)§ Three-Wire	Four-Wire	Maximum System Voltages	References
Low voltage*		*Single-Phase Systems*		127 or 127/254	IEEE Std 100-1972 defines low-voltage system as "less than 750 V." It is now being proposed as "maximum rms ac voltage of 1000 V."
	(120)	120/240			
		Three-Phase Systems	208Y/120	220	
		(240)	240/120	254	
		480	480Y/277	508	
		(600)		635	
Medium voltage†		(2400)		2 540	These voltages are listed in ANSI C84.1-1970 but are not identified by voltage class. Note that additional voltages in this class are listed in ANSI C84.1-1970 but are not included here because they are primarily oriented to electric-utility system practice.
		4160		4 400	
		(4800)		5 080	
		(6900)		7 260	
			12 470Y/7200	13 200	
			13 200Y/7620	13 970	
		13 800	(13 800Y/7970)	14 520	
		(23 000)		24 340	
			24 940Y/14 400	26 400	
		(34 500)	34 500Y/19 920	36 500	
		(46 000)		48 300	
		69 000		72 500	
High voltage‡		115 000		121 000	ANSI C84.1-1970 identifies these as "higher voltage three-phase systems" (as well as 46 and 69 kV).
		138 000		145 000	
		(161 000)		169 000	
		230 000		242 000	
Extra-high voltage#		345 000		362 000	ANSI C92.2-1974.
		500 000		550 000	
		735 000—765 000		800 000	

NOTES: (1) Voltage class designations were approved for use in documents published by the IEEE Industrial and Commercial Power Systems Committee at its Committee Meeting on May 16, 1973 (and on June 6, 1974 and January 28, 1975); approved by IEEE Standards Board September 4, 1975.

(2) Limits for the 600 V nominal system voltage established by ANSI C84.1a-1973.

(3) The maximum and minimum voltage ranges for system voltages from 120 to 239 kV are given in ANSI C84.1-1970, Voltage Ratings for Electric Power Systems and Equipment (60 Hz).

(4) It is appropriate to observe that the voltage class of "medium voltage" has had long acceptance in industrial practice. In contrast, the term "medium voltage" does not as easily apply to electric-utility practice because of the traditional use of overlapping voltage areas of transmission and distribution [as defined in IEEE Std 100-1972, Dictionary of Electrical and Electronics Terms (ANSI C42.100-1972)].

(5) The preferred maximum voltages for three-phase system voltages from 345 to 735-765 kV are given in ANSI C92.2-1974, Preferred Voltage Ratings for Alternating-Current Electrical Systems and Equipment Operating at Voltages above 240 Kilovolts Nominal. When ANSI C92.2-1974 superseded ANSI C92.2-1967, the specific identification of the nominal system voltages for the extra-high-voltage systems was withdrawn. They are now only identified as "typical 'nominal' voltages." (The limits for the extra-high-voltage systems which appear in ANSI C84.1-1970, page 9, are superseded by ANSI C92.2-1974.)

*low-voltage system (electric power). An electric system having a maximum rms ac voltage of 1000 V or less.

†medium-voltage system (electric power for industrial and commercial systems only). An electric system having a maximum rms ac voltage above 1000 V or 72 500 V.

‡high-voltage system (electric power for industrial and commercial systems only). An electric system having a maximum rms ac voltage above 72 500 V to 242 000 V.

#extra-high-voltage system (electric power). An electric system having a maximum rms ac voltage above 242 000 V to 800 000 V.

§ System voltages shown without parentheses are preferred. See ANSI C84.1-1970, Section 2.2.

Table 2 (Cont'd)

(b) Nominal System Voltages

Standard Nominal System Voltages*	Associated Nonstandard Nominal System Voltages
Low-Voltage Systems	
(120)	110, 115, 125
120/240	110/220, 115/230, 125/250
208Y/120	216Y/125
240/120	
(240)	230, 250
480Y/277	460Y/265
480	440, 460
(600)	550, 575
Primary Distribution Voltage Systems	
(2400)	2200, 2300
(4160Y/2400)	
4160	4000
(4800)	4600
(6900)	6600, 7200
(8320Y/4800)	
(12 000Y/6930)	11 000, 11 500
12 470Y/7200	
13 200Y/7620	
(13 800Y/7970)	
13 800	14 400
(20 780Y/12 000)	
(22 860Y/13 200)	
(23 000)	
24 940Y/14 400	
34 500Y/19 920	
(34 500)	33 000
Transmission Voltage Systems	
(46 000)	
69 000	66 000
115 000	110 000, 120 000
138 000	132 000
(161 000)	154 000
230 000	220 000

*System voltages shown without parentheses are preferred.

47

This led to overvoltage on equipment connected close to the supply, so the utilization equipment rating was also raised to 110 V. As generator sizes increased and distribution and transmission systems developed, an effort to keep transformer ratios in round numbers lead to a series of utilization voltages of 110, 220, 440, and 550 V, a series of primary distribution voltages of 2200, 4400, 6600, and 13 200 V, and a series of transmission voltages of 22 000, 33 000, 44 000, 66 000, 110 000, 132 000, and 220 000 V.

As a result of the effort to maintain the supply voltage slightly above the utilization voltage, the supply voltages were raised again to multiples of 115 V, which resulted in a new series of utilization voltages of 115, 230, 460, and 575 V, a new series of primary distribution voltages of 2300, 4600, 6900, and 13 800 V, and a new series of transmission voltages of 23 000, 34 500, 46 000, 69 000, 115 000, 138 000, and 230 000 V.

As a result of continued problems with the operation of voltage-sensitive lighting equipment and voltage insensitive motors on the same system, and the development of the 208Y/120 V network system, the supply voltages were raised again to multiples of 120 V. This resulted in a new series of utilization voltages of 120, 208Y/120, 240, 480, and 600 V, and a new series of primary distribution voltages of 2400, 4160Y/2400, 4800, 12 000, and 12 470Y/7200 V. However, most of the existing primary distribution voltages continued in use and no 120 V multiple voltages developed at the transmission level.

3.2.7 *Standard Nominal System Voltages in the United States.* The nominal system voltages listed in the center of Table 2(a) and in the left-hand column of Table 2(b) are officially designated as "standard nominal system voltages" in the United States by ANSI C84.1-1970. In addition, those shown without parentheses are designated as preferred standards to provide a long-range plan for reducing the multiplicity of voltages.

In the case of utilization voltages of 600 V and below, the associated nominal system voltages in the right-hand column are obsolete and should not be used. Manufacturers are encouraged to design utilization equipment to provide acceptable performance within the utilization voltage tolerance limits specified in the standard where possible. Some numbers listed in the right-hand column are used in equipment ratings, but these should not be confused with the numbers designating the nominal system voltage on which the equipment is designed to operate.

In the case of primary distribution voltages, the numbers in the right-hand column may designate an older system in which the voltage tolerance limits are maintained at a different level than the standard nominal system voltage, and special consideration must be given to the distribution transformer ratios, taps, and tap settings. See Sections 3.2.8 and 3.3.2.

3.2.8 *Use of Distribution Transformer Taps to Shift Utilization Voltage Spread Band.* Except for small distribution transformers in sizes of 50 to 100 kVA or less, power and distribution transformers are normally provided with four taps on the primary winding in 2½ percent steps. These taps permit the transformer ratio to be changed to raise or lower the secondary voltage spread band to provide a closer fit to the tolerance limits of the utilization equipment. There are two situations requiring the use of taps.

(1) Where the primary voltage spread band is above or below the limits required to provide a satisfactory secondary voltage spread band. This occurs under two conditions.

(a) The primary voltage has a nominal value which is slightly different from the transformer primary nameplate rating. For example, if a 13 200–480 V transformer is connected to a nominal 13 800 V system, the nominal secondary voltage would be (13 800/13 200) × 480 = 502 V. However, if the 13 800 V system were connected to the plus 5 percent tap of the 13 200–480 V transformer at 13 860 V, the nominal secondary voltage would be (13 860/13 800) × 480 = 482 V, which is practically the same as would be obtained from a transformer having the proper ratio of 13 800–480 V.

(b) The primary voltage spread is in the upper or lower portion of the tolerance limits provided in ANSI C84.1-1970. For example, a 13 200–480 V transformer is connected to a 13 200 V primary distribution system close to

Table 3
Tolerance Limits for Lighting Circuits
from Table 1, Range A, in Volts

Nominal System Voltage (volts)	Transformer Tap	Minimum Utilization Voltage (volts)	Maximum Utilization Voltage (volts)
480Y/277	Normal	440Y/254	500Y/288
468Y/270	Plus 2½%	429Y/248	488Y/281
456Y/263	Plus 5%	418Y/241	475Y/274

Table 4
Tolerance Limits for Low-Voltage
Three-Phase Motors, in Volts

Motor Rating (volts)	− 10 Percent	+ 10 Percent
460	414	506
440	396	484

Table 5
Tolerance Limits for Low-Voltage
Standard Florescent Lamp Ballasts, in Volts

Ballast Rating (volts)	− 10 Percent	+ 10 Percent
277	249	305
265	238	292

the substation so that the primary voltage spread band falls in the upper half of the tolerance zone for range A in the standard, or 13 200 to 13 860 V. This would result in a nominal secondary voltage under no-load conditions of 480 to 504 V. By setting the transformer on the plus 2½ percent tap at 13 530 V, the secondary voltage would be lowered 2½ percent to a range of 468 to 491 V. This would materially reduce the overvoltage on utilization equipment.

(2) Adjusting the utilization voltage spread band to provide a closer fit to the tolerance limits of the utilization equipment. For example, Table 3 shows the shift in the utilization voltage spread band for the plus 2½ and plus 5 percent taps as compared to the utilization voltage tolerance limits for range A of ANSI C84.1-1970 for the 480 V system. Table 4 shows the voltage tolerance limits of standard 460 and 440 V three-phase induction motors. Table 5 shows the tolerance limits for standard 277 and 265 V fluorescent lamp ballasts. A study of these three tables shows that a tap setting of "normal" will provide the best fit with the tolerance limits of the 460 V motor and the 277 V ballast, but a setting on·the plus 5 percent tap

will provide the best fit for the 440 V motor and the 265 V ballast. For buildings having appreciable numbers of both ratings of motors and ballasts, a setting on the plus 2½ percent tap may provide the best compromise.

Note that these examples assume that the tolerance limits of the supply and utilization voltages are within the tolerance limits specified in ANSI C84.1-1970. This may not be true, so the actual voltages should be measured, preferably with a seven-day chart, to obtain readings during the night and over weekends when maximum voltages occur. These actual voltages can then be used to prepare a voltage profile similar to Fig 9 to check the proposed transformer ratios and tap settings.

Where a plant has not yet been built so that actual voltages cannot be measured, the supplying utility should be requested to provide the expected spread band for the supply voltage, preferably supported by a seven-day graphic chart from the nearest available location. If the plant furnishes the distribution transformers, recommendations should also be obtained from the supplying utility on the transformer ratios, taps, and tap settings. With this information a voltage profile can be prepared to check the expected voltage spread at the utilization equipment.

Where the supplying utility offers a voltage over 600 V which differs from the standard nominal voltages listed in ANSI C84.1-1970, the supplying utility should be asked to furnish the expected tolerance limits of the supply voltage, preferably supported by a seven-day chart from a nearby location, and the recommended distribution transformer ratio and tap settings to be used to obtain a satisfactory utilization voltage range. With this information, a voltage profile for the supply voltage and utilization voltage limits can be constructed for comparison with the tolerance limits of utilization equipment. If the supply voltage offered by the utility is one of the associated nominal system voltages listed in Table 2, the taps on a standard distribution transformer will generally be sufficient to adjust the distribution transformer ratio to provide a satisfactory utilization voltage range.

Since the taps are on the primary side of the transformer, raising the tap setting by using the 2½ percent plus tap increases the transformer ratio by 2½ percent and lowers the secondary voltage spread band by 2½ percent minus the voltage drop in the transformer. Taps only serve to move the secondary voltage spread band up or down in the steps of the taps. They cannot correct for excessive spread in the supply voltage or excessive drop in the plant wiring system. If the voltage spread band at the utilization equipment falls outside the satisfactory operating range of the equipment, then action must be taken to improve voltage conditions by others means (see Section 3.7).

In general, transformers should be selected with the same primary nameplate voltage rating as the nominal voltage of the primary supply system, and the same secondary voltage rating as the nominal voltage of the secondary system. Taps should be provided at plus 2½ and plus 5 percent and at minus 2½ and minus 5 percent to allow for adjustment in either direction.

3.3 Voltage Selection

3.3.1 *Selection of Low-Voltage Utilization Voltages.* The preferred utilization voltage for industrial plants is 480Y/277 V. Three-phase power and other 480 V load is connected directly to the system at 480 V, and gaseous discharge lighting is connected phase to neutral at 277 V. Small dry-type transformers rated 480—208Y/120 V are used to provide 120 V, single phase, for convenience outlets, and 208 V, single phase and three phase, for small tools and other machinery. Where requirements are limited to 120 or 240 V, single phase, a 480—120/240 V single-phase transformer may be used.

For small industrial plants supplied at utilization voltage by a single distribution transformer, the choice of voltages is limited to those the utility will supply. However, most utilities will supply most of the standard nominal voltages listed in ANSI C84.1-1970 with the exception of 600 V, although all voltages supplied may not be available at every location.

The built-up downtown areas of most large cities are supplied from secondary networks. Originally only 208Y/120 V was available, but most utilities now provide spot networks at 480Y/277 V for large installations.

3.3.2 *Utility Service Supplied from a Medium-Voltage Primary Distribution Line.* When an industrial plant becomes too large to be supplied at utilization voltage from a single distribution transformer normally furnished by the supplying utility and located outdoors, a tap from the primary distribution line is brought into the plant as a primary service to supply distribution transformers, generally dry type or askarel-insulated type suitable for indoor installation. Generally these distribution transformers are combined with primary and secondary switching and protective equipment to become unit substations. They are designated as primary unit substations when the secondary voltage is over 1000 V and secondary unit substations when the secondary voltage is 1000 V and below. Primary distribution may also be used to supply industrial plants involving more than one building. In this case the primary distribution line may be run overhead or underground between buildings and may supply distribution transformers located outside the building or unit substations inside the building.

Originally primary distribution voltages were limited to the range from 2400 through 14 400 V, but the increase in load densities in recent years has forced utilities to limit expansion of primary distribution voltages below 15 000 V and to begin converting transmission voltages in the range from 15 000 to 50 000 V (sometimes called subtransmission voltages) to primary distribution. ANSI C84.1-1970 provides tolerance limits for primary supply voltages up through 34 500 V. ANSI C57.12.20-1974, Requirements for Overhead-Type Distribution Transformers 67 000 Volts and Below; 500 kVA and Smaller, lists overhead distribution transformers for primary voltages up through 69 000 V.

In case an industrial plant, supplied at utilization voltage from a single primary distribution transformer, contemplates an expansion which cannot be supplied from the existing transformer, a changeover to primary distribution will be required unless a separate supply to the new addition is permitted by the local electrical code enforcing authority and the higher cost resulting from separate bills from the utility is acceptable. In any case, the proposed expansion must be discussed with the supplying utility to determine whether the expansion can be supplied from the existing primary distribution system or whether the entire load can be transferred to another system. Any utility charges and the plant costs associated with the changes must be clearly established.

In general, primary distribution voltages between 15 000 and 25 000 V can be brought into a plant and handled like the lower voltages. Primary distribution voltages from 25 000 through 35 000 V will require at least a preliminary economic study to determine whether they can be brought into the plant or transformed to a lower primary distribution voltage. Voltages above 35 000 V will require transformation to a lower voltage.

In most cases, plants with loads of less than 10 000 kVA will find that 4160 V is the most economical plant primary distribution voltage and plants with loads over 20 000 kVA will find 13 800 V the most economical considering only the cost of the plant wiring and transformers. If the utility supplies a voltage in the range from 12 000 to 15 000 V, a transformation down to 4160 V at plant expense cannot normally be justified. For loads of 10 000 to 20 000 kVA, an economic study including consideration of the costs of future expansion must be made to determine the most economical primary distribution voltage.

Where overhead lines are permissible on plant property, an overhead primary distribution system may be built around the outside of the building or to separate buildings to supply utility-type outdoor equipment and transformers. This system is especially economical at voltages over 15 000 V. However, any plans for such an installation must be approved by the insurance company providing fire insurance because of rules governing the installation of oil-filled transformers close to buildings, and with the local electrical code enforcing authority because some jurisdictions consider each transformer as a separate service.

Utility primary distribution systems are almost always solidly grounded wye systems, and this must be considered in the design of the plant primary distribution system.

3.3.3 *Utility Service Supplied from Medium-Voltage or High-Voltage Transmission Lines.* Voltages on transmission lines used to supply large industrial plants range from 23 000 through 230 000 V. There is an overlap with primary distribution system voltages in the range from 23 000 through 69 000 V with voltages of 34 500 V and below tending to fall into the category of regulated primary distribution voltages and voltages above 34 500 V tending to fall into the category of unregulated transmission lines. The transmission voltage will be limited to those voltages the utility has available in the area. A substation is required to step the transmission voltage down to a primary distribution voltage to supply the distribution transformers in the plant.

(1) *The Substation is Supplied by the Industrial Plant.* Most utilities have a low rate for service from unregulated transmission lines which requires the plant to provide the substation. This permits the plant designer to select the primary distribution voltage but requires the plant personnel to assume the operation and maintenance of the substation. The substation designer should obtain from the supplying utility the voltage spread on the transmission line, and recommendations on the substation transformer ratio, tap provisions, and tap setting, and whether regulation should be provided.

With this information a voltage profile similar to Fig 9 is obtained using the actual values for the spread band of the transmission line and the estimated maximum values for the voltage drops in the substation transformer, primary distribution system, distribution transformers, and secondary distribution system to obtain the voltage spread at the utilization equipment. If this voltage spread is not within satisfactory limits, then regulations must be provided in the substation preferably by equipping the substation transformer or transformers with tap changing under load.

For plants supplied at 13 800 V the distribution transformers or secondary unit substations should have a ratio of 13 800—480Y/277 V with two 2½ percent taps, plus and minus. Where a number of medium-size motors in the few hundred horsepower range are used, a distribution transformer stepping down to 4160 or 2400 V may be more economical than supplying these motors from the 480 V system.

For plants supplied at 4160 V distribution transformers or secondary unit substations should have a ratio of 4160—480Y/277 V with two 2½ percent taps, plus and minus. Medium-size motors of a few hundred horsepower may economically be connected directly to the 4160 V system, preferably from a separate primary distribution circuit, to avoid the lighting dip which might occur when the large motors are started if distribution transformers supplying lighting equipment were connected to the same circuit.

(2) *The Substation is Supplied by the Utility.* Most utilities have a rate for power purchased at the primary distribution voltage which is higher than the rate for service at transmission voltage because the utility provides the substation. The choice of the primary distribution voltage is limited to those supplied by the particular utility, but the utility will be responsible for keeping the limits specified for service voltages in ANSI C84.1-1970. The utility should be requested to provide recommendations for the ratio of the distribution transformers or secondary unit substations, provisions for taps, and the tap settings. With this information a voltage profile similar to Fig 9 can be constructed using the estimated maximum values for the voltage drops in the primary distribu-

tion system, the transformers, and the secondary distribution system to make sure that the utilization voltages fall within satisfactory limits.

3.4 Voltage Ratings for Low-Voltage Utilization Equipment. Utilization equipment is defined as electric equipment which uses electric power by converting it into some other form of energy such as light, heat, or mechanical motion. Every item of utilization equipment must have, among other things, a nameplate listing the nominal supply voltage for which the equipment is designed. With one major exception, most utilization equipment carries a nameplate rating which is the same as the voltage system on which it is to be used, that is, equipment to be used on 120 V systems is rated 120 V (except for a few small appliances rated 117 or 118 V), for 208 V systems, 208 V, and so on. The major exception is motors and equipment containing motors. These are also about the only utilization equipment used on systems over 600 V. Single-phase motors for use on 120 V systems have been rated 115 V for many years. Single-phase motors for use on 208 V single-phase systems are rated 200 V, and for 240 V single-phase systems they are rated 230 V.

Prior to the late 1960s, low-voltage three-phase motors were rated 220 V for use on both 208 and 240 V systems, 440 V for use on 480 V systems, and 550 V for use on 600 V systems. The reason was that most three-phase motors were used in large industrial plants where relatively long circuits resulted in voltages considerably below nominal at the ends of the circuits. Also utility supply systems had limited capacity and low voltages were common during heavy load periods. As a result, the average voltage applied to three-phase motors approximated the 220, 440, and 550 V nameplate ratings.

In recent years supplying utilities have made extensive changes to higher distribution voltages. Increased load density has resulted in shorter primary distribution systems. Distribution transformers have been moved inside buildings to be closer to the load. Lower impedance wiring systems have been used in the secondary distribution system. Capacitors have been used to improve power factors. All of these changes

Table 6
Nameplate Voltage Ratings of Standard Induction Motors

Nominal System Voltage	Nameplate Voltage
Single-phase motors	
120	115
240	230
Three-phase motors	
208	200
240	230
480	460
600	575
2400	2300
4160	4000
4800	4600
6900	6600
13 800	13 200

have contributed to reducing the voltage drop in the distribution system which raised the voltage applied to utilization equipment. By the mid 1960s surveys indicated that the average voltage supplied to 440 V motors on 480 V systems was 460 V, and there were increasing numbers of complaints of overvoltages as high as 500 V during light-load periods.

At about the same time the Motor and Generator Committee of the National Electrical Manufacturers Association (NEMA) decided that the improvements in motor design and insulation systems would allow a reduction of two frame sizes for standard induction motors rated 600 V and below. However, the motor voltage tolerance would be limited to ± 10 percent of the nameplate rating. As a result the nameplate voltage rating of the new motor designated as the T frame motor was raised from the 220/440 V rating of the U frame motor to 230/460 V. Subsequently a motor rated 200 V for use on 208 V systems was added to the program. Table 6 shows the nameplate voltage ratings of standard induction motors as specified in NEMA MG1-1972.

The question has been raised why the confusion between equipment ratings and system nominal voltage cannot be eliminated by making the nameplate rating of utilization equip-

ment the same as the nominal voltage of the system on which the equipment is to be used. However, manufacturers say that the performance guarantee for utilization equipment is based on the nameplate rating and not the system nominal voltage. For utilization equipment such as motors where the performance peaks in the middle of the tolerance range of the equipment, better performance can be obtained over the tolerance range specified in ANSI C84.1-1970 by selecting a nameplate rating closer to the middle of this tolerance range.

3.5 Effect of Voltage Variations on Low-Voltage and Medium-Voltage Utilization Equipment

3.5.1 *General Effects.* When the voltage at the terminals of utilization equipment deviates from the value on the nameplate of the equipment, the performance and the operating life of the equipment are affected. The effect may be minor or serious depending on the characteristics of the equipment and the amount of the voltage deviation from the nameplate rating. Generally, performance conforms to the utilization voltage limits specified in ANSI C84.1-1970 and C84.1a-1973, but it may vary for specific items of voltage-sensitive equipment. In addition, closer voltage control may be required for precise operations.

3.5.2 *Induction Motors.* The variation in characteristics as a function of the applied voltage is given in Table 7. Motor voltages below nameplate rating result in reduced starting torque and increased full-load temperature rise. Motor voltages above nameplate rating result in increased torque, increased starting current, and decreased power factor. The increased starting torque will increase the accelerating forces on couplings and driven equipment. Increased starting current causes greater voltage drop in the supply circuit and increases the voltage dip on lamps and other equipment. In general, voltages slightly above nameplate rating have less detrimental effect on motor performance than voltages slightly below nameplate rating.

3.5.3 *Synchronous Motors.* Synchronous motors are affected in the same manner as induction motors, except that the speed remains constant (unless the frequency changes) and the maximum or pull-out torque varies directly

Table 7
General Effect of Voltage Variations on Induction-Motor Characteristics
(a) U-Frame Motors

		Voltage Variation	
Characteristic	Function of Voltage	90 Percent Voltage	110 Percent Voltage
Starting and maximum running torque	(Voltage)^2	Decrease 19%	Increase 21%
Synchronous speed	Constant	No change	No change
Percent slip	1/(Voltage)^2	Increase 23%	Decrease 17%
Full-load speed	Synchronous speed slip	Decrease 1½%	Increase 1%
Efficiency			
Full load	—	Decrease 2%	Increase ½-1%
¾ load	—	Practically no change	Practically no change
½ load	—	Increase 1-2%	Decrease 1-2%
Power factor			
Full load	—	Increase 1%	Decrease 3%
¾ load	—	Increase 2-3%	Decrease 4%
½ load	—	Increase 4-5%	Decrease 5-6%
Full-load current	—	Increase 11%	Decrease 7%
Starting current	Voltage	Decrease 10-12%	Increase 10-12%
Temperature rise, full load	—	Increase 6-7°C	Decrease 1-2°C
Maximum overload capacity	(Voltage)^2	Decrease 19%	Increase 21%
Magnetic noise—no load in particular	—	Decrease slightly	Increase slightly

with the voltage if the field voltage remains constant as in the case where the field is supplied by a generator on the same shaft with the motor. If the field voltage varies with the line voltage as in the case of a static rectifier source, then the maximum or pull-out torque varies as the square of the voltage.

3.5.4 *Incandescent Lamps.* The light output and life of incandescent filament lamps are critically affected by the impressed voltage. The variation of life and light output with voltage is given in Table 8. The figures for 125 and 130 V lamps are also included because these ratings are useful in signs and other locations where long life is more important than light output.

3.5.5 *Fluorescent Lamps.* Fluorescent lamps, unlike incandescent lamps, operate satisfactorily over a range of ± 10 percent of the ballast nameplate voltage rating. Light output varies approximately in direct proportion to the ap-

plied voltage. Thus a 1 percent increase in applied voltage will increase the light output by 1 percent and, conversely, a decrease of 1 percent in the applied voltage will reduce the light output by 1 percent. The life of fluorescent lamps is affected less by voltage variation than that of incandescent lamps.

The voltage-sensitive component of the fluorescent fixture is the ballast, a small reactor or transformer which supplies the starting and operating voltages to the lamp and limits the lamp current to design values. These ballasts may overheat when subjected to above normal voltage and operating temperature, and ballasts with integral thermal protection may be required. See NEC, Article 410.

3.5.6 *High-Intensity Discharge Lamps (Mercury, Sodium, and Metal Halide).* Mercury lamps using the conventional unregulated ballast will have a 30 percent decrease in light output for a 10 percent decrease in terminal

Table 7 (Cont'd)

(b) T-Frame Motors

Characteristic	Function of Voltage	Voltage Variation	
		90 Percent Voltage*	110 Percent Votlage*
Starting and maximum running torque	(Voltage)^2	Decrease 19%	Increase 21%
Percent slip	1/(Voltage)^2	Increase 20-30%	Decrease 15-20%
Full-load speed	Synchronous speed slip	Slight decrease	Slight increase
Efficiency			
Full load	—	Decrease 0-2%	Decrease 0-3%
¾ load	—	Practically no change	No change to slight decrease
½ load	—	Increase 0-1%	Decrease 0-5%
Power factor			
Full load	—	Increase 1-7%	Decrease 5-15%
¾ load	—	Increase 2-7%	Decrease 5-15%
½ load	—	Increase 3-10%	Decrease 10-20%
Full-load current	—	Increase 5-10%	Slight decrease to 5% increase
Starting current	Voltage	Decrease ≈ 10%	Increase ≈ 10%
Temperature rise, full load	—	Increase 10-15%	Increase 2-15%
Maximum overload capacity	(Voltage)^2	Decrease 19%	Increase 21%
Magnetic noise — no load in particular	—	Slight decrease	Slight increase

*There may be wide variations depending upon type of motor, such as drip proof (DP) or totally enclosed fan cooled (TEFC), and horsepower rating, with the smaller ratings showing the greater variations. Some data will vary according to manufacturers.

Table 8
Effect of Voltage Variations on Incandescent Lamps

Applied Voltage (volts)	Lamp Rating					
	120 V		125 V		130 V	
	Percent Life	Percent Light	Percent Life	Percent Light	Percent Life	Percent Light
105	575	64	880	55	—	—
110	310	74	525	65	880	57
115	175	87	295	76	500	66
120	100	100	170	88	280	76
125	58	118	100	100	165	88
130	34	132	59	113	100	100

voltage. If a constant wattage ballast is used the decrease in light output for a 10 percent decrease in terminal voltage will be about 2 percent.

Mercury lamps require 4 to 8 minutes to suitably vaporize the mercury in the lamp and reach full brilliancy. At about 20 percent undervoltage, the mercury arc will be extinguished and the lamp cannot be restarted until the mercury condenses; this takes from 4 to 8 minutes unless the lamps have special cooling controls. The lamp life is related inversely to the number of starts so that, if low-voltage conditions require repeated starting, lamp life will be reduced. Excessively high voltage raises the arc temperature which could damage the glass enclosure if the temperature approaches the glass softening point.

Sodium and metal halide lamps have similar characteristics to mercury lamps although the starting and operating voltages may be somewhat different so that the lamps and ballasts may not be interchangeable. See the manufactuers' catalogs for detailed information.

3.5.7 *Infrared Heating Processes.* Although the filaments in the lamps used in these installations are of the resistance type, the energy output does not vary with the square of the voltage because the resistance varies at the same time. The energy output does vary roughly as some power of the voltage slightly less than the square, however. Voltage variations can produce unwanted changes in the process heat available unless thermostatic control or other regulating means are used.

3.5.8 *Resistance Heating Devices.* The energy input and, therefore, the heat output of resistance heaters varies, in general, with the square of the impressed voltage. Thus a 10 percent drop in voltage will cause a drop of 19 percent in heat output. This, however, holds true only for an operating range over which the resistance remains essentially constant.

3.5.9 *Electron Tubes.* The current-carrying ability or emission of all electron tubes is affected seriously by voltage deviation from rating. The cathode life curve indicates that the life is reduced by half for each 5 percent increase in cathode voltage. This is due to the re-

duced life of the heater element and to the higher rate of evaporation of the active material from the surface of the cathode. It is extremely important that the cathode voltage be kept near rating on electron tubes for satisfactory service. In many cases this will necessitate a regulated power source. This may be located at or within the equipment, and often consists of a regulating transformer having constant output voltage or current.

3.5.10 *Capacitors.* The reactive power output of capacitors varies with the square of the impressed voltage. A drop of 10 percent in the supply voltage, therefore, reduces the reactive power output by 19 percent, and where the user has made a sizable investment in capacitors for power factor improvement, he loses the benefit of almost 20 percent of this investment.

3.5.11 *Solenoid-Operated Devices.* The pull of alternating-current solenoids varies approximately as the square of the voltage. In general, solenoids are designed to operate satisfactorily on 10 percent overvoltage and 15 percent undervoltage.

3.5.12 *Solid-State Equipment.* Thyristors, transistors, and other solid-state devices have no thermionic heaters. Thus they are not nearly as sensitive to long-time voltage variations as the electron tube components they are largely replacing. Internal voltage regulators are frequently provided for sensitive equipment such that it is independent of supply system regulation. This equipment as well as power solid-state equipment is, however, generally limited regarding peak reverse voltage, since it can be adversely affected by abnormal voltages of even microsecond duration. An individual study of the maximum voltage of the equipment, including surge characteristics, is necessary to determine the effect of maximum system voltage or whether abnormally low voltage will result in malfunction.

3.6 Voltage Drop Considerations in Locating the Low-Voltage Secondary Distribution System Power Source. One of the major factors in the design of the secondary distribution system is the location of the power source as close as possible to the center of the load. This applies in every

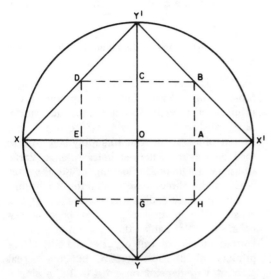

Fig 10
Effect of Secondary Distribution System
Power Source Location on Area which
Can Be Supplied under Specified
Voltage Drop Limits

Table 9
Areas that Can Be Supplied for
Specific Voltage Drops and Voltages at
Various Secondary Distribution System
Power Source Locations

| Nominal System Voltage (volts) | Distance (feet) | | | |
| | 5 % Voltage Drop | | 2½% Voltage Drop | |
	0X	0A	0X	0A
120/240	360	180	180	90
208	312	156	156	78
240	360	180	180	90
480	720	360	360	180

case, from a service drop from a distribution transformer on the street to a distribution transformer located outside the building or a secondary unit substation located inside the building. Frequently building esthetics or available space require the secondary distribution system power supply to be installed in a corner of a building without regard to what this adds to the cost of the building wiring to keep the voltage drop within satisfactory limits.

Fig 10 [1] shows that if a power supply is located in the center of a horizontal floor area at point 0, the area that can be supplied from circuits run radially from point 0 with specified circuit constants and voltage drop would be the area enclosed by the circle of radius 0—X. However, conduit systems are run in rectangular coordinates, so with this restriction, the area which can be supplied is reduced to the square X—Y—X'—Y' when the conduit system is run parallel to the axes X—X' and Y—Y'. But the limits of the square are not parallel to the conduit system. Thus to fit the conduit system into a square building with walls parallel to the conduit system, the area must be reduced to F—H—B—D.

If the supply point is moved to the center of one side of the building, which is a frequent situation when the transformer is placed outside the building, the area which can be served with the specified voltage drop and specified circuit constants is E—A—B—D. If the supply station is moved to a corner of the building, a frequent location for buildings supplied from the rear or from the street, the area is reduced to 0—A—B—C.

Every effort should be made to place the secondary distribution system supply point as close as possible to the center of the load area. Note that this study is based on a horizontal wiring system and any vertical components must be deducted to establish the limits of the horizontal area which can be supplied.

Using an average value of 30 ft per volt drop for a fully loaded conductor, which is a good average figure for the conductor sizes normally used for feeders, the distances in Fig 10 for 5 and 2½ percent voltage drops are shown in Table 9. For a distributed load the distances will be approximately twice the values shown.

3.7 Improvement of Voltage Conditions. Poor equipment performance, overheating, nuisance tripping of overcurrent protective devices, and excessive burnouts are signs of unsatisfactory voltage. Abnormally low voltage occurs at the end of long circuits. Abnormally high voltage occurs at the beginning of circuits close to the source of supply, especially under lightly loaded conditions such as at night and over weekends.

In cases of abnormally low voltage, the first step is to make a load survey to measure the current taken by the affected equipment, the current in the circuit supplying the equipment, and the current being supplied by the secondary unit substation under peak-load conditions to make sure that the abnormally low voltage is not due to overloaded equipment. If the abnormally low voltage is due to overload, then corrective action must be taken to relieve the overloaded equipment.

If overload is ruled out or if the utilization voltage is excessively high, a voltage survey should be made, preferably by using graphic voltmeters, to determine the voltage spread at the utilization equipment under all load conditions and the voltage spread at the utility supply. This survey can be compared with ANSI C84.1-1970 to determine if the unsatisfactory voltage is caused by the plant distribution system or the utility supply. If the utility supply exceeds the tolerance limits specified in ANSI C84.1-1970, the utility should be notified. If the industrial plant is supplied at a transmission voltage and furnishes the distribution substation, the operation of the voltage regulators should be checked.

If excessively low voltage is caused by excessive voltage drop in the plant wiring (over 5 percent), then additional circuits must be run in parallel with the affected circuits, or to supply portions of the affected equipment in order to reduce the voltage drop to suitable values. If the load power factor is low, capacitors may be installed to improve the power factor and reduce the voltage drop. Where the excessively low voltage affects a large area, the best solution may be to go to primary distribution if the building is supplied from a single distribution transformer, or to install an additional distribution transformer in the center of the affected

area if the plant has primary distribution. Plants wired at 208Y/120 or 240 V may be changed over economically to 480Y/277 V if an appreciable portion of the wiring system is rated 600 V and motors are dual rated 220 × 440 or 230 × 460 V.

3.8 Phase-Voltage Unbalance in Three-Phase Systems

3.8.1 *Causes of Phase-Voltage Unbalance.* Most utilities use four-wire grounded-wye primary distribution systems so that single-phase distribution transformers can be connected phase to neutral to supply single-phase load such as residences and street lights. Variations in single-phase loading cause the currents in the three-phase conductors to be different, producing different voltage drops and causing the phase voltages to become unbalanced. Normally the maximum phase-voltage unbalance will occur at the end of the primary distribution system, but the actual amount will depend on how well the single-phase loads are balanced between the phases on the system.

Perfect balance can never be maintained because the loads are continually changing, causing the phase-voltage unbalance to vary continually. Blown fuses on three-phase capacitor banks will also unbalance the load and cause phase-voltage unbalance.

Industrial plants make extensive use of 480Y/277 V utilization voltage to supply lighting loads connected phase to neutral. Proper balancing of single-phase loads among the three phases on both branch circuits and feeders is necessary to keep the load unbalance and the corresponding phase-voltage unbalance within reasonable limits.

3.8.2 *Measurement of Phase-Voltage Unbalance* [2]. The simplest method of expressing the phase-voltage unbalance is to measure the voltages in each of the three phases:

$$\text{phase-voltage unbalance} = \frac{\text{maximum deviation from average phase voltage}}{\text{average phase voltage}}$$

The amount of voltage unbalance is better expressed in symmetrical components as the

Table 10
Effect of Phase-Voltage Unbalance on Motor Temperature Rise

Motor Type	Load	Percent Voltage Unbalance	Percent Added Heating	Insulation System Class	Temperature Rise (°C)
U frame	Rated	0	0	A	60
	Rated	2	8	A	65
	Rated	3½	25	A	75
T frame	Rated	0	0	B	80
	Rated	2	8	B	86.4
	Rated	3½	25	B	100

negative sequence component of the voltage:

$$\text{voltage-unbalance factor} = \frac{\text{negative-sequence voltage}}{\text{positive-sequence voltage}}$$

3.8.3 *Effect of Phase-Voltage Unbalance.* When unbalanced phase voltages are applied to three-phase motors, the phase-voltage unbalance causes additional negative-sequence currents to circulate in the motor, increasing the heat losses primarily in the rotor. The most severe condition occurs when one phase is opened and the motor runs on single-phase power. A recent paper [3] published a table showing the increased temperature rise which occurs for specified values of phase-voltage unbalance for both U frame and T frame motors (Table 10). A later paper [4] provides a more comprehensive review of the problem.

Although there will generally be an increase in the motor load current when the phase voltages are unbalanced, the increase is insufficient to indicate the actual temperature rise which occurs because standard current-responsive thermal or magnetic overload devices only provide a trip characteristic that correlates with the motor thermal damage due to normal overload current (positive sequence) and not negative-sequence current.

All motors are sensitive to phase-voltage unbalance, but sealed compressor motors used in air conditioners are most susceptible to this condition. These motors operate with higher current densities in the windings because of the added cooling effect of the refrigerant. Thus the same percent increase in the heat loss due to circulating currents, caused by phase-voltage unbalance, will have a greater effect on the sealed compressor motor than it will on a standard air-cooled motor.

Since the windings in sealed compressor motors are inaccessible, they are normally protected by thermally operated switches embedded in the windings, set to open and disconnect the motor when the winding temperature exceeds the set value. The motor cannot be restarted until the winding has cooled down to the point at which the thermal switch will reclose.

When a motor trips out, the first step in determining the cause is to check the running current after it has been restarted to make sure that the motor is not overloaded. The next step is to measure the three phase voltages to determine the amount of phase-voltage unbalance. Table 10 indicates that where the phase-voltage unbalance exceeds 2 percent, the motor is likely to become overheated if it is operating close to full load.

Some electronic equipment such as computers may also be affected by phase-voltage unbalance of more than 2 or 2½ percent. The equipment manufacturer can supply the necessary information.

In general, single-phase loads should not be connected to three-phase circuits suppling equipment sensitive to phase-voltage unbalance. A separate circuit should be used to supply this equipment.

3.9 Voltage Dips and Flicker. The previous discussion has covered the relatively slow changes in voltage associated with steady-state voltage spreads and tolerance limits. However, certain types of utilization equipment such as motors have a high initial inrush current when turned on and impose a heavy load at a low power factor for a very short time. This sudden increase in the current flowing to the load causes a momentary increase in the voltage drop along the distribution system and a corresponding reduction in the voltage at the utilization equipment. A voltage drop of more than ¼ percent will cause a noticeable reduction in the light output of an incandescent lamp and a less noticeable reduction in the light output of gaseous discharge lighting equipment.

In general the starting current of a standard motor averages about 5 times the full-load running current. The approximate values for all alternating-current motors over ½ hp are indicated by a code letter on the nameplate of the motor. The values indicated by these code letters are given in NEMA MG1-1972 and also in Article 430 of the NEC.

A motor requires about 1 kVA for each motor horsepower in normal operation; so the starting current of the average motor will be about 5 kVA for each motor horsepower. When the motor rating in horsepower approaches 5 percent of the secondary unit substation transformer capacity in kilovolt-amperes, the motor starting apparent power approaches 25 percent of the transformer capacity which, with a transformer impedance voltage of 6 to 7 percent, will result in a noticeable voltage dip on the order of 1 percent.

In addition, a similar voltage dip will occur in the wiring between the secondary unit substation and the motor when starting a motor with a full-load current which is on the order of 5 percent of the rated current of the circuit. This will result in a full-load voltage drop on the order of 4 or 5 percent. However, the voltage drop is distributed along the circuit so that maximum dip occurs only when the motor and the affected equipment are located at the far end of the circuit. As the motor is moved from the far end to the beginning of the circuit, the voltage drop in the circuit approaches zero. As the affected equipment is moved from the far end to the beginning of the circuit, the voltage dip remains constant up to the point of connection of the motor and then decreases to zero as the equipment connection approaches the beginning of the circuit.

The total voltage dip is the sum of the dip in the secondary unit substation transformer and the secondary circuit. In the case of very large motors of several hundred to a few thousand horsepower, the impedance of the supply system must be considered.

Where loads are turned on and off rapidly as in the case of resistance welders, or fluctuate rapidly as in the case of arc furnaces, the rapid fluctuations in the light output of incandescent lamps, and to a lesser extent, gaseous discharge lamps, is called flicker. When flicker continues over an appreciable period, voltage variations as low as ½ percent may be objectionable. If utilization equipment involving rapidly fluctuating loads is on the order of 10 percent of the capacity of the secondary unit substation transformer and the secondary circuit, accurate calculations should be made using the actual load currents and system impedances to determine the effect on lighting equipment.

Fig 11 shows a chart used by many utilities to determine whether voltage fluctuations will cause objectionable fluctuations in the light output of incandescent lamps. The border line of irritation curve starts with a voltage change of 1½ percent at a frequency of 20 fluctuations per second, which is the maximum frequency at which individual fluctuations can be detected, decreases to a minimum of ½ percent at 6 fluctuations per second, and then increases to 6 percent for 1 fluctuation per hour. The range between permissible flicker and objectionable flicker is due to the fact that some people are bothered more than others. Also, the effect of flicker depends upon lighting intensity and working conditions.

In using this curve, the purpose for which the lighting is provided must be considered. For example, lighting used for close work such as drafting requires flicker limits approaching the border line of visibility curve. For general area lighting such as storage areas, the flicker limits may approach the border line of irritation curve. Note that the effect of voltage dips de-

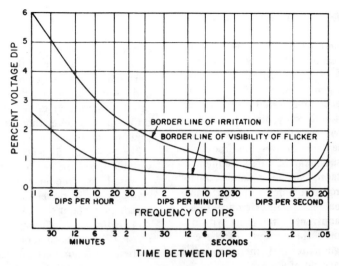

Fig 11
Effect of Recurrent Voltage Dips on Incandescent Lamps

pends on the frequency of occurrence. An occasional dip even though quite large is rarely objectionable.

When objectionable flicker occurs, either the load causing the flicker must be reduced or eliminated, or the capacity of the supply system must be increased to reduce the voltage drop caused by the fluctuating load. In large plants, flicker-producing equipment should be segregated on separate transformers and feeders so as not to disturb flicker-sensitive equipment.

Objectionable dips in the supply voltage from the utility should be reported to the utility for correction.

3.10 Harmonics

3.10.1 *Nature of Harmonics.* Harmonics are integral multiples of the fundamental frequency. For example, for 60 Hz power systems, the second harmonic would be 2 × 60 or 120 Hz and the third harmonic would be 3 × 60 or 180 Hz.

Harmonics are caused by devices which change the shape of the normal sine wave of voltage or current in synchronism with the 60 Hz supply. In general those include three-phase devices in which the three phase coils are not exactly symmetrical, and single- and three-phase loads in which the load impedance

changes during the voltage wave produce a distorted current wave such as the magnetizing current in a coil with an iron core. It can be shown that a distorted wave must be made up of a fundamental and harmonics of various frequencies and magnitudes.

Inductive reactance varies directly as the frequency so that the current in an inductive circuit is reduced in proportion to the frequency for a given harmonic voltage. Conversely, capacitive reactance varies inversely as the frequency so that the current in a capacitive circuit is increased in proportion to the frequency for a given harmonic voltage. If the inductive reactance and the capacitive reactance in a series circuit are the same, they will cancel each other, and a given harmonic voltage will cause a large current to flow limited only by the resistance of the circuit. This condition is called resonance, and is more likely to occur at the higher harmonic frequencies.

3.10.2 *Characteristics of Harmonics.* The harmonic content and magnitude existing in any power system is largely unpredictable and effects will vary widely in different parts of the same system because of the different effects of different frequencies. Since the distorted wave is in the supply system, harmonic effects may occur at any point on the system where the distorted wave exists; this is not limited to the immediate vicinity of the harmonic-producing device. Where power is converted to direct current or some other frequency, harmonics will exist in any distorted alternating component of the converted power.

Harmonics may be transferred from one circuit or system to another by direct connection or by inductive or capacitive coupling. Since 60 Hz harmonics are in the low-frequency audio range, the transfer of these frequencies into communication, signaling, and control circuits employing frequencies in the same range may cause objectional interference. In addition, harmonic currents circulating within a power circuit reduce the capacity of the current-carrying equipment and increase losses without providing any useful work.

3.10.3 *Harmonic-Producing Equipment*

(1) *Arc Equipment.* Arc furnaces and arc welders have a changing load characteristic during each half-cycle which demands harmonic currents from the supply system. Normally these do not cause very much trouble unless the supply conductors are in close proximity to communication and control circuits or there are large capacitor banks on the system.

(2) *Gaseous-Discharge Lamps.* Fluorescent and mercury lamps produce small arcs and, in combination with the ballast, produce harmonics, particularly the third. Experience shows that the third-harmonic current may be as high as 30 percent of the fundamental in the phase conductors and up to 90 percent in the neutral where the third harmonics from each phase add directly since they are displaced one third of a cycle. This is why the NEC requires a full neutral for circuits supplying this type of load.

(3) *Rectifiers.* Half-wave rectifiers which suppress alternate half-cycles of current generate both even- and odd-numbered harmonics. Full-wave rectifiers tend to eliminate the even-numbered harmonics and usually diminish the magnitude of the odd-numbered harmonics.

The major producer of harmonics is the controlled rectifier which chops the alternating-current wave, particularly near the peak of the cycle. Since the wave shapes of both the input and the output depend upon the control setting

to start rectification, both the shape of the input and output waves and hence the frequency and magnitude of the harmonics will vary with the setting of the control. Large rectifiers which are frequently supplied from six- and twelve-phase transformer connections to produce smoother direct current will produce different harmonics than those supplied from a three-phase system.

Phase-controlled rectifiers used to provide variable-speed drives for direct-current motors, or used as frequency changers to provide variable-speed drives for alternating-current motors, are major sources of harmonics.

(4) *Rotating Machinery.* Normally the three phase coils of both motors and generators are sufficiently symmetrical that any harmonic voltages generated from lack of symmetry are too small to cause any interference. The nonlinear characteristics of the stator iron can produce appreciable harmonics, especially at high flux densities.

(5) *Induction Heaters.* Induction heaters use 60 Hz or higher frequency power to induce circulating currents in metals to heat the metal. Harmonics are generated by the interaction of the magnetic fields caused by the current in the induction heating coil and the circulating currents in the metal being heated. Large induction heating furnaces may create objectionable harmonics.

(6) *Capacitors.* Capacitors do not generate harmonics. However, the reduced reactance of the capacitor to the higher frequencies magnifies the harmonic current in the circuit containing the capacitors. In cases of resonance, this magnification may be very large. High harmonic currents may overheat the capacitors. In addition, the high currents may induce interference with communication, signal, and control circuits.

Special capacitors prescribed by equipment manufacturers are required to perform satisfactorily under actual operating conditions. Thus the manufacturer must be furnished the harmonic voltage content of the power supply to determine the correct type to be used.

3.10.4 *Reduction of Harmonic Effects.* Where harmonic interference exists, the regular measures of increasing the separation between the power and communication conductors and the

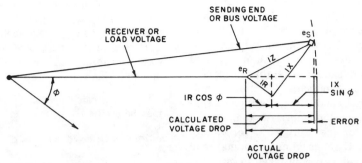

Fig 12
Phasor Diagram of Voltage Relations for Voltage-Drop Calculations

use of shielded communication conductors should be considered. Where capacitor banks magnify the harmonic current, the capacitors should be changed to suitable types or removed. Where resonant conditions exist, the capacitor bank should be changed in size to shift the resonant point to another frequency. Where harmonics pass from a power system to a communications, signal, or control circuit through a direct connection such as a power supply, filters may be required to suppress or short-circuit the harmonic frequencies.

During preliminary meetings with the supplying utility the anticipated harmonic analysis of the power supply should be determined. This information coupled with that provided by the manufacturer of any equipment to be installed that may generate a voltage distortion (along with appropriate safety factors) can be used to govern the specifications or the application of other equipment that may be exposed to the harmonic voltage condition.

3.11 Calculation of Voltage Drops. Building wiring designers must have a working knowledge of voltage drop calculations, not only to meet NEC requirements, but also to ensure that the voltage applied to utilization equipment is maintained within proper limits. Due to the phasor relationships between voltage and current and resistance and reactance, voltage drop calculations require a working knowledge of trigonometry, especially for making exact computations. Fortunately most voltage drop calculations are based on assumed limiting conditions and approximate formulas are adequate.

3.11.1 *General Mathematical Formulas.* The phasor relationships between the voltage at the beginning of a circuit, the voltage drop in the circuit, and the voltage at the end of the circuit are shown in Fig 12.

The approximate formula for the voltage drop is

$$V = IR \cos \phi + IX \sin \phi$$

where

V = Voltage drop in circuit, line to neutral

I = Current flowing in conductor

R = Line resistance for one conductor, in ohms

X = Line reactance for one conductor, in ohms

ϕ = Angle whose cosine is the load power factor

$\cos \phi$ = Load power factor, in decimals

$\sin \phi$ = Load reactive factor, in decimals

The voltage drop V obtained from this formula is the voltage drop in one conductor, one way, commonly called the line-to-neutral voltage drop. The reason for using the line-to-neutral voltage is to permit the line-to-line voltage to be computed by multiplying by the following constants:

Voltage System	Multiply By
Single-phase	2
Three-phase	1.732

In using this formula, the line current I is generally the maximum or assumed load current-carrying capacity of the conductor.

The resistance R is the alternating-current resistance of the particular conductor used and the particular type of raceway in which it is installed as obtained from the manufacturer. It depends on the size of the conductor measured in American wire gauge (AWG) for smaller conductors and in thousands of circular mils (kcmil) for larger conductors, the type of conductor (copper or aluminum), the temperature of the conductor (normally 75°C for average loading and 90°C for maximum loading), and whether the conductor is installed in magnetic (steel) or nonmagnetic (aluminum or nonmetallic) raceway. The resistance opposes the flow of current and causes the heating of the conductor.

The reactance X is obtained from the manufacturer. It depends on the size and material of the conductor, whether the raceway is magnetic or nonmagnetic, and on the spacing between the conductors of the circuit. The spacing is fixed for multiconductor cable, but may vary with single-conductor cables so that an average value must be used. Reactance occurs because the alternating current flowing in the conductor causes a magnetic field to build up and collapse around each conductor in synchronism with the alternating current. This magnetic field, as it builds up and falls radially, cuts across the conductor itself and the other conductors of the circuit, causing a voltage to be induced into each in the same way that current flowing in the primary of a transformer induces a voltage in the secondary of the transformer. Since the induced voltage is proportional to the rate of change of the magnetic field which is maximum when the current passes through zero, the induced voltage will be a maximum when the current passes through zero or, in vector terminology, lags the current wave by 90 degrees.

ϕ is the angle between the load voltage and the load current and is obtained by looking up the power factor expressed as a decimal (1 or less) in the cosine section of a trigonometric table or slide rule.

Cos ϕ is the power factor of the load expressed in decimals and may be used directly in the computation of IR cos ϕ.

Sin ϕ is obtained by looking up the angle ϕ in a trigonometric table of sines, or obtaining it from a calculator. By convention, sin ϕ is positive for lagging power factor loads and negative for leading power factor loads.

IR cos ϕ is the resistance component of the voltage drop and IX sin ϕ is the reactive component of the voltage drop.

For exact calculations, the following formula may be used:

$$e_R = e_S + IR \cos \phi + IX \sin \phi$$
$$- \sqrt{e_S^2 - (IX \cos \phi - IR \sin \phi)^2}$$

where the symbols correspond to those in Fig 12.

3.11.2 *Cable Voltage Drop.* Voltage drop tables and charts are sufficiently accurate to determine the approximate voltage drop for most problems. Table 11 contains four sections giving the three-phase line-to-line voltage drop for 10 000 circuit A · ft for copper and aluminum conductors in both magnetic and nonmagnetic conduit. The figures are for single-conductor cables operating at 60°C. However, the figures are reasonably accurate up to a conductor temperature of 75°C and for multiple-conductor cable. Although the length of cable runs over 600 V is generally too short to produce a significant voltage drop, Table 11 may be used to obtain approximate values. For borderline cases the exact values obtained from the manufacturer for the particular cable should be used. The resistance is the same for the same wire size, regardless of the voltage, but the thickness of the insulation is increased at the higher voltages, which increases the conductor spacing resulting in increased reactance causing increasing errors at the lower power factors. For the same reason, Table 11 cannot be used for open-wire or other installations such as trays where there is appreciable spacing between the individual phase conductors.

In using Table 11, the normal procedure is to look up the voltage drop for 10 000 A · ft and multiply this value by the ratio of the actual number of ampere-feet to 10 000. Note that the distance in feet is the distance from the source to the load.

Example 1. 500 kcmil copper conductor in steel (magnetic) conduit; circuit length 200 ft;

Table 11
Three-Phase Line-to-Line Voltage Drop for 600 V Single-Conductor Cable per 10 000 A · ft 60°C Conductor Temperature, 60 Hz

Load Power Factor Lagging	Wire Size (AWG or kcmil)																						
	1000	900	800	750	700	600	500	400	350	300	250	4/0	3/0	2/0	1/0	1	2	4	6	8*	10*	12*	14*
Section 1: Copper Conductors in Magnetic Conduit																							
1.00	0.28	0.31	0.34	0.35	0.37	0.42	0.50	0.60	0.68	0.78	0.92	1.1	1.4	1.7	2.1	2.6	3.4	5.3	8.4	13	21	33	53
0.95	0.50	0.52	0.55	0.57	0.59	0.64	0.71	0.81	0.88	1.0	1.1	1.3	1.5	1.9	2.3	2.8	3.5	5.3	8.2	13	20	32	50
0.90	0.57	0.59	0.62	0.64	0.66	0.71	0.78	0.88	0.95	1.1	1.2	1.3	1.6	1.9	2.3	2.8	3.4	5.2	8.0	12	19	30	48
0.80	0.66	0.68	0.71	0.73	0.74	0.80	0.85	0.95	1.0	1.1	1.2	1.4	1.6	1.9	2.3	2.6	3.2	4.8	7.3	11	17	27	43
0.70	0.71	0.73	0.76	0.78	0.80	0.83	0.88	0.97	1.0	1.1	1.2	1.3	1.5	1.8	2.1	2.5	3.0	4.4	6.6	9.9	15	24	38
Section 2: Copper Conductors in Nonmagnetic Conduit																							
1.00	0.23	0.26	0.28	0.29	0.33	0.38	0.45	0.55	0.62	0.73	0.88	1.0	1.3	1.6	2.1	2.6	3.3	5.3	8.4	13	21	33	53
0.95	0.40	0.43	0.45	0.47	0.50	0.54	0.62	0.71	0.80	0.92	1.0	1.1	1.5	1.8	2.2	2.7	3.4	5.3	8.2	13	20	32	50
0.90	0.47	0.48	0.52	0.54	0.55	0.59	0.68	0.76	0.85	0.95	1.1	1.1	1.5	1.8	2.2	2.7	3.3	5.1	7.9	12	19	30	48
0.80	0.54	0.55	0.57	0.59	0.62	0.66	0.73	0.81	0.88	0.97	1.1	1.1	1.4	1.7	2.1	2.5	3.1	4.7	7.2	11	17	27	43
0.70	0.57	0.59	0.62	0.64	0.66	0.69	0.74	0.83	0.88	0.97	1.1	1.1	1.4	1.6	2.0	2.4	2.8	4.3	6.4	9.7	15	24	38
Section 3: Aluminum Conductors in Magnetic Conduit																							
1.00	0.42	0.45	0.49	0.52	0.55	0.63	0.74	0.91	1.0	1.2	1.4	1.7	2.1	2.6	3.3	4.2	5.2	8.4	13	21	33	52	—
0.95	0.62	0.65	0.70	0.73	0.76	0.83	0.94	1.1	1.2	1.4	1.6	1.8	2.3	2.7	3.4	4.2	5.3	8.2	13	20	32	50	—
0.90	0.69	0.72	0.76	0.79	0.82	0.88	0.99	1.2	1.3	1.4	1.6	1.9	2.3	2.7	3.4	4.1	5.1	7.9	12	19	30	48	—
0.80	0.76	0.80	0.83	0.85	0.88	0.95	1.0	1.2	1.3	1.4	1.6	1.8	2.2	2.6	3.2	3.9	4.7	7.3	11	17	27	43	—
0.70	0.80	0.83	0.87	0.89	0.92	0.98	1.1	1.2	1.3	1.4	1.6	1.7	2.1	2.4	2.9	3.6	4.3	6.5	10	15	24	37	—
Section 4: Aluminum Conductors in Nonmagnetic Conduit																							
1.00	0.36	0.39	0.44	0.47	0.51	0.59	0.70	0.88	1.0	1.2	1.4	1.7	2.1	2.6	3.3	4.2	5.2	8.4	13	21	33	52	—
0.95	0.52	0.56	0.60	0.63	0.67	0.74	0.85	1.0	1.1	1.3	1.5	1.8	2.2	2.7	3.4	4.2	5.0	8.2	13	20	32	50	—
0.90	0.57	0.61	0.65	0.68	0.71	0.79	0.89	1.1	1.2	1.3	1.5	1.8	2.2	2.6	3.3	4.1	5.0	7.9	12	19	30	48	—
0.80	0.63	0.66	0.71	0.73	0.76	0.83	0.92	1.1	1.2	1.3	1.5	1.7	2.1	2.5	3.1	3.8	4.6	7.2	11	17	27	42	—
0.70	0.66	0.69	0.73	0.75	0.78	0.83	0.92	1.1	1.2	1.3	1.4	1.6	1.7	2.3	2.8	3.4	4.2	6.4	9.9	15	24	37	—

*Solid Conductor. Other conductors are stranded.

To convert voltage drop to	Multiply by
Single phase, three wire, line to line	1.18
Single phase, three wire, line to neutral	0.577
Three phase, line to neutral	0.577

load 300 A at 80 percent power factor. What is the voltage drop?

Using Section 1 of Table 11, the intersection between 500 kcmil and 80 percent power factor gives a voltage drop of 0.85 V for 10 000 A · ft.

$$200 \text{ ft} \times 300 \text{ A} = 60\ 000 \text{ circuit A} \cdot \text{ft}$$

$$(60\ 000/10\ 000) \times 0.85 = 6 \times 0.85 = 5.1 \text{ V drop}$$

voltage drop, phase to neutral = 0.577×5.1

$$= 2.9 \text{ V}$$

Example 2. No 12 aluminum conductor in aluminum (nonmagnetic) conduit; circuit length 200 ft; load 10 A at 70 percent power factor. What is the voltage drop?

Using Section 4 of Table 11, the intersection between No 12 aluminum conductor and 0.70 power factor is 37 V for 10 000 A · ft.

$$200 \text{ ft} \times 10 \text{ A} = 2000 \text{ circuit A} \cdot \text{ft}$$

$$\text{voltage drop} = (2000/10\ 000) \times 37$$

$$= 7.4 \text{ V}$$

Example 3. Determine the wire size in Example 2 to limit the voltage drop to 3 V.

The voltage drop in 10 000 A · ft would be

$$(10\ 000/2\ 000) \times 3 = 15 \text{ V}$$

Using Section 4 of Table 11, move along the 0.70 power factor line to find the voltage drop not greater than 15 V. No 8 aluminum has a voltage drop of 15 V for 10 000 A · ft, so it is the smallest aluminum conductor in aluminum conduit which could be used to carry 10 A for 200 ft with a voltage drop of not more than 3 V, line to line.

3.11.3 *Busway Voltage Drop.* See Chapter 11 for busway voltage drop tables and related information.

3.11.4 *Transformer Voltage Drop.* Voltage-drop curves in Figs 13 and 14 may be used to determine the approximate voltage drop in single-phase and three-phase 60 Hz liquid-filled, self-cooled, and dry transformers. The voltage drop through a single-phase transformer is found by entering the chart at a kilovolt-ampere rating three times that of the single-phase transformer. Fig 13 covers transformers in the following ranges:

(1) *Single-Phase*
250—500 kVA, 8.6—15 kV insulation classes
833—1250 kVA, 5—25 kV insulation classes
(2) *Three-Phase*
225—750 kVA, 8.6—15 kV insulation classes
1000—10 000 kVA, 5—25 kV insulation classes

An example of the use of the chart is given in the following.

Example. Find the voltage drop in a 2000 kVA three-phase 60 Hz transformer rated 4160—480 V. The load is 1500 kVA at 0.85 power factor.

Solution. Enter the chart on the horizontal scale at 2000 kVA, extend a line vertically to its intersection with the 0.85 power factor curve. Extend a line horizontally from this point to the left to its intersection with the vertical scale. This point on the vertical scale gives the percent voltage drop for rated load. Multiply this value by the ratio of actual load to rated load:

$$\text{percent drop at rated load} = 3.67$$

$$\text{percent drop at 1500 kVA} = 3.67 \times \frac{1500}{2000}$$

$$= 2.75$$

$$\text{actual voltage drop} = 2.75 \text{ percent} \times 480$$

$$= 13.2 \text{ V}$$

Fig 14 applies to the 34.5 kV insulation class power transformer in ratings from 1500 to 10 000 kVA. These curves can be used to determine the voltage drop for transformers in the 46 and 69 kV insulation classes by using appropriate multipliers at all power factors except unity.

To correct for 46 kV, multiply the percent voltage drop obtained from the chart by 1.065 and for 69 kV, multiply by 1.15.

3.11.5 *Motor-Starting Voltage Drop.* It is characteristic of most alternating-current motors that the current which they draw on starting is much higher than their normal running current. Synchronous and squirrel-cage induction motors started on full voltage may draw a current as high as seven or eight times their full-load running current. This sud-

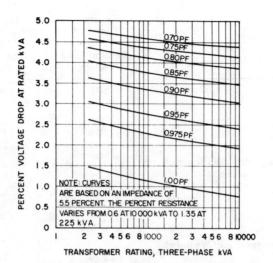

**Fig 13
Approximate Voltage Drop Curves for Three-Phase
Transformers, 225—10 000 kVA, 5—25 kV**

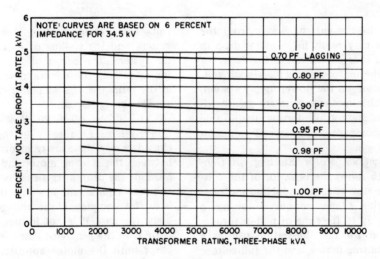

**Fig 14
Approximate Voltage Drop Curves for Three-Phase
Transformers, 1500—10 000 kVA, 34.5 kV**

Table 12
Comparison of Motor-Starting Methods

Type of Starter (Settings Given Are the More Common for Each Type)	Motor Terminal Voltage (Percent Line Voltage)	Starting Torque (Percent Full-Voltage Starting Torque)	Line-Current (Percent Full-Voltage Starting Current)
Full-voltage starter	100	100	100
Autotransformer			
80-percent tap	80	64	68
65-percent tap	65	42	46
50-percent tap	50	25	30
Resistor starter, single step (adjusted for motor voltage to be 80 percent of line voltage)	80	64	80
Reactor			
50-percent tap	50	25	50
45-percent tap	45	20	45
37.5-percent tap	37.5	14	37.5
Part-winding starter (low-speed motors only)			
75-percent winding	100	75	75
50-percent winding	100	50	50

NOTE: For a line voltage not equal to the motor rated voltage multiply all values in the first and last columns by the ratio (actual voltage)/(motor rated voltage). Multiply all values in the second column by the ratio $[(\text{actual voltage})/(\text{motor rated voltage})]^2$.

den increase in the current drawn from the power system may result in excessive drop in voltage unless it is considered in the design of the system. The motor-starting load in kilovolt-amperes, imposed on the power supply system, and the available motor torque are greatly affected by the method of starting used. Table 12 gives a comparison of several common methods.

3.11.6 *Effect of Motor Starting on Generators.* Fig 15 shows the behavior of the voltage of a generator when an induction motor is started. Starting a synchronous motor has a similar effect up to the time of pull-in torque. The case used for this illustration utilizes a full-voltage starting device, and the full-voltage motor starting load in kilovolt-amperes is about 100 percent of the generator rating. It is assumed for curves A and B that the generator is provided with an automatic voltage regulator.

The minimum voltage of the generator as shown in Fig 15 is an important quantity because it is a determining factor affecting under-

voltage devices and contactors connected to the system and the stalling of motors running on the system. The curves of Fig 16 can be used for estimating the minimum voltage occurring at the terminals of a generator supply power to a motor being started.

3.11.7 *Effect of Motor Starting on Distribution System.* Frequently in the case of purchased power, there are transformers and cables between the starting motor and the generator. Most of the drop in this case is within the distribution equipment. When all the voltage drop is in this equipment, the voltage falls immediately (because it is not influenced by a regulator as in the generator case) and does not recover until the motor approaches full speed. Since the transformer is usually the largest single impedance in the distribution system, it takes almost the total drop. Fig 17 has been plotted in terms of motor starting load in kilovolt-amperes which would be drawn if rated transformer secondary voltage were maintained.

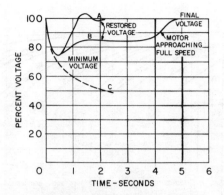

Motor-starting kVA = 100 percent of
generator rating
A — No initial load on generator
B — 50 percent initial load on generator
C — No regulator

Fig 15
Typical Generator-Voltage Behavior Due to
Full-Voltage Starting of a Motor

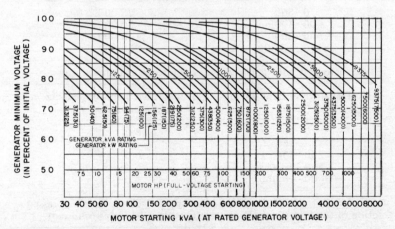

NOTES: (1) The scale of motor horsepower is based on the starting current being equal to approximately 5.5 times normal.

(2) If there is no initial load, the voltage regulator will restore voltage to 100 percent after dip to values given by curves.

(3) Initial load, if any, is assumed to be of constant-current type.

(4) Generator characteristics are assumed as follows: (a) Generators rated 1000 kVA or less: Performance factor $k = 10$; transient reactance $X_d' = 25$ percent; synchronous reactance $X_d = 120$ percent. (b) Generators rated above 1000 kVA: Characteristics for 3600 r/min turbine generators.

Fig 16
Minimum Generator Voltage Due to Full-Voltage Starting of a Motor

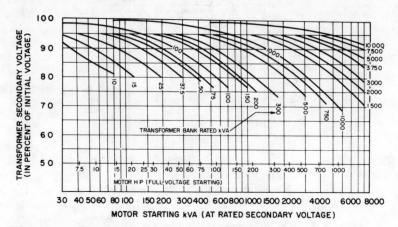

NOTES: (1) The scale of motor horsepower is based on the starting current being equal to approximately 5.5 times normal.

(2) Short-circuit capacity of primary supply is assumed to be as follows:

Transformer Bank Load (kVA)	Primary Short-Circuit Capacity (kVA)
0—300	25 000
500—1000	50 000
1500—3000	100 000
3760—10 000	250 000

(3) Transformer impedances are assumed to be as follows:

Transformer Bank Load (kVA)	Transformer Bank Impedance (percent)
10—50	3.0
75—150	4.0
200—500	5.0
750—2000	5.5
3000—10 000	6.0

(4) Representative values of primary system voltage drop, expressed as a fraction of total drop, for the assumed conditions, are as follows:

Transformer Bank Load (kVA)	System Drop / Total Drop
100	0.09
1000	0.25
10 000	0.44

Fig 17
Approximate Voltage Drop in a Transformer Due to
Full-Voltage Starting of a Motor

3.12 Standards References. The following standards publications were used as references in preparing this chapter.

ANSI C57.12.20-1974, Requirements for Overhead-Type Distribution Transformers 67 000 Volts and Below; 500 kVA and Smaller

ANSI C84.1-1970, Voltage Ratings for Electric Power Systems and Equipment (60 Hz), including Supplement C84.1a-1973

ANSI C92.2-1974, Preferred Voltage Ratings for Alternating-Current Electrical Systems and Equipment Operating at Voltages above 230 Kilovolts Nominal

IEEE Std 100-1972, Dictionary of Electrical and Electronics Terms (ANSI C42.100-1972)

NEMA MG1-1972, Motors and Generators

NFPA No 70, National Electrical Code (1975), (ANSI C1-1975)

3.13 References and Bibliography
3.13.1 *References*

[1] MICHAEL, D.T. Proposed Design Standard for the Voltage Drop in Building Wiring for Low-Voltage Systems. *IEEE Transactions on Industry and General Applications*, vol IGA-4, Jan/Feb 1968, pp 30-32.

[2] *Electric Utility Engineering Reference Book*; vol 3, *Distribution Systems*. Trafford, PA: Westinghouse Electric Corporation, 1965.

[3] ARNOLD, R.E. NEMA Suggested Standards for Future Design of AC Integral Horsepower Motors. *IEEE Transactions on Industry and General Applications*, vol IGA-6, Mar/Apr 1970, pp 110-114.

[4] LINDERS, J.R. Effects of Power Supply Variations on AC Motor Characteristics. *Conference Record, 1971 6th Annual Meeting of the IEEE Industry and General Applications Group*, IEEE 71 C1-IGA, pp 1055-1068.

3.13.2 *Bibliography*

[5] BRERETON, D.S., and MICHAEL, D.T. Significance of Proposed Changes in AC System Voltage Nomenclature for Industrial and Commercial Power Systems: I — Low-Voltage Systems. *IEEE Transactions on Industry and General Applications*, vol IGA-3, Nov/Dec 1967, pp 504-513.

[6] BRERETON, D.S., and MICHAEL, D.T. Significance of Proposed Changes in AC System Voltage Nomenclature for Industrial and Commercial Power Systems: II — Medium-Voltage Systems. *IEEE Transactions on Industry and General Applications*, vol IGA-3, Nov/Dec 1967, pp 514-520.

[7] BRERETON, D.S., and MICHAEL, D.T. Developing a New Voltage Standard for Industrial and Commercial Power Systems. *Proceedings of the American Power Conference*, vol 30, 1968, pp 733-751.

4. System and Equipment Protection

4.1 Introduction

4.1.1 *Purpose.*
The system and equipment protective devices guard the power system from the ever present threat of damage caused by overcurrents and transient overvoltages that can result in equipment loss and system failure. This chapter presents the principles of adequate system and equipment protection by introducing the many protective devices and their applications, special problems, system conditions associated with transient overvoltages, sound engineering techniques of protective device application and coordination, and suggested maintenance and testing procedures for circuit interrupting and protective devices.

4.1.2 *Considering Plant Operation* [1].
Industrial plants vary greatly in the complexity of their electric distribution systems. A small plant may have a simple radial design with low-voltage fuse protection only, whereas a large plant complex may incorporate an intricate network of medium- and low-voltage distribution substations, uninterruptible power sources, and in-plant generation required to operate in parallel with or isolated from local utility networks. At an early design stage, the plant engineering representatives should meet with the local power company to review and resolve the requirements of both the plant and the utility.

The need for higher production from industrial plants has created demands for greater industrial power system reliability. Trends to network systems and parallel operation with utilities have produced sources having extremely high overcurrents during fault conditions and the development of new equipment standards.

The high costs of power distribution equipment and the time required to repair or replace damaged equipment such as transformers, cable, high-voltage circuit breakers, etc, make it imperative that serious consideration be given to system protection design.

The losses associated with an electrical service interruption due to equipment or system failures vary widely with different types of industries. For example, a service interruption in a machining operation means loss of production, loss of tooling, and loss from damaged products. Likewise, a service interruption in a chemical plant can cause loss of product and create major clean-up and restart problems. To avoid a disorderly shutdown which can be both hazardous and costly, it may be necessary to tolerate a short-time overload condition and the associated reduction in life expectancy of the affected electric apparatus. Other industries such as refineries, paper mills, automotive plants, textile mills, steel mills, and food processing plants are similarly affected, and losses can represent a substantial expense.

For some types of loads involving continuous processes and complex automation, even a momentary dip in voltage is as serious as a complete service interruption, whereas other types of loads can tolerate an interruption. Thus the nature of the industrial operation is a major consideration in determining the degree of protection that can be justified.

4.1.3 *Equipment Capabilities.* In addition to meeting the demands of overall system performance as dictated by the nature of the load, the protective devices must operate in conjunction with the associated circuit interrupters so as to afford protection to other power system equipment components. Transformers, cable, busway, circuit breakers, and other switching apparatus all have short-circuit withstand limits as established by the National Electrical Manufacturers Association and the American National Standards Institute.

When a fault condition occurs these devices, although perfectly intact, may be connected in series in the same circuit and subjected to severe thermal and magnetic stresses accompanying the passage of high-magnitude short-circuit current through their conducting parts. An important function of the system protective devices is to initiate operation of the circuit interrupter responsible for isolating the fault so that the other equipment connected in the same circuit is not stressed beyond safe limits. Otherwise the initial fault condition can affect far more than the specific circuit to be isolated, and a widespread outage can result.

Where the system may be subjected to transient overvoltages, it is essential to understand the nature and effects of such disturbances. For all predictable abnormalities, equipment damage can be minimized through the proper application of surge protection methods compatible with the overvoltage withstand capabilities of the exposed apparatus.

The design engineer must examine the performance capability of all the individual system equipment components and not just the process sensitivity to local outages when justifying the protection to be applied. System and equipment protection is one of the most important items in the process of system planning, and enough time must be allowed in the early stages of a system design to properly investigate the selection and application of the protective devices.

4.1.4 *Importance of Responsible Planning.* Some industrial plants, because of their size or the nature of their operations, are able to maintain electrical engineering staffs capable of the design, installation, and maintenance of an efficient protective system, while other plants will probably find it more economical to engage competent engineering advice and services from consultants. This work is specialized and often very complex, and it is neither safe nor fair to the operating engineer to expect him to do it as a sideline. Modern computerized methods of calculating fault currents on complex systems are available from consulting firms and manufacturers. These provide accurate information essential for making decisions relative to the protection design in a short period of time (see Chapter 5) [2].

Protection in an electric system is a form of insurance. It pays nothing as long as there is no fault or other emergency, but when a fault occurs it can be credited with reducing the extent and duration of the interruption, the hazards of property damage, and personnel injury. Economically, the premium paid for this insurance must be balanced against the cost of repairs and lost production. Protection, well integrated with the class of service desired, may reduce capital investment by eliminating the need for equipment reserves in the industrial plant or utility supply system.

While the protective devices are the watchmen of the power system, the industrial electrical engineer must be the custodian of the protection system. Adequate and regular maintenance and testing must be done, as well as reanalysis of the protection scheme when major system changes occur. Integrity of the system protection and, thereby, the system performance requires a continuing effort if it is to be preserved.

4.2 Analysis of System Behavior and Protection Needs

4.2.1 *Nature of the Problem.* It would be neither practical nor economical to build a fault-proof power system. Consequently, modern systems are designed to provide reasonable insulation, clearances, etc, but a certain number of faults must be tolerated during the life of the system. Even with the best design possible, materials deteriorate and the likelihood of faults increases with age. Every system is subject to short circuits and ground faults that should be removed quickly, and a knowledge of the effect of faults on system voltages and currents

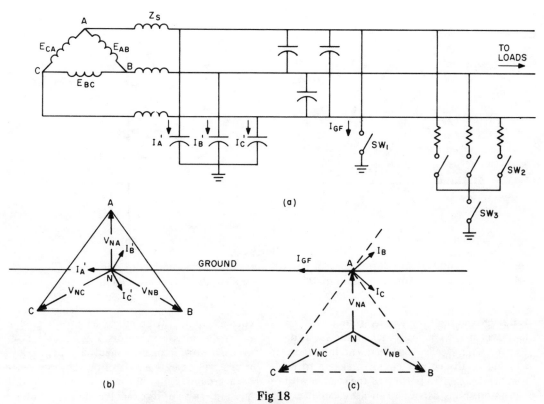

Fig 18
Analysis of Steady-State Conditions on an Ungrounded System
Before and After the Occurrence of a Ground Fault
(a) Typical Ungrounded System Conditions. (b) Normal Distribution of
Voltage and Current Phasors prior to Fault Application. (c) Distribution of
Voltage and Current Phasors after Applying a Ground Fault on Phase A (by Closing SW_1)

is necessary to design suitable protection, since these quantities are used to develop a protective plan. A reliable protective system is not only properly designed and maintained, but it must not be unnecessarily complicated or it can cause more problems than it solves in the form of undesired or incorrect operations.

Operating records show that the majority of electric circuit faults originate as a line-to-ground failure. Protective devices must detect three-phase, phase-to-phase, double-phase-to-ground, as well as single-phase-to-ground short circuits. There are two general classifications of three-phase systems, (1) ungrounded systems, and (2) grounded systems where one conductor, generally the neutral, is grounded either solidly or through an impedance. Both classifi-

cations of systems are subject to all the aforementioned types of faults, but the severity of those faults involving ground depends to a large extent on the method of system grounding and the magnitude of the grounding impedance [3].

4.2.2 *Grounded and Ungrounded Systems.* The general subject of system grounding [see IEEE Std 142-1972, Grounding of Industrial and Commercial Power Systems (ANSI C114.1-1973)] is treated from the viewpoint of system design in Chapter 6, and it is necessary to observe here only the effect on basic relaying methods of the choice between a grounded and an ungrounded system.

In grounded systems, phase-to-ground faults produce currents of sufficient magnitude to op-

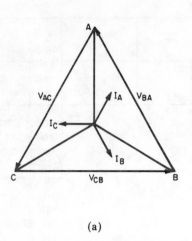

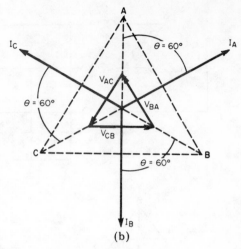

(a)

(b)

Fig 19
Voltage and Current Phasor Relationships for a Balanced Fault Condition
(a) Normal Conditions with Load Applied. (b) Three-Phase Fault

erate ground-fault-responsive overcurrent relays, which automatically detect the fault, determine which feeder has failed, and initiate the tripping of the correct circuit breakers to deenergize the faulted portion of the system without interrupting service to healthy circuits. If the system neutral is grounded through a properly chosen impedance, the value of the ground-fault current can be restricted to a level which will avoid extensive damage at the point of the fault, yet be adequate for ground-fault relaying. In addition, the voltage dip caused by the flow of short-circuit current will be materially reduced.

In ungrounded systems, as shown in Fig 18(a), phase-to-ground faults produce relatively insignificant values of fault current. In a small isolated-neutral industrial installation, the ground-fault current for a single line-to-ground fault may be well under 1 A, while the largest plant, containing miles of cable to provide electrostatic capacitance to ground, may produce not more than 20 A of ground-fault current. These currents usually are not of sufficient magnitude for the operation of overcurrent relaying to locate and remove such faults, not only because of the extreme sensitivity of the relays that would be required, but also because of the complexity of the flow pattern resulting from the fact that the "source" of the ground

current is the distributed capacitance to ground of the unfaulted conductors. It is possible, however, to provide phase-to-neutral voltage relays which will operate an alarm on the occurrence of a ground fault, but which cannot provide any indication of its exact location. The voltage and current distribution for normal operation and a single-line-to-ground fault (phase A) condition for an ungrounded system are shown in Fig 18(b) and (c), respectively.

The one advantage of an ungrounded system lies in the possibility of maintaining service on the entire system, including the faulted section, until the fault can be located and the equipment shut down for repair. Against this advantage must be balanced such disadvantages as the impossibility of relaying the fault automatically, the difficulty of locating the fault, the continuation of burning and the escalation of damage at the point of the fault, the long-continued overstressing of the insulation of the unfaulted phases (1.73 times operating voltage in the case of solid ground faults and perhaps much more in the case of intermittent ground faults), and the hazard of multiple ground faults and transient overvoltages.

4.2.3 *Distortion of Phase Voltages and Currents during Faults.* Balanced three-phase faults do not cause voltage distortion or current unbalance. Fig 19 shows the balanced conditions, both before and after a fault is applied on a system having an X/R ratio of approximately 1.7, which corresponds to an angle of 60 degrees between phase voltage and current or a 50 percent fault-circuit power factor. This condition would be realized by closing all three poles of switch SW_2 in Fig 18(a). Other types of faults, such as phase to phase, single-phase to ground, and two-phase to ground, cause distorted voltages and unbalanced currents. The voltage distortion is greatest at the fault and minimum at the generator or source.

Currents and voltages which exist during a fault vary widely for different systems, depending on type and location of the fault and the impedance of the system grounding connection. The vector diagrams of Fig 20 show voltage and current relations which exist for different types of faults on a solidly grounded system in which the currents lag the voltages by 60 degrees. Load currents are not included.

These diagrams are typical of the fault conditions which cause protective devices to operate. The characteristics of the voltage distortion which accompanies a fault are used to enable special types of relays to discriminate between different types of faults having otherwise similar current conditions. Some of these special devices will be discussed in further detail later in this chapter. The distortion can be greater or less than that shown, depending on the impedance of the fault and its distance from the relay. The small voltage drop shown between the faulted phases represents the fault impedance (arc) voltage drop plus the voltage drop in the system conductors due to the flow of fault current between the relay and the fault point.

4.2.4 *Analytical Restraints.* The one-line diagram commonly used to represent three-phase systems is a very useful analytical tool when its limitations are properly observed. Its validity is limited to symmetrical three-phase loading of symmetrical systems.

One-line diagrams, for example, provide no means for properly representing the effect of single-phase loads on the operation of the system or the protective devices. Likewise, the influence of surge protection equipment acting independently in any of the three phases cannot be correctly evaluated.

In the case of a line-to-ground fault on a three-phase system, the opening of that phase protector alone alters the system symmetry. Examination of a more precise three-phase diagram reveals that it is still possible for current to flow through the remaining phases via the line-to-line connected paths at the load apparatus, and then to ground at the fault point.

Determining how much current continues to flow to the ground fault after opening of one phase protector is a complex problem due to the alteration in system symmetry and introduction of additional variable impedances. However, it is generally a much lower value than the initial line-to-ground fault current and, hence, it can take the remaining protectors some time to sense and clear the circuit. Substantial heat with associated damage can be generated at the fault point before complete isolation of the circuit is accomplished.

By simultaneously opening all phases, the three-phase symmetry is not altered and the condition described above need not be given consideration.

4.2.5 *Practical Limits of Protection.* When the industrial power system is in normal operation, all parts should have some form of automatic protection. However, some fault possibilities may be legitimately considered as too improbable to justify the cost of specific protection. Before accepting a risk on this basis alone, the magnitude of the probable damage should also be seriously considered. Too much protection might be provided for failures which occur frequently but cause only minor difficulties, while rare but serious causes of trouble might be neglected. For example, internal transformer failures rarely occur, but the consequences may be very serious since such faults can cause fires and endanger personnel and equipment.

Most systems have some flexibility in the manner in which circuits are connected. The various possible arrangements should be considered in planning the protection system so as not to leave some emergency operating condition without protection. Some types of systems have so many possible operating combinations that protection cannot be applied to operate properly for all conditions. In such cases the operating connections for which the protection is inadequate should be avoided.

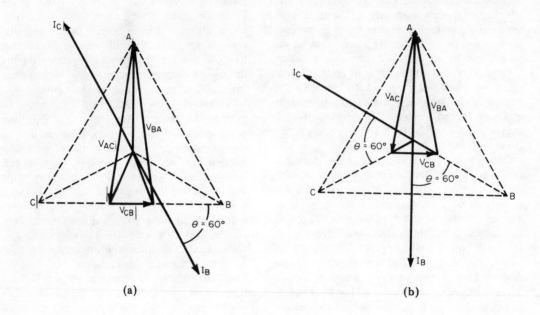

(a)

(b)

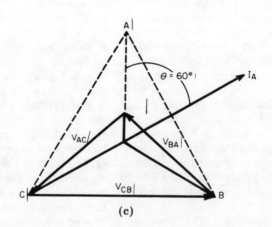

(c)

Fig 20
Voltage and Current Phasor Relationships for
Various Unbalanced Fault Conditions. System X/R = 1.7
(a) Phase-to-Phase Fault Between Phases B and C on Ungrounded System
[Close Poles B and C of SW$_2$ in Fig 18(a)]
(b) Two-Phase-to-Ground Fault Between Phases B and C and Ground on Grounded System
[Close SW$_3$ and Poles B and C of SW$_2$ in Fig 18(a)]
(c) Phase-to-Ground Fault Between Phase A and Ground on Grounded System

4.3 Protective Devices and Their Applications

4.3.1 *General Discussion.* The power system protective devices provide the intelligence and initiate the action which enables circuit switching equipment to respond to abnormal or dangerous system conditions. Normally, relays control power circuit breakers rated above 600 V and current-responsive self-contained elements operate multipole low-voltage circuit breakers to isolate circuits experiencing overcurrents on any leg. Similarly, fuses and single-pole interrupters function either alone or in combination with other suitable means to properly provide isolation of faulted or overloaded circuits. In other cases special types of relays that respond to abnormal electric system conditions may cause circuit breakers or other switching devices to disconnect defective equipment from the remainder of the system.

Following is a brief description of the types and chacteristics of relays and other protective devices most commonly used in industrial plant power systems along with some brief application considerations. A table of ANSI relay device numbers referenced with the respective device functions, as given in ANSI C37.2-1970, Manual and Automatic Station Control, Supervisory, and Associated Telemetering Equipments, appears in the Appendix.

4.3.2 *Overcurrent Relays.* The most common relay for short-circuit protection of the industrial power system is the overcurrent relay. The overcurrent relays used in the industry are mostly of the electromagnetic attraction, induction, and solid-state types. Relays with bimetallic elements used for thermal overload protection are discussed in Section 4.3.16. The simplest overcurrent relay using the electromagnetic attraction principle is the solenoid type. The basic elements of this relay are a solenoid wound around an iron core and steel plunger or armature which moves inside the solenoid and supports the moving contacts. Other electromagnetic-attraction-type relays have hinged armatures or clappers of different shapes. Fig 21 shows an overcurrent relay.

The construction of the induction-type overcurrent relay is similar to a watthour meter since it consists of an electromagnet and a movable armature which is usually a metal disk on a vertical shaft restrained by a coiled spring. The

**Fig 21
Typical Electromagnetic Overcurrent Relay**

relay contacts are operated by the movable armature (Fig 22).

The pickup or operating current for all overcurrent relays is adjustable. When the current through the relay coil exceeds a given setting, the relay contacts close and initiate the circuit breaker tripping operation. The relay usually operates on current from the secondary of a current transformer.

If the current operates the relay without intentional time delay, the protection is called instantaneous overcurrent protection. When the overcurrent is of a transient nature such as caused by the starting of a motor or some sudden overload of brief duration, the circuit breaker should not open. For this reason most overcurrent relays are equipped with a time delay which permits a current several times in excess of the relay setting to persist for a limited period of time without closing the contacts. If a relay operates faster as current increases, it is said to have an inverse-time characteristic. Overcurrent relays are available with inverse, very inverse, and extremely inverse time characteristics

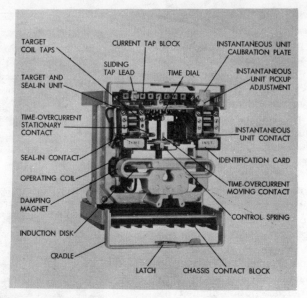

Fig 22
Induction-Disk Overcurrent Relay with Instantaneous Attachment
(Relay Removed from Drawout Case)

to fit the requirements of the particular application. There are also definite minimum-time overcurrent relays having an operating time that is practically independent of the magnitude of current after a certain current value is reached. Induction overcurrent relays have a provision for variation of the time adjustment and permit change of operating time for a given current. This adjustment is called the time lever or time dial setting of the relay.

Fig 23 shows the family of time—current operating curves available with a typical inverse-time overcurrent relay. Similar curves are published for other overcurrent relays having different time-delay characteristics. As is apparent, it is possible to adjust the operating time of relays. This is important since they are normally used to "selectively" trip circuit breakers which operate in series on the same system circuit. With increasing current values, the relay operating time will decrease in an inverse manner down to a certain minimum value. Fig 24 shows the characteristic curves of inverse (A), very inverse (B), and extremely inverse (C) time relays when set on their minimum and maximum time

dial positions. It also shows the characteristics of the instantaneous element (D) that is usually supplied in these relays [4].

4.3.3 *Overcurrent Relays with Voltage Restraint or Voltage Control.* A short circuit on an electric system is always accompanied by a corresponding voltage dip, whereas an overload will cause only a moderate voltage drop. Therefore a voltage-restrained or voltage-controlled overcurrent relay is able to distinguish between overload and fault conditions. A voltage-restrained overcurrent relay is subject to two opposing torques, an operating torque due to current and a restraining torque due to voltage. As such the overcurrent required to operate the relay is higher at normal voltage than it is at reduced voltage. A voltage-controlled overcurrent relay operates by virtue of current torque only, the application of which is controlled by another relay element set to operate at some predetermined value of voltage. Such relay characteristics are useful where it is necessary to set the relay close to or below load current, while retaining certainty that it will not operate improperly on normal load current.

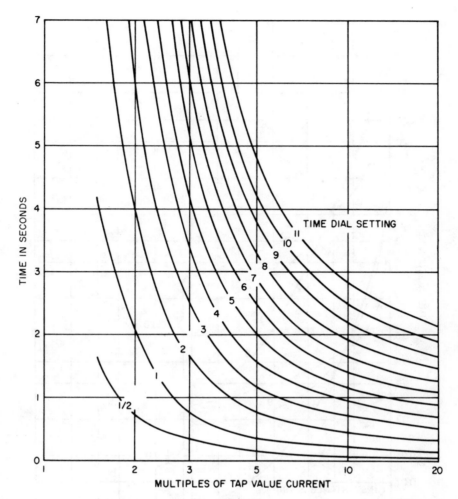

Fig 23
Time—Current Characteristics of a Typical Inverse-Time Overcurrent Relay

4.3.4 *Directional Relays*

4.3.4.1 *Directional Overcurrent Relay.* Directional overcurrent relays consist of a typical overcurrent unit and a directional unit which are combined to operate jointly for a predetermined phase-angle and magnitude of current. In the directional unit the current in one coil is compared in phase-angle position with a voltage or current in another coil of that unit. The reference current or voltage is called the polarization. Such a relay operates only for current flow to a fault in one direction and will be insensitive to current flow in the opposite direction. The overcurrent unit of the directional overcurrent relay is practically the same as for the usual overcurrent relay and has similar definite-minimum-time, inverse, and very inverse time—current characteristics. The directional overcurrent relays can be supplied with voltage restraint on the overcurrent element.

The latest type of directional relay is usually "directionally controlled," that is, the overcurrent unit is inert until the directional unit detects the current in the tripping direction and releases or activates the overcurrent unit. Many directional relays are equipped with instan-

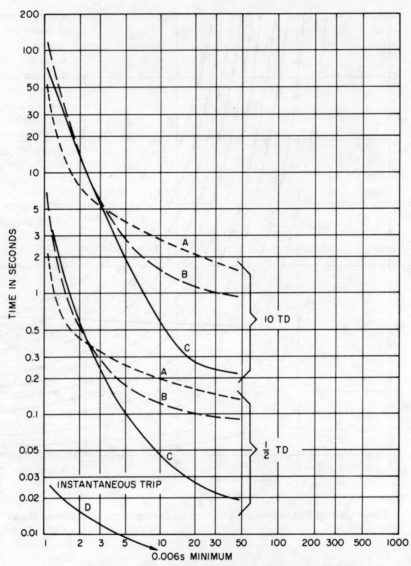

A Inverse
B Very Inverse
C Extremely Inverse
D Instantaneous
TD Relay Time Dial Setting

Fig 24
Typical Relay Time—Current Characteristics

taneous elements which in some cases operate nondirectionally, and unless it is possible to determine the direction of the fault by magnitude alone, the nondirectional instantaneous tripping feature should not be used.

4.3.4.2 *Directional Ground Relay.* The grounded-neutral industrial power system consisting of parallel circuits or loops may use directional ground relays which are generally constructed in the same manner as the directional overcurrent relays used in the phase leads. In order to properly sense the direction of fault current flow, they require a polarizing source which may be either potential or current as the situation requires. Obtaining a suitable polarizing source requires special consideration of the system conditions during faults involving ground and a unique application of auxiliary devices.

4.3.4.3 *Directional Power Relay.* The directional power relay is in principle a single-phase or three-phase contact-making wattmeter and operates at a predetermined value of power. It is often used as a directional overpower relay set to operate if excess energy flows out of an industrial plant power system into the utility power system. Under certain conditions it may also be useful as an underpower relay to separate the two systems if the power flow drops below a predetermined value. Care should be used in the application of single-phase watt relays because at certain power factors they may cause a false trip operation.

4.3.5 *Differential Relays.* All the previously described relays have the common characteristic of adjustable settings to operate at a given value of some electrical quantity such as current, voltage, frequency, power, or a combination of current and voltage or current and phase angle. There are other fault-protection relays which function by virtue of continually comparing two or more currents [Fig 25(a)]. Fault conditions will cause a change of these compared values with reference to each other and the resulting "differential" current can be used to operate the relay. However, current transformers have a small error in ratio and phase angle between the primary and secondary currents, depending upon variations in manufacture, the magnitude of current, and the connected secondary burden. These errors will cause a differential current to flow even when the primary currents are balanced. The error currents may become proportionately larger during fault conditions, especially when there is a direct-current component present in the fault current. The differential relays, of course, must not operate for the maximum error current which can flow for a fault condition external to the protected zone. To provide this feature, the percentage-type relay [device 87, Fig 25(a)] has been developed which has special restraint windings to prevent improper operation due to the error currents on heavy "through" fault conditions while providing very sensitive detection of low-magnitude faults inside the differentially protected zone [5].

4.3.5.1 *Differential Protection of Motors and Generators.* Using the connection arrangement illustrated at the left of Fig 25(a), overcurrent relays can be used for the differential protection of a motor or generator. As long as the current flowing into each winding of the motor is equal to the current flowing out of the same winding, no net current will flow through the relay operating coil (ignoring current transformer error currents). Any leakage of fault current to other phases or to ground will upset this balance and send differential current through the operating winding. When this current is greater than the pickup of the relay, its contacts close to activate tripping of the circuit breaker and disconnect the faulted apparatus.

Since no means is provided with this scheme to prevent false operation on current transformer error current, the overcurrent relays have to be set so that they will not operate on the maximum error current which can flow in the relay during an external fault. This results in a substantial sacrifice of sensitivity for low-magnitude internal faults.

Another form of motor differential protection involves a special routing of the machine phase and neutral leads of each leg through a common "window-type" current transformer as shown in Fig 25(b). Under normal conditions the magnetizing flux produced by the phase and neutral currents adds to zero and no output is produced for the instantaneous relay. A fault in any winding will result in the fault current by-passing the current transformer causing a

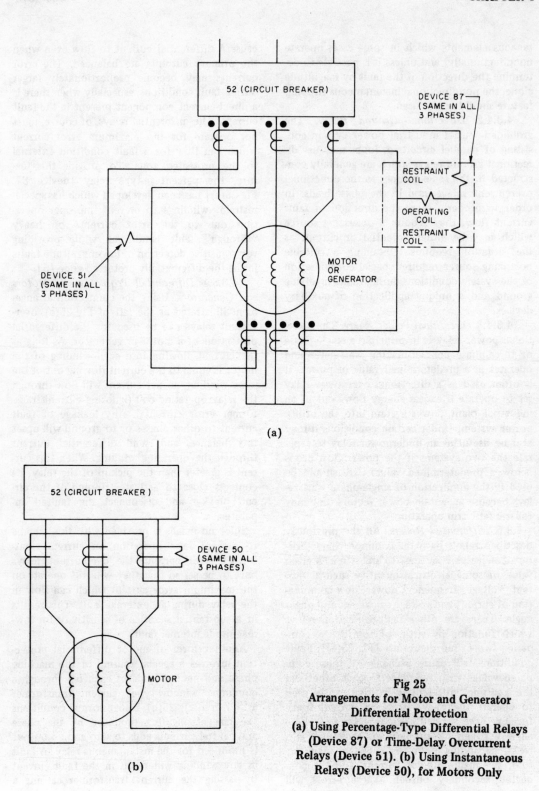

(a)

(b)

**Fig 25
Arrangements for Motor and Generator
Differential Protection
(a) Using Percentage-Type Differential Relays
(Device 87) or Time-Delay Overcurrent
Relays (Device 51). (b) Using Instantaneous
Relays (Device 50), for Motors Only**

differential current (and flux) which will in turn produce an output signal. The single relay and current transformer employed per phase in this scheme is less expensive, although additional machine terminal box space for neutral conductor cabling is required.

4.3.5.2 *Differential Protection of a Two-Winding Transformer Bank.* When differential relays are used for transformer protection, the inherent characteristics of power transformers introduce a number of problems which do not exist in the generators and motors. If the current transformer secondary currents on the two sides of the transformer differ in magnitude by more than the range provided by the relay taps, the relay currents can be altered by means of auxiliary current transformers or current-balancing autotransformers. If the high-voltage and low-voltage line currents are not in phase due to a delta—wye connection in the transformer, the secondary currents can be brought into phase by connecting the current transformers in delta on the wye side, and in wye on the delta side. The differential output signals of the current transformers are subject to the same errors as discussed above for generators. In addition, a significant trip current signal can be observed at the relay input due to primary magnetizing inrush current which occurs upon transformer energization. This is why ordinary overcurrent relays cannot be given sensitive settings, and induction-type percentage differential relays are used instead. Additional protection can be afforded by the harmonic restraint type relay which distinguishes between magnetizing inrush current and internal fault current without reducing sensitivity.

4.3.5.3 *Differential Protection of Buses.* Large industrial power system buses often have sectionalizing circuit breakers so that a fault in one of the bus sections can be isolated without involving the remaining sections. Each of the bus sections or in some cases the whole bus (if not sectionalized) can be provided with differential relay protection, which in case of an internal fault isolates the bus section involved.

Differential bus protection distinguishes between internal and external fault by comparing the magnitudes of the currents flowing in and out of the protected bus. The major differences between bus protection and generator or transformer protection are in the number of circuits in the protected zone and in the magnitudes of currents involved in the various circuits. Fig 26 shows phase and ground differential protection of an eight-circuit bus using overcurrent relays. This method, of course, is subject to the same disadvantages discussed in the preceding paragraphs. Several more acceptable types of bus protective relays are used, including the percentage differential relay, the linear coupler, and the differential voltage relay.

(1) *Percentage Differential Relay.* Where the number of circuits connected to the bus is relatively small, relays using the percentage differential principle similar to the transformer differential relay may be used. The problem of application of percentage differential relays for bus protection, however, increases with the number of circuits connected to the bus. All current transformers supplying the relays must have identical ratios and characteristics. Variations in the characteristics of the current transformers, particularly the saturation phenomena under short-circuit conditions, present the greatest problem to this type of protection and often limit it to applications where only a limited number of feeders are present.

Various relay-desensitizing schemes are used to avoid operation from the inrush of magnetizing current when switching transformers. The harmonic restraint type has features which distinguish between magnetizing inrush current and internal fault current. Another type utilizes external timing relays and shunting resistors during the switching interval.

(2) *Linear Coupler* [6]. The linear-coupler bus-protection scheme eliminates the difficulty due to differences in the characteristics of iron-core current transformers by using air-core mutual inductances. Since it does not contain any iron in its magnetic circuit, the linear coupler is free of any direct-current or alternating-current saturation. The linear couplers of the different circuits are connected in series and produce voltages that are directly proportional to the currents in the circuits. For normal conditions, or for external faults, the sum of the voltages produced by linear couplers is zero. During internal (bus) faults, however, this voltage is no longer zero and operates a sensitive

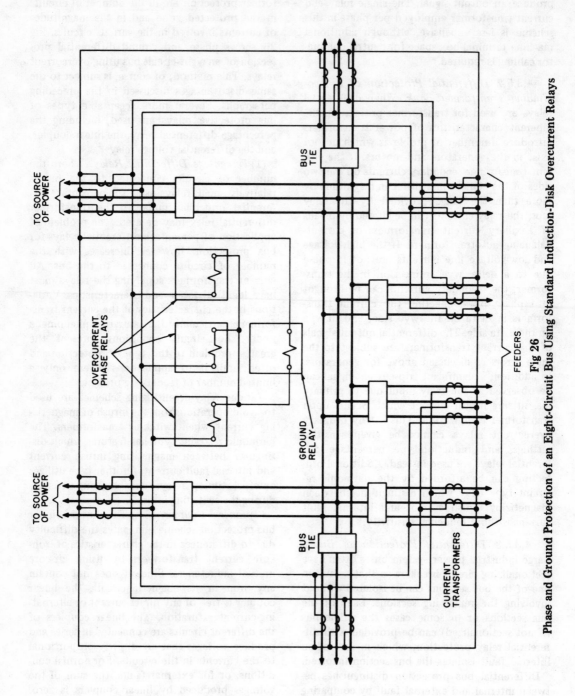

Fig 26

Phase and Ground Protection of an Eight-Circuit Bus Using Standard Induction-Disk Overcurrent Relays

relay to trip all circuit breakers to clear the bus fault.

(3) *Differential Voltage Relay.* Another method of bus protection is the use of differential voltage relays. This scheme uses through-type iron-core current transformers. The problem of current-transformer saturation is overcome by using a voltage-responsive (high-impedance) operating coil in the relay.

Bus protection using linear couplers or differential voltage relays is not limited as to number of source and load feeders, and in general is faster in operation than protection using the percentage differential principle. It should be noted that linear couplers or current transformers used for differential voltage relays cannot be used for other purposes. Separate current transformers are required for line relaying and metering.

4.3.6 *Current Balance Relay.* The principle of a differential relay as applied to rotating machinery protection requires that the current transformers be available at both ends of the phase windings to permit the comparison between the current magnitudes at these ends. In some instances, particularly in smaller units, it may not be possible to justify the cost of installing these current transformers or of bringing out the extra winding terminals to make installation of differential relays possible. In such cases phase-balance current-comparison relays can provide an acceptable substitute for differential protection. A negative-sequence current relay is a more sensitive device which also detects unbalanced phase currents. In applying these relays it is assumed that under normal conditions the phase currents in the three-phase supply to the equipment and the corresponding output signals from each phase current transformer are balanced. Should the fault occur in the motor or generator involving one or two phases or should an open circuit develop in any of the phases, the currents will become unbalanced and the relay will operate. In addition to protecting against winding faults, the phase-balance current relay affords protection against damage to the motor or generator due to single-phase operation. This type of protection is not provided for by the usual differential relays. Another current-balance type of differential protection for motors, which is both simple

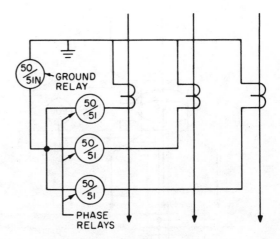

Fig 27
Standard Arrangement for Residually
Connected Ground Relay

and relatively inexpensive, is provided by the use of a single current transformer zero-sequence relay scheme and is discussed further in Section 4.3.7.2.

4.3.7 *Ground-Fault Relaying* [7], [8]

4.3.7.1 *Residually Connected Protection.* Where the industrial power system neutral is intentionally grounded and ground-fault current can flow in the conductors, ground relaying may be used to provide improved protection. This is often an overcurrent relay connected in the common lead of the wye-connected secondaries of three line-current transformers. Fig 27 shows the typical current transformer and relay connections for this application. The ground relay can be set to pick up at a much lower current value than the phase relays because there is no neutral current with normal balanced load current.

Overcurrent relays used for ground-fault protection are generally the same as those used for phase-fault protection, except that a more sensitive range of minimum operating current values is possible since they see only fault currents. Relays with inverse, very inverse, and extremely inverse time characteristics, as well as instantaneous relays, are all applicable for ground-fault relays. Precaution must be used, however, in applying this type of residually connected ground relay, since it is subject to

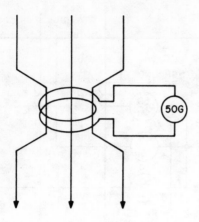

Fig 28
Relay and Current-Transformer Connection for
Zero-Sequence Ground Relay

nuisance operation due to error currents arising from current transformer saturation and unmatched characteristics in the manner described for differential relays. Often the optimum speed and sensitivity of a residual ground relay must be compromised because of this.

4.3.7.2 *Zero-Sequence Relay.* An improved type of ground-fault protection can be obtained by a zero-sequence relay scheme in which a single window-type current transformer is mounted so as to encircle all three phase conductors as illustrated in Fig 28. On four-wire systems with possible unbalanced line-to-neutral loads, the neutral conductor must also pass through the current transformer window. A ground fault produces a zero-sequence current in the current transformer secondary which operates an overcurrent relay in control of circuit breaker tripping. Since only one current transformer is employed in this method of sensing ground faults, the relaying is not subject to current-transformer errors due to ratio mismatch or saturation effects. As a result, very sensitive instantaneous tripping can be used on branch circuits. Only circuit faults involving ground will produce a current in the current-transformer secondary to operate the relay, as the vectorial sum of normal load or three-phase fault currents will be zero. This scheme is widely applied on 5 and 15 kV systems and is also used on large low-voltage systems for improved protection. It is also often used as an economical alternative to differential protection for large motors on grounded systems.

4.3.7.3 *Neutral Relaying.* Ground-fault protection of some type provides improved protection and is now required by NFPA No 70, National Electrical Code (1975), (ANSI C1-1975), (NEC), in many cases on wye-connected solidly grounded services from 150 through 600 V. Transformer neutral relaying with a current transformer connected to the neutral grounding circuit provides a convenient low-cost method of obtaining ground-fault tripping. On a three-phase three-wire grounded system (that is, no line-to-neutral loads) the ground-fault pickup can be set at the minimum value available, unless it causes loss of the necessary selectivity. In this case, if a longer operating time delay does not correct the problem, a higher setting can be used to allow downstream devices to clear the fault. When no ground-fault relaying is present on downstream circuit interrupters, the pickup of the source ground-fault protection may have to be set as high as the phase protection of the largest feeder if selective operation is to be accomplished. Provided ground-fault protection is applied throughout the system, coordination can be easily achieved at very low settings as the ground-fault element will not sense normal load currents, but will only pick up along with the downstream device when there is a ground fault. In four-wire solidly grounded systems, 60 Hz and third-harmonic unbalanced load currents may flow in the neutral, and thus residually connected ground relaying requires an additional current transformer to balance the residual signal of the normal line-to-line neutral load currents. On such systems, neutral grounding circuit relaying or zero-sequence type of protection can be used with the load-carrying neutral brought through the current transformer along with the phase conductors, thereby avoiding the possibility of nuisance tripping due to the inaccuracies associated with the residual connection.

For three-phase three-wire systems of 480 V and above supplied from wye-connected transformers, where the neutral is not run to permit connection of phase-to-neutral load, high-resistance grounding may be used to limit the ground-fault current to values which will not cause substantial additional damage to the equipment at the point where the ground fault occurs. Such protection is frequently applied to critical circuits supplying equipment such as ventilating or cooling systems in chemical plants where an unplanned power interruption could create a dangerous situation. In these cases the ground-fault detector operates an alarm so that the operation can be shut down in an orderly manner or a transfer made to an emergency power supply before the faulted circuit is disconnected.

4.3.8 *Synchronism-Check and Synchronizing Relays.* The synchronism-check relay is used to verify when two alternating-current circuits are within the desired limits of frequency and voltage phase angle to permit them to operate in parallel. These relays should be employed on switching applications on systems known to be normally paralleled at some other location so that they are only checking that the two sources have not become electrically separated or displaced by an unacceptable phase angle. The synchronizing relay, on the other hand, monitors two separate systems that are to be paralleled, automatically initiating switching as a function of the phase-angle displacement, frequency difference (beat frequency), and voltage deviation as well as the operating time of the switching equipment, to accomplish interconnection when conditions are acceptable. An example of this application would be where a plant generates its own power with a parallel-operated tie with a utility system. The utility end of the tie line must have synchronizing relays that will check conditions on both systems prior to paralleling and initiate the interconnection so as to avoid any possibility of tying with the industrial plant generators out of phase.

4.3.9 *Pilot-Wire Relays* [9], [10]. The relaying of tie lines, either between the industrial system and the utility system or between major load centers within the industrial system, often presents a special problem. Such lines must be capable of carrying maximum emergency load currents for any length of time, and they must be removable from service quickly should a fault occur. A type of differential relaying called pilot-wire relaying responds very quickly to faults in the protected line, clearing the fault promptly and minimizing line damage and disturbance to the system. Yet it is normally unresponsive to load currents and to currents flowing to faults in other lines and equipment. The various types of pilot-wire relaying schemes all operate on the principle of comparing the conditions at the terminals of the protected line, the relays being connected to operate if the comparison indicates a fault in the line. The information necessary for this comparison is transmitted between terminals over a pilot-wire circuit, hence the designation of this type of relaying. Because, like all differential schemes, it is completely and inherently balanced within itself and completely selective, the pilot-wire relay scheme does not provide protection for faults of the adjacent station bus or beyond it.

4.3.10 *Voltage Relays.* Voltage relays function at predetermined values of voltage, which may be overvoltage, undervoltage, a combination of both, voltage unbalance (comparing two sources of voltage), reverse phase voltage, and excess negative-sequence voltage (that is, single phasing of a three-phase system). Plunger-type, induction-type, or solid-state-type relays are available. Adjustments of pickup or dropout voltage and operation timing is usually provided in these relays. Plunger-type relays are usually instantaneous in operation, although bellows, dash pots, or other delay means can be provided. The time-delay feature is often required in order that transient voltage disturbances will not cause nuisance relay operation. Some typical voltage relay applications are as follows:

(1) *Over- or Undervoltage Relays*
    (a) Capacitor switching control
    (b) Alternating- and direct-current overvoltage protection for generators
    (c) Automatic transfer of power supplies
    (d) Load shedding on undervoltage
    (e) Undervoltage protection for motors
(2) *Voltage Balance Relays.* Blocking the operation of a voltage-controlled current relay when a potential transformer fuse blows
(3) *Reverse-Phase Voltage Relays*
    (a) Detection of reverse phase connections of interconnecting circuits, transformers, motors, or generators

(b) Prevention of any attempt to start a motor with one phase of the system open

(4) *Negative-Sequence Voltage Relays.* Detection of single phasing, damaging phase voltage unbalance, and reversal of phase rotation of supply for protection of rotating equipment.

4.3.11 *Distance Relays* [9], [11]. Distance relays comprise a family of relays which measure voltage and current, and the ratio is expressed in terms of impedance. Typically this impedance is an electrical measure of the distance along a transmission line from the relay location to a fault. The impedance can also represent the equivalent impedance of a generator or large synchronous motor when a distance relay is used for loss-of-field protection.

The measuring element is usually instantaneous in action, with time delay provided by a timer element so that the delay, after operation of a given measuring element, is constant. In a typical transmission-line application three measuring elements are provided. The first operates only for faults within the primary-protection zone of the line and trips the circuit breaker without intentional time delay. The second element operates on faults not only in the primary protection zone, but also in one adjacent or backup protection zone, and initiates tripping after a short time delay. The third element is set to include a still more remote zone and to trip after a longer time delay. These relays have their greatest usefulness in applications where selective "stepped" operation of circuit breakers in series is essential, where changes in operating conditions cause wide variations in magnitudes of fault current, and where load currents may be large enough, in comparison with fault currents, to make overcurrent relaying undesirable.

The three main types of distance relay and their usual applications are as follows.

(1) *Impedance Type.* Phase-fault relaying for moderate-length lines.

(2) *Mho Type.* Phase-fault relaying for long lines or where severe synchronizing power surges may occur. Generator or large synchronous motor loss-of-field relaying.

(3) *Reactance Type.* Ground-fault relaying and phase-fault relaying on very short lines and lines of such physical design that high values of fault arc resistance are expected to occur and affect relay "reach," and on systems where severe synchronizing power surge are not a factor.

4.3.12 *Phase-Sequence or Reverse-Phase Relays.* Reversal of the phase rotation of a motor may result in costly damage to machines, long shutdown, and lost production. Important motors are frequently equipped with phase-sequence or reverse-phase relay protection. If this relay is connected to a suitable potential source, it will close its contacts whenever the phase rotation is in the opposite direction. It also can be made sensitive to unbalanced voltage or undervoltage conditions (see Section 4.3.10).

4.3.13 *Frequency Relays* [12]. Frequency relays sense under- or overfrequency conditions during system disturbances. Most frequency relays have provision for adjustment of operating frequency and voltage. The speed of operation depends on the deviation of the actual frequency from the relay setting. Some frequency relays operate instantaneously if the frequency deviates from the set value. Others are actuated by the rate at which the frequency is changing. The usual application of this type of relay is to selectively drop system load based on the frequency decrement in order to restore normal system stability.

4.3.14 *Temperature-Sensitive Relays.* Temperature-sensitive relays usually operate in conjunction with temperature-detecting devices such as resistance temperature detectors or thermocouples located in the equipment to be protected and are used for protection against overheating of large motors (above 1500 hp), generator stator windings, and large transformer windings.

For generators and large motors several temperature detectors are usually embedded in the stator windings and the hottest (by test) reading detector is connected into the temperature relay bridge circuit. The bridge circuit is balanced at this temperature, and an increase in winding temperature will increase the resistance of the detector, unbalance the bridge circuit, and cause relay operation. Transformer temperature relays operate in a similar manner from detecting devices set in winding "hot spot" areas. Some relays are provided with a

$10°C$ differential feature which will prevent reenergizing of the equipment until the winding temperature has dropped $10°C$.

4.3.15 *Pressure-Sensitive Relays.* Pressure-sensitive relays used in power systems respond either to the rate of rise of gas pressure (sudden pressure relay) or to a slow accumulation of gas (gas-detector relay), or a combination of both. Such relays are valuable supplements to differential or other forms of relaying on power, regulating, and rectifier transformers.

A sudden rise in the gas pressure above the liquid insulating medium in a liquid-filled transformer indicates that a major internal fault has occurred. The "sudden-pressure" relay will respond quickly to this condition and isolate the faulted transformer. Slow accumulation of gas (in conservator tank-type transformers) indicates the presence of a minor fault such as loose contacts, grounded parts, short-circuited turns, leakage of air into the tank, etc. The gas-detector relay will respond to this condition and either sound an alarm or isolate the faulted transformer.

4.3.16 *Replica-Type Temperature Relays.* Thermally activated relays respond to heat generated by current flow in excess of a certain predetermined value. The input to the relay is normally the output of the current transformer whose ratio must be carefully selected to match the available relay ratings. Many varied types are available, the most common being the bimetal strip and the melting alloy types. The relay must be checked for variations in operating characteristics as a function of ambient temperature.

Since the operating characteristics of this thermal "replica"-type relay closely match general-purpose motor heating curves in the light and medium overload areas, they are used almost exclusively for overload protection of motors up to 1500 hp.

4.3.17 *Auxiliary Relays.* Auxiliary relays are used in protection schemes whenever a protective device cannot in itself provide all the functions necessary for satisfactory protection. This type of relay is available with a wide range of coil ratings, contact arrangements, and tripping functions, each suited for a particular applica-

tion. Some of the most common applications of auxiliary relays are circuit breaker lockout, circuit breaker latching, targeting, multiplication of contacts, timing, circuit supervision, and alarming.

4.3.18 *Direct-Acting Trip Devices for Low-Voltage Circuit Breakers*

4.3.18.1 *Electromechanical Trip Devices.* All low-voltage circuit breakers which employ electromechanical trip devices have a heavy copper coil on each pole of the circuit breaker that is an integral part of the trip unit capable of carrying the full-load current. Magnetizing force produced by the current passing through the coil acts on an armature to overcome dashpot restraint to trip the circuit breaker and provide overload protection. Short-circuit protection is provided by magnetic forces suddenly overcoming spring restraint. A separate adjustable unit is required for each circuit breaker frame size.

Long-time overcurrent, short-time overcurrent, and instantaneous tripping are available on trip devices which can include any combination of the three forms of protection and with adjustable characteristics. Although current transformers and relays have been applied to low-voltage circuit breakers, electromechanical series overcurrent devices for many years have been the basic form of protection for all types of low-voltage switchgear. However, they do have some inherent disadvantages. The trip point will vary depending on age and severity of duty, and they have a limited calibration range. Because the trip characteristic curve of electromechanical devices has a very inverse shape with a somewhat unpredictable broad operating band (Fig 29), coordination of tripping with other devices is difficult.

4.3.18.2 *Static Trip Devices.* In contrast to electromechanical devices, static trip devices operate from a low-current signal generated by current sensors or current transformers in each phase. Output from the sensors is fed into a static trip unit which evaluates the incoming signal with respect to its calibration setpoints and acts to trip the circuit breaker if preset values are exceeded. In addition to phase protection, the static trip device is available with ground-fault trip protection.

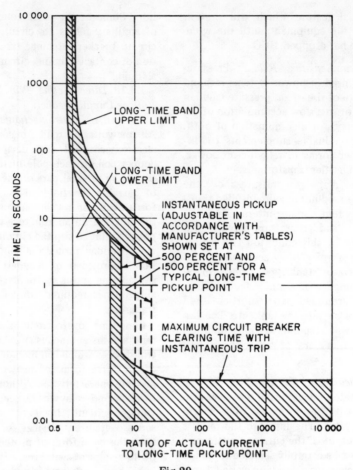

Fig 29
Typical Time—Current Plot for Electromechanical Trip Devices

Static trip devices are more accessible on the circuit breaker than electromechanical trip devices and are much easier to calibrate since low values of currents can be fed through the device to simulate the effect of an actual fault-current signal. Special care or provisions are sometimes necessary to guarantee predictable operation when applying static trip devices to loads having other than the pure sinusoidal current wave shapes. Vibration, temperature, altitude, and duty cycle have virtually no effect on the calibration of static trip devices. Excellent reliability, therefore, is generally possible. The most important advantage of static trip devices is the shape of the trip characteristic curve, which is essentially a straight line throughout its work-

ing portion (Fig 30), having a very narrow and predictable operating band.

4.3.19 *Fuses* [13]—[15]. [See also IEEE Std 313-1971, Relays and Relay Systems Associated with Electric Power Apparatus (ANSI C37.90-1971).] A high-voltage fuse is defined by IEEE Std 100-1972, Dictionary of Electrical and Electronics Terms (ANSI C42.100-1972), as "an overcurrent protective device with a circuit-opening fusible part that is heated and severed by the passage of overcurrent through it." From this definition it can be seen that a fuse is intended to be responsive to current and provide protection against system overcurrent conditions.

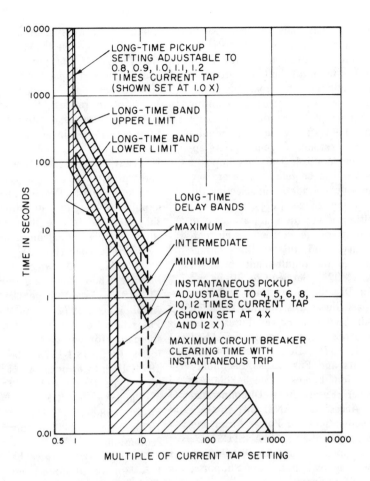

Fig 30
Typical Time—Current Plot for Solid-State Trip Devices

4.3.19.1 *Power Fuses (Over 600 V).* Most power fuses comply with the NEMA E rating which has the following requirements. (1) All fuses must carry their current rating continuously. (2) Fuses rated 100 A or less shall melt in 300 s at an rms current within the range of 200 to 240 percent of the continuous rating of the fuse element. (3) Fuses rated above 100 A shall melt in 600 s at an rms current within the range of 220 to 264 percent of the continuous rating of the fuse element.

(1) *Current-Limiting Power Fuses* [13], [15], [16]. This fuse is designed so that the melting of the fuse element introduces a high arc resistance into the circuit in advance of the pro-

spective peak current of the first half-cycle. This restricts the short-circuit current to a lower value than would occur without the current-limiting effect of the fuse. Current-limiting fuses were primarily developed for use in conjunction with motor starter contactors to limit the short-circuit current to a value within the withstand capabilities of the contactors so as to obtain a high interrupting capacity starter on systems above 600 V. They are now, however, widely used in high-capacity industrial systems wherever it is necessary to limit the short-circuit current for the protection of series-connected equipment components. Typical applications are protection of potential transformers and protection of small loads on high-capacity circuits. The time—current characteristic of current-limiting power fuses is near vertical (Fig 31), which makes them difficult to coordinate with overcurrent relays on their load side.

The current-forcing action of current-limiting fuses during interruption produces transient overvoltage on the system which may require the application of suitable surge-protective apparatus for proper control. The duty imposed on surge arrestors can be relatively severe and should be carefully considered in selecting the equipment to be applied (see [17]).

Current-limiting power fuses are available in various frequency, voltage, continuous current-carrying capacity, and interrupting ratings which conform to the requirements of ANSI C37.40-1969 (R 1974), Service Conditions and Definitions for Distribution Cutouts and Fuse Links, Secondary Fuses, Distribution Enclosed Single-Pole Air Switches, Power Fuses, Fuse Disconnecting Switches, and Accessories; ANSI C37.41-1969 (R 1974), Design Tests for Distribution Cutouts and Fuse Links, Secondary Fuses, Distribution Enclosed Single-Pole Air Switches, Power Fuses, Fuse Disconnecting Switches, and Accessories; ANSI C37.46-1969 (R 1974), Specifications for Power Fuses and Fuse Disconnecting Switches; and ANSI C37.47-1969 (R 1974), Specifications for Distribution Fuse Disconnecting Switches, Fuse Supports, and Current-Limiting Fuses.

(2) *Expulsion-Type Fuses.* This type of fuse is generally used in distribution system cutouts or disconnect switches. To interrupt a fault current, an arc-confining tube with a deionizing fiber liner and fusible element is employed. Arc interruption is accomplished by the rapid production of pressurized gases within the fuse tube which extinguishes the arc by expulsion from the open end or ends of the fuse.

Enclosed, open, and open-link types of expulsion fuses are available for use as cutouts. Enclosed cutouts have terminals, fuse clips, and fuse holders mounted completely within an insulating enclosure. Open cutouts have these parts completely exposed. Open-link cutouts have no integral fuseholders and the arc-confining tube is incorporated as part of the fuse link.

Fused cutouts and disconnect switches are used outdoors for the protection of industrial plant distribution systems and applications such as line-fault and overload protection of distribution feeder circuits, transformer primary fault protection, and capacitor-bank fault protection.

Since gases are released rapidly during the interruption process, the operation of expulsion-type fuses is comparatively noisy. When they are applied in an enclosure such as a disconnect switch, special care must be taken to vent any ionized gases that might be released and which would cause a flashover between internal live parts. Despite these disadvantages, expulsion-type fuses are often applied because they have an inverse time—current characteristic which is more compatible with standard overcurrent relays (Fig 31).

4.3.19.2 *Low-Voltage Fuses (600 V and Below).* These fuses are covered by NEMA FU1-1972, Low-Voltage Cartridge Fuses, by ANSI C97.1-1972, Low-Voltage Cartridge Fuses 600 Volts or Less, and in the following standards for Safety by Underwriters Laboratories, Inc: 198.1-1973, Class H Fuses; 198.2-1974, High-Interrupting-Capacity Current-Limiting Type Fuses; 198.3-1974, High-Interrupting Capacity Class K Fuses; 198.4-1974, Class R Fuses; 198.5-1973, Plug Fuses; and 198.6-1975, Fuses for Supplementary Overcurrent Protection.

Plug fuses are of three basic types, all rated 125 V or less to ground and up to 30 A maximum. Although they have no interrupting rating, they are subjected to one alternating-current short-circuit test with an available current of 10 000 A. The three types are Edison base with no time delay and all ratings interchangeable, Edison base with a time delay and interchangeable ratings, and type S base available in three noninterchangeable current ranges: 0—15 A, 16—20 A, and 21—30 A. These last two types normally have a time-delay characteristic of at least 12 s at 200 percent of their rating, although time-delay plug fuses are no longer required by the NEC.

Cartridge fuses may be either renewable or nonrenewable. Nonrenewable fuses are factory assembled and must be replaced after operating. Renewable fuses can be disassembled and the fusible element replaced. Renewal elements are usually designed to give a greater time delay than ordinary nonrenewable fuses, and in some

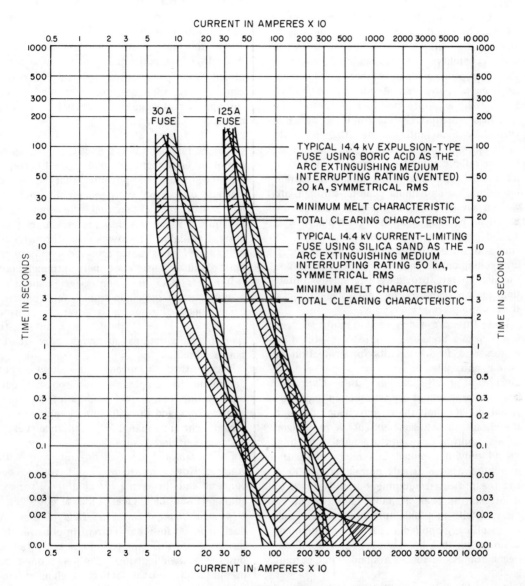

Fig 31
Time—Current Characteristic Curves Showing the Difference Between
Boric-Acid Expulsion-Type and Current-Limiting Fuses

designs the delay on moderate overcurrents is considerable. Generally the renewable type fuse is not available in the higher interrupting ratings.

(1) *Current-Limiting Fuses.* Current-limiting fuses are intended for use in circuits where available short-circuit current is beyond the withstand capability of downstream equipment or the interrupting rating of ordinary fuses or standard circuit breakers. According to NEMA FU1-1972, "an alternating-current-limiting fuse is a fuse which safely interrupts all available currents within its interrupting rating and, within its current-limiting range, limits the clearing time at rated voltage to an interval equal to or less than the first major or symmetrical current loop duration and limits peak let-through current to a value less than the peak current that would be possible with the fuse replaced with a solid conductor of the same impedance." A current-limiting fuse, therefore, places a definite ceiling on the peak let-through current and thermal energy, providing equipment protection against damage from excessive magnetic stresses and thermal energy.

These fuses are widely used in motor starters, fused circuit breakers, and fused switches of motors and feeder circuits for protection of busway and cable.

(2) *Non-Current-Limiting Fuses (Class H).* These fuses interrupt overcurrents up to 10 000 A but do not limit the current which flows in the circuit to the same extent as recognized current-limiting fuses. As a general rule, they should only be applied in circuits where the maximum available fault current is 10 000 A and the protected equipment is fully rated to withstand the peak available fault current associated with this fault duty, unless such fuses are specifically applied as part of an equipment combination that has been type-tested and designed for use at higher available fault current levels.

(3) *Let-Through Considerations.* Fig 32 illustrates typical current-limiting fuse operating characteristics during a high-fault-current interruption. In applications involving high available fault currents, the operating characteristic of the current-limiting fuse limits the actual current which is allowed to flow through the circuit to a level substantially less than the "pro-spective" maximum. The peak let-through current of a current-limiting fuse is the instantaneous peak value of current through the fuse during fuse opening. The let-through I^2t of a fuse is a measure of the thermal energy developed throughout the entire circuit during clearing of the fault. Both values are important in evaluating fuse performance and can be determined from peak let-through and I^2t curves supplied by the fuse manufacturer. A let-through current value considerably less than the available fault current will greatly reduce the magnetic stresses (which increase as the square of the current) and thus reduce fault damage in protected equipment. In some cases it becomes possible to use components (that is, motor starters, disconnect switches, circuit breakers, and bus duct) in the system which have fault capabilities much less than the maximum fault current available. The low peak let-through current and I^2t levels can be achieved with current-limiting fuses because of their extremely fast (often less than one quarter-cycle) speed of response when subjected to high fault current.

The speed of response is governed by fuse design. For highest speed, silver links with special configurations surrounded by quartz sand are used.

Peak let-through current values alone cannot determine the comparable effectiveness of current-limiting fuses. The product of the total clearing time and the effective value of the let-through current squared I^2t, or thermal energy, must be considered as well.

The melting I^2t of a fuse does not vary with voltage. However, arcing I^2t is voltage dependent and the arcing I^2t at 480 V, for example, will not be as great as that at 600 V.

(4) *Dual-Element or Time-Delay Fuses.* A dual-element fuse has current-responsive elements of two different fusing characteristics in series in a single cartridge. The fuse is one-time in operation, and the fast-acting element protects against short-circuit currents in much the same way as an ordinary fuse. The time-delay element permits short-duration overloads, but melts if these overloads are sustained. The most important application for these fuses is motor and transformer protection. They do not open on motor starting or transformer magnetizing inrush currents, but protect the motor and

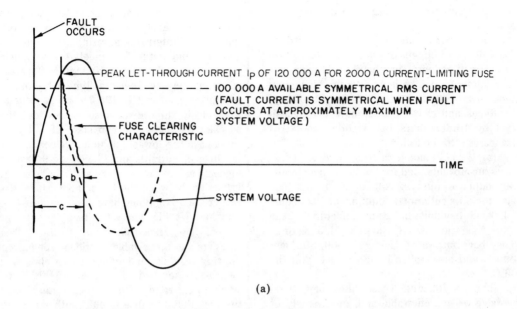

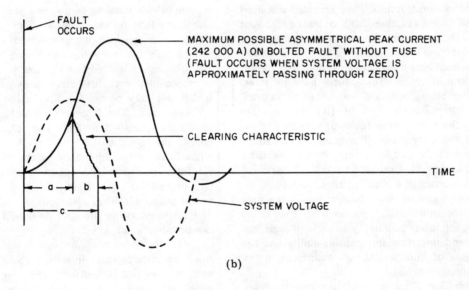

a — Melting Time
b — Arcing Time
c — Total Clearing Time

Fig 32
Typical Current-Limitation Characteristics Showing Peak
Let-Through and Maximum Prospective Fault Current as a Function of
the Time of Fault Occurrence (100 kA Available Symmetrical rms Current)
(a) Fault Occurring at Peak Voltage. (b) Fault Occurring at Zero Voltage

branch circuits from damage by sustained over-loads.

(5) *Fuse Standards (Cartridge).* Cartridge fuses differ in dimensions according to voltage and current ratings. They have ferrule contacts in ratings of 60 A or less and knife-blade contacts in larger ratings. Cartridge fuses of varying types and characteristics have been classified by Underwriters Laboratories, Inc (UL) standards into the following classes.

(a) *Miscellaneous cartridge fuses.* These fuses are not intended for use in branch circuits but rather for use in control circuits, special electronic or automotive equipment, etc. To be UL listed they must not be manufactured in the same dimensions as UL classes G, H, J, K, or L. They have ratings of 125, 250, 300, 500, and 600 V and are applied in accordance with the NEC.

(b) *Class H fuses.* These fuses may be renewable or nonrenewable and are generally of zinc-link construction. They are rated and sized up to 600 A at either 250 V or less or 600 V or less. They may or may not be of dual-element construction, but if labeled as "time delay" they must have a minimum delay of 10 s at 500 percent of rating. These fuses have no interrupting rating, but have been tested on an available alternating current of 10 000 A and are generally used where fault currents do not exceed this magnitude. The label shows neither the class letter nor any interrupting rating. Class H fuses are often referred to as NEC fuses, and the nonrenewable class H fuse is sometimes referred to as a one-time fuse, although this may describe other fuse types as well. Renewable fuses are losing popularity because of their limited and uncertain interrupting ability and the dangers of "double-linking" or insecurely fastening renewal links.

(c) *Class K high-interrupting-capacity fuses.* These fuses are manufactured in identical physical sizes as class H fuses, with which they are interchangeable. However, they have been tested at various high available fault-current levels up to their maximum labeled ratings, which may be either 50 000, 100 000, 200 000 A rms. Thus fuses in this class are referred to as high-interrupting-capacity fuses. Class K fuses must meet specified maximum values of instantaneous peak let-through currents and I^2t ener-

gy let-through maximum values for each physical case size. Each fuse label bears the alternating-current interrupting rating, the class and subclass, and if the fuse meets the time-delay requirement of at least 10 s at 500 percent rating, it may be labeled as time-delay or with the letter D. Per UL 198.3 class K fuses are available in three distinct subclasses identified as class K1, class K5, and class K9. Class K1 fuses have the lowest maximum values for peak let-through currents and I^2t, class K9 have the highest maximum values, and the class K5 values are between the K1 and K9 values. Most earlier class K9 types have been modified and now have class K5 characteristics.

(d) *Class R current-limiting fuses.* These fuses are a nonrenewable cartridge-type current-limiting fuse also manufactured to class H dimensional standards, having a 200 000 A rms symmetrical rating. The R designation signifies the fact that the fuse is built with a rejection feature which consists of notches provided in either the fuse ferrule or blade, depending on the size involved. Equipment rated and approved only for use with fuses having the current-limiting characteristics of class R fuses is then provided with fuse attachment hardware which will only permit the installation of the notched fuses. The fuses are available in two subclasses identified as RK1 and RK5 which denote the fact that the fuses have let-through characteristics corresponding to class K1 and K5 fuses, respectively. Since equipment which is approved for class R service is always rated to withstand the higher let-through conditions of RK5 fuses, either the RK1 or the RK5 fuse can be safely applied.

(e) *Class J current-limiting fuses.* These fuses are manufactured in ratings up to 600 A and in specified dimensions per UL 198.2 which are noninterchangeable with class H and class K fuses. They are labeled as current-limiting. There is no 250 V or less rating; all are labeled 600 V or less and may be used only in fuseholders of suitable class J dimensions. Each case size has specified maximum values of peak let-through current values and maximum thermal (I^2t) values. Fuses having a time-delay characteristic are available in class J dimensional sizes although they do not meet the UL standards for current-limiting performance.

Class J fuses have a 200 000 A rms interrupting rating.

(f) *Class L current-limiting fuses.* These are the only UL labeled fuses available with current ratings in excess of 600 A. Per UL 198.2 their ratings range from 601 to 6000 A rms, all at 600 V or less. There is no 250 V size. Class L fuses have an interrupting rating of 200 000 A rms and will safely interrupt any overcurrent up to this value. Like class J fuses, each case size or mounting dimension (mounting holes drilled in the blades) has a maximum allowable peak let-through current and I^2t value.

4.3.19.3 *Fuse-Selection Considerations.* For each fuse classification the corresponding UL standards specify the following design features which are particularly important to fuse application: current rating, voltage rating, frequency rating, interrupting rating, maximum peak let-through current, and maximum clearing thermal energy I^2t. Standards also specify maximum opening times at certain overload values such as 135 and 200 percent of rating and, for time-delay qualification, minimum opening time at a specific overload percentage. Within these parameters and from various other overcurrent test data manufacturers construct time—current curves. Normally such curves are based on rms available currents (only above 0.01 s) and show the average melting time. However, some manufacturers' curves may show minimum melting time, maximum clearing time, or "virtual" time. Caution should be exercised in the use of such curves to be certain that equivalent characteristics are being compared.

A fuse must be selected for voltage, current-carrying capacity, and interrupting rating. When fuses must be coordinated with other fuses or circuit breakers, the time—current characteristic curves, peak let-through curves, and I^2t curves may be useful. The load characteristics will dictate the time-delay performance required of the fuse. If fuses are applied in series in a circuit, it is essential for short-circuit coordination to verify that the clearing I^2t of the downstream fuse during a fault will be less than the melting I^2t of the upstream fuse. Fuse manufacturers publish fuse-ratio tables that provide listings of fuses which are known to operate selectively. Use of these tables permits coordination without the need for detailed analysis provided the fuses being applied are all of the same manufacturer.

When coordinating an upstream circuit breaker with a downstream fuse, the let-through energy of the fuse (clearing I^2t) must be less than the required amount to release the circuit breaker trip latch mechanism. This is not easily accomplished with many types of circuit breakers. Critical operation occurs in the region for currents greater than the circuit breaker instantaneous trip device pickup or periods of time less than 0.01 s, even though a normal time—current plot would suggest that selective performance exists. Similar problems exist when attempting to coordinate a downstream circuit breaker with an upstream fuse. The clearing time of the circuit breaker can often exceed the minimum melting time of the fuse. Overload coordination for low-magnitude or moderate faults can be established with standard time—current curve overlays (for details see IEEE Std 242-1975, Protection and Coordination of Industrial and Commercial Power Systems).

For protection of a downstream circuit breaker with an upstream fuse during high fault currents, the peak let-through current of the fuse must be compatible with the momentary withstand rating of the circuit breaker. Manufacturers' tables for the selection of fuses to protect circuit breakers are an easy solution providing that such tables are based upon current styles, types, and classes of fuses and circuit breakers.

Although the proper selection of a fuse to protect a circuit breaker, starter, or cable circuit will generally prevent equipment failure during a fault condition, some apparatus design practices allow damage to bimetals, contacts, and other parts. Unless the combination has been specifically tested and rated as a unit, the application of a fuse of a given interrupting rating in a switch or other fusible device does not confer that rating on the equipment involved. A switch, for example, might not withstand the let-through energy of a current-limiting fuse during certain fault-current conditions. When a combination rating is not available, the rating of the fuse or the device, whichever is the less, should be used.

The voltage rating of a fuse should be selected as equal to or higher than the nominal system voltage on which it is used. When applying a high-voltage current-limiting fuse of a given voltage rating on a circuit of a lower voltage rating, consideration must be given to the magnitude and effect of overvoltages that will be induced due to the zero current forcing action of the fuse during the interruption of high-magnitude fault currents. A low-voltage fuse of any labeled voltage rating will always perform satisfactorily on lower service voltages. This is not a problem at 600 V or less.

The current rating of a fuse should be selected so that it clears only on a fault or an overload and not on current inrush. Ambient temperatures and types of enclosures affect fuse performance and must be considered. Fuse manufacturers must be requested to supply correction factors for unusual ambient temperatures.

4.4 Performance Limitations

4.4.1 *Load Current and Voltage Wave Shape.* The published operating characteristics of all protective relays and trip devices are based on an essentially pure sinusoidal wave shape of current and voltage. Many industrial loads are of such a nature as to produce harmonics in the system current and voltage. This condition is aggravated by the presence of any distribution equipment in the system with nonlinear electrical characteristics. As a result it is important to understand the nature of the expected system current and voltage as well as the effect that wave-shape distortion might have on the protective devices being applied.

4.3.2 *Instrument Transformers* [11]. If a protective relay is to operate predictably and reliably, it must receive information that accurately represents conditions that exist on the power system from the circuit instrument transformers. Since current and potential transformers become significantly nonlinear devices under certain conditions, they may not produce an output precisely representative of power system conditions either in wave shape or magnitude. The exact extent of any distortion is a function of the input signal level and transformer burden (total connected impedance) as well as the actual design (accuracy class) of the

instrument transformer being applied. Potential transformer performance characteristics are classified by ANSI C57.13-1968, Requirements for Instrument Transformers. This same standard provides separate classifications for current transformers with regard to burden capability and accuracy for both metering and relaying service. The burden and output requirements of all instrument transformers should be carefully checked against their rating for any relay application to verify that proper relay operation will result. In some cases where a larger burden is encountered or the expected level of fault current is higher than that encompassed by the ANSI C57.13-1968 rating structure, it may be necessary to obtain the current transformer saturation curve from the manufacturer in order to analytically establish acceptable performance.

4.5 Principles of Protective Relay Application

[18]–[20]. [See also IEEE Std 273-1967, Protective Relay Applications to Power Transformers (ANSI C37.91-1972).] Fault-protection relaying can be classified into two groups, primary relaying which should function first in removing faulted equipment from the system, and backup relaying which functions only when primary relaying fails.

To illustrate the areas of protection associated with primary relaying, Fig 33 shows the various areas together with circuit breakers which feed each electric element of the system. Note that it is possible to disconnect any piece of faulted equipment by opening one or more circuit breakers. For example, when a fault occurs on the incoming line L_1 in Fig 33, the fault is within a specific area of protection (area A) and should be cleared by the primary relays which operate circuit breakers 1 and 2. Likewise a fault on bus 1 is within a specific area of protection, area B, and should be cleared by the primary relaying actuating circuit breakers 2, 3, and 4. If circuit breaker 2 fails to open and the faulted equipment remains connected to the system, the backup protection provided by circuit breaker 1 and its relays must be depended upon to clear the fault.

Fig 33 illustrates the basic principles of primary relaying, with separate areas of protection established around each system element

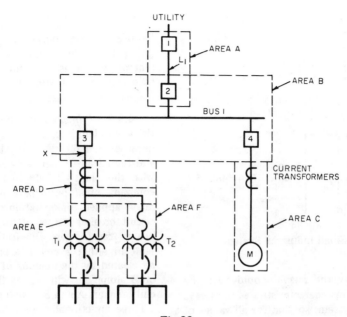

Fig 33
One-Line Diagram Illustrating Zones of Protection

that can be isolated by a separate interrupting device. Any equipment failure occurring within a given area will cause tripping of all circuit breakers supplying power to that area.

To assure that all faults within a given zone will operate the relays of that zone, the current transformers associated with that zone should be placed on the line side of each circuit breaker so that the circuit breaker itself is a part of two adjacent zones. This is known as "overlapping." Sometimes it is necessary to locate both sets of current transformers on the same side of the circuit breaker. In radial circuits the consequences of this lack of overlap are not usually very serious. For example, a fault at X on the load side of circuit breaker 3 in Fig 33 could be cleared by the opening of circuit breaker 3 if there were any way to cause it to open circuit breaker 3. Since the fault is between the circuit breaker and the current transformers, the relays of circuit breaker 3 will not see it, and circuit breaker 2 will have to open and consequently interrupt the other load on the bus. When the current transformers are located immediately at the load bushings of the circuit breaker, the amount of circuit exposed to this problem is minimized. The consequences of lack of overlap become more serious in the case of tie circuit breakers between differentially protected buses and bus feeds protected by differential or pilot-wire relaying.

In applying relays to industrial systems, safety, simplicity, reliability, maintenance, and the degree of selectivity required must be considered. Before attempting to design a protective relaying plan, the various elements that make up the distribution system together with the operating requirements must be examined.

4.5.1 *Typical Small-Plant Relay Systems* [21]. One of the simplest industrial power systems consists of a single service entrance circuit breaker and one distribution transformer stepping the utility's primary distribution voltage down to utilization voltage, as illustrated in Fig 34. There would undoubtedly be several circuits on the secondary side of the transformer, protected by either fuses or circuit breakers.

Protection for the circuit between the incoming line and the circuit breakers on the transformer secondary would normally consist of conventional induction-disk-type overcurrent

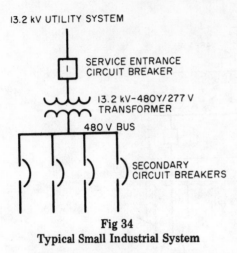

Fig 34
Typical Small Industrial System

relays. Preferably the relays should have the same time—current characteristics as the relays on the utility system, so that for all values of fault current the local service entrance circuit breaker can be programmed to trip before the utility supply line circuit breaker. The phase relays should also have instantaneous elements to promptly clear high-current faults.

This simple system provides both primary and backup relay protection. For instance, a fault on a secondary feeder should be cleared by the secondary protective device; however, if this device should fail to trip, the primary relays will trip circuit breaker 1.

This simple industrial system can be expanded by tapping the primary feeder and providing fuse protection on the primary of each distribution transformer as shown in Fig 35.

An additional step or area of protection is included over the simpler system shown in Fig 34. All secondary feeder faults should be cleared by the secondary circuit breakers as before, while faults within the transformer should now be cleared by the transformer primary fuses. The fuses may also act as backup protection for the faults which are not cleared by the secondary feeder protective devices. The primary feeder faults will, as before, be cleared by circuit breaker 1. This circuit breaker will in turn act as backup protection for the transformer primary fuses.

4.5.2 *Protective Relaying for a Large Industrial Plant Power System* [2], [4], [22], [23].

As an electric system becomes larger, the number of sequential steps of relaying also increases, giving rise to the need for a protective relaying scheme which is inherently selective within each zone of protection. Fig 36 shows the main connections of a large system.

4.5.2.1 *Primary Protection.* The relay selectivity problem is of great concern to the utility because their 69 kV supply lines are paralleled and their transformers are connected in parallel with the plant's local generation. The utility company must participate in the selection of relays applied for operation of either incoming circuit breaker in case of trouble in the 69 kV bus or transformers. Due to the 69 kV bus tie a fault in either a bus or a transformer cannot be cleared by the opening of circuit breaker A or B alone but will require the opening of circuit breakers A and B as well as C or D.

Three directionally controlled overcurrent relays (device 67) must be installed for circuit breakers C and D and connected to trip for current flow toward the respective 69 kV transformer. Directionally controlled overcurrent relays are suggested because their sensitivity is not limited by the flow of load current in the normal or nontrip direction. Three overcurrent relays having inverse-time characteristics must be installed at circuit breaker positions A and B as backup protection for faults that may occur on or immediately adjacent to the 69 kV buses. The connection of these overcurrent relays (device 51), shown in Fig 36 as being energized from the output of paralleled current transformers located at the incoming 69 kV lines and bus tie, provides the advantage of isolating only the faulted bus section in a shorter time than would be possible if individual circuit breaker relays were used.

The next zones of protection are the 13.8 kV buses 1 and 2. Fault currents are relatively high for any equipment failure on or near the main 13.8 kV buses. For this reason a differential protective relay scheme (device 87B) is recommended for each bus. Differential relaying is instantaneous in operation and is inherently selective within itself. Without such relaying, high-current bus faults must be cleared by proper operation of overcurrent devices on the several sources. This usually results in long-time

13.2 kV UTILITY SYSTEM

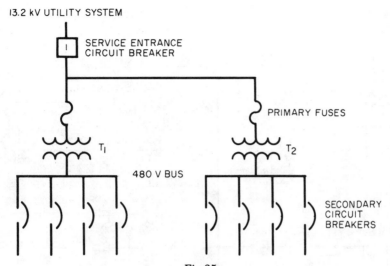

SERVICE ENTRANCE
CIRCUIT BREAKER

PRIMARY FUSES

T_1

T_2

480 V BUS

SECONDARY
CIRCUIT
BREAKERS

Fig 35
System of Fig 34 Expanded by Addition of a Transformer
and Associated Secondary Circuits

clearing since the overcurrent devices have pickup and time settings determined by other than bus fault considerations. General practice is to use separate current transformers with the same ratio and output characteristics for the differential relay scheme.

A multicontact auxiliary relay (device 86B) is used with the differential relays to trip all the circuit breakers connected to the bus whenever bus trouble occurs. To realize maximum sensitivity, the time-delay ground relays (device 51N) at the 69—13.8 kV source transformers are connected to the output of current transformers measuring the current in the neutral connection to ground. The 87TN relay is differentially connected to provide sensitive tripping on faults between the transformer secondary and the 13.8 kV main circuit breaker. Unlike the time-delay relays 51N-1 and 51N-2, this relay does not have to be set to coordinate with other downstream ground-fault relays.

Superior protection for the cable tie between buses 2 and 3 is provided by pilot-wire differential relays (device 87L). In addition to being instantaneous in operation, pilot-wire schemes are inherently selective within themselves and require only two pilot wires if the proper relays are used. Backup protection provided by

overcurrent relays should be installed at both ends of the tie line.

Nondirectional relays can be applied at circuit breaker M. At circuit breaker N directional relays are more advantageous since the 10 MVA generator represents a fault source at bus 3.

Separate current transformers are used for the pilot-wire differential relaying to provide reliability and flexibility in the application of other protective devices.

The 9000 hp 13.8 kV synchronous motor is provided with a reactor-type reduced-voltage starting arrangement using metalclad switchgear. Overload protection is provided by a thermal relay (device 49) whose sensor is a resistance temperature detector (RTD) imbedded in the stator windings. This relay can be used to either trip or alarm. Internal fault protection is provided by the differential relay scheme (device 87M). Backup fault protection and locked-rotor protection is provided by an overcurrent relay (device 51/50) applied in all three phases. Undervoltage and reverse phase rotation protection is provided by the voltage-sensitive relay (device 47) connected to the main bus potential transformers.

Ground-fault protection is provided by the instantaneous zero-sequence current relay (de-

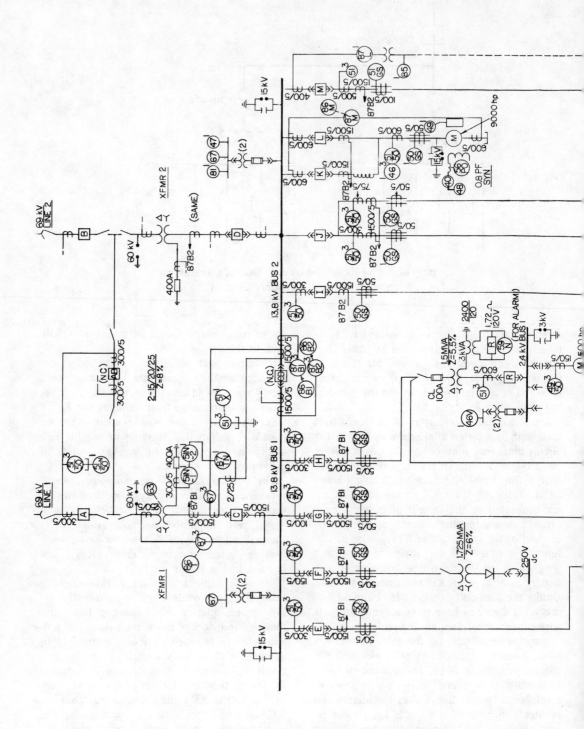

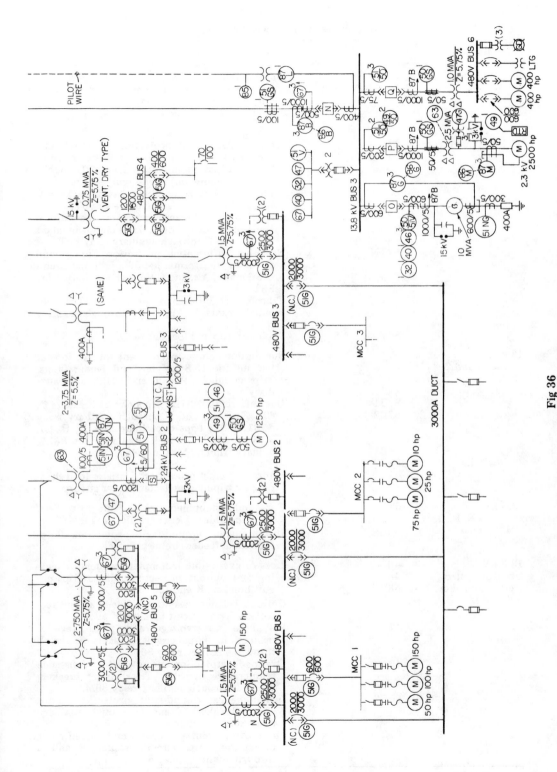

Fig 36

One-Line Diagram Showing Protection for Typical Large Industrial Plant System

Protective Device Legend for Fig 36

Location	Device	Description
69 kV supply lines	51/50 51N/50N	Summation phase and ground overcurrent protection for 69 kV bus and backup for transformer differential relaying. Trips circuit-breakers A, AB, and C through auxiliary relay 86T.
15 MVA main transformers	51N-1 51N-2	Backup ground-fault protection for transformer secondary 13.8 kV bus and feeder circuits. 51N-1 trips tie circuit breaker CD; 51N-2 (after time interval) trips circuit breaker.
	63	Sudden pressure relay. Trips circuit breakers A, AB, and C through auxiliary relay 86T.
	67	Directional phase overcurrent as backup to transformer differential. Trips circuit breakers A, AB, and C through auxiliary relay 86T.
	87TN	Sensitive differential protection for ground faults in transformer secondary. Trips circuit breakers A, AB, and C through auxiliary relay 86T.
	87T	Transformer differential protection. Trips circuit breakers A, AB, and C through auxiliary relay 86T.
	86T	Auxiliary trip and lockout relay.
13.8 kV buses 1 and 2	51	Summation phase overcurrent protection as backup for 13.8 kV bus and feeder faults. Trips circuit breakers C and CD through auxiliary relay 51X.
	87B1 87B2	Bus differential. 87B1 trips circuit breakers C, CD, E, F, G, and H through auxiliary relay 86B1. 87B2 trips circuit breakers D, CD, I, J, K, L and M through auxiliary relay 86B2.
	86B1 86B2	Auxiliary trip and lockout relay.
	81	Underfrequency protection. Initiates load shedding by triping preselected feeder circuits.
13.8 kV feeders E, F, G, H, I, and J	51/50	Time and instantaneous phase-fault protection. Trips individual feeder circuit breaker.
	50GS	Instantaneous ground-fault protection. Trips individual feeder circuit breaker.
13.8 kV motor control circuit breakers K and L	40 48 56PO	Loss of excitation, incomplete sequence checking and pullout protective relays. Trips circuit breakers K and L.
	46	Current balance relay for single-phase protectection. Trips circuit breakers K and L.
	47	Polyphase undervoltage and phase reversal protection. Trips circuit breakers K and L through auxiliary relay 86M.
	49	Overload protection using stator resistance temperature detector. Trips circuit breakers K and L through auxiliary relay 86M.
	50GS	Instantaneous ground-fault protection. Trips circuit breakers K and L through auxiliary relay 86M.
	51/50	Phase overcurrent protection and locked rotor protection. Trips circuit breakers K and L through auxiliary relay 86M.

Protective Device Legend for Fig 36 (Cont'd)

Location	Device	Description
13.8 kV motor control circuit breakers K and L (cont'd)	87M	Motor differential protection. Trips circuit breakers K and L through auxiliary relay 86M.
	86M	Auxiliary trip and lockout relay.
13.8 kV tie line circuit breaker M	51GS	Sensitive ground-fault protection as backup to pilot-wire relaying and backup for bus and feeder ground faults at 13.8 kV bus 3. Trips circuit breaker M.
	51	Phase overcurrent protection as backup to pilot-wire relaying and backup for bus and feeder faults at 13.8 kV buses. Trips circuit breaker M.
	87L	Line differential protection for phase and ground faults using pilot wire. Trips circuit breaker M.
	85	Pilot wire monitoring relay to alarm for open, short-circuited, or grounded pilot wire.
3.75 MVA transformer and 2.4 kV buses 2 and 3	51	Summation phase overcurrent protection for 2.4 kV bus faults and backup protection for feeder faults. Trips circuit breakers S and ST through auxiliary relay 51X.
	51N-1 51N-2	Ground-fault protection for transformer secondary and bus and backup for feeder ground faults. 51N-1 trips tie circuit breaker ST; 51N-2 (after time interval) trips circuit breaker S.
	63	Sudden pressure relay on transformer. Trips circuit breakers S and H.
	67	Directional phase overcurrent protection for transformer faults and 13.8 kV line faults. Trips circuit breaker S.
	87TN	Sensitive differential protection for ground faults in transformer secondary. Trips circuit breakers S and H.
	46	Current balance relay for single-phase protection. Trips contactor.
	47	Polyphase undervoltage and phase sequence protection. Trips circuit breakers S and ST through auxiliary relay 51X.
	49	Replica-type thermal overload protection. Trips contactor.
	50GS	Instantaneous ground-fault protection. Trips contactor.
	51	Overcurrent relay for motor locked rotor protection. Trips contactor.
1.5 MVA transformer and 2.4 kV bus 1	49/50	Replica-type thermal overload protection including instantaneous element for short-circuit protection. Trips contactor.
	51	Phase overcurrent protection. Trips circuit breaker R.
	59N	Sensitive voltage detection of ground faults for high-resistance grounded system. Initiates alarm signal.

Protective Device Legend for Fig 36 (Cont'd)

Location	Device	Description
13.8 kV tie line, circuit breaker N at 13.8 kV bus 3	67	Directional phase overcurrent protection as backup to pilot wire and backup for bus and feeder faults at 13.8 kV buses. Trips circuit breaker N.
	87L	Line differential protection for phase and ground faults using pilot wire. Trips circuit breaker N.
	85	Pilot-wire monitoring relay to initiate alarm for open, short-circuited, or grounded pilot wire.
	51G	Sensitive ground-fault protection as backup to pilot-wire relaying and backup for bus and feeder faults at 13.8 kV buses. Trips circuit breaker N.
13.8 kV bus 3	87B	Bus differential. Trips circuit breakers N, O, P, and Q through auxiliary relay 86B.
	86B	Auxiliary trip and lockout relay.
10 MVA generator	32	Reverse power or antimotoring protection. Trips circuit breaker O.
	40	Loss of excitation protection. Alarm and subsequent trip of circuit breaker O.
	46	Negative-sequence overcurrent protection for generator due to external unbalanced fault. Trips circuit breaker O.
	51V	Backup overcurrent protection for external three-phase faults. Trips circuit breaker O.
	51NG	Ground-fault protection for generator and backup for differential relays and feeder ground relays. Trips circuit breaker O and field circuit breaker through 86G.
	87G	Generator differential protection. Trips circuit breaker O and field circuit breaker through 86G.
	86G	Auxiliary trip and lockout relay.
2.5 MVA transformer and motor at 13.8 kV bus 3	47	Polyphase undervoltage and phase sequence protection. Trips circuit breaker P.
	49	Temperature overload protection using stator resistance temperature detectors. Initiates alarm.
	49/50	Replica-type thermal overload protection including instantaneous element for short-circuit protection. Trips circuit breaker P.
	50GS	Instantaneous ground-fault protection for 13.8 kV transformer primary. Trips circuit breaker P.
	51/50	Phase overcurrent and locked rotor protection. Trips circuit breaker P.
	63	Sudden pressure relay, transformer mounted. Trips circuit breaker P.
	87M	Motor differential relay. Trips circuit breaker P through auxiliary relay 86M.
	86M	Auxiliary trip and lockout relay.
1.0 MVA transformer at 13.8 kV bus 3	50GS	Instantaneous ground-fault protection. Trips circuit breaker Q.
	51/50	Time and instantaneous phase-fault protection. Trips circuit breaker Q.
480 V transformer secondary circuit breakers	67	Directional phase overcurrent protection for transformer faults and 13.8 kV line faults. Trips 480 V circuit breaker.

vice 50GS). The current-balance relay (device 46) protects the motor against damage from excessive rotor heating caused by single phasing or another unbalanced voltage condition.

Surge protection is provided by the surge arrester and surge capacitor combination, which is located as close as possible to the motor terminals. The motor rotor starting winding can be damaged by excessive current due to loss of excitation or suddenly applied loads which cause the motor to pull out of step. Rotor damage could also result from excessive time for the motor to reach synchronous speed and lock into step. To protect against damage from these causes, loss of excitation (device 40), pull-out (device 56PO), and incomplete sequence (device 48) relays should be provided.

In addition to the protection against (1) internal faults, (2) sustained overloads, (3) undervoltage, and (4) voltage surges, generators must be protected from overheating caused by external unbalanced low-magnitude faults and from being driven as motors (antimotoring) when the prime mover is a steam turbine and can be damaged by such operation. The overload protection must be capable of detecting an external fault-current condition which corresponds to the minimum level of generator contribution. For the 10 MVA generator connected to bus 3, internal fault detection is provided by the percentage differential relay (devide 87G), surge protection by the surge arresters and capacitors at the generator terminals, ground-fault-current limitation is accomplished by the 400 A resistor in the generator neutral, and ground-fault detection is afforded by the overcurrent relay (device 51NG). Loss-of-excitation protection is provided by device 40, external unbalanced fault-current (negative-sequence) protection by device 46, antimotoring protection by device 32, and backup overcurrent protection by device 51V/50.

It is good practice for transformers of the size shown on the incoming service, where a circuit breaker is used on both the primary and secondary sides, to install transformer percentage differential relays with inverse characteristic overcurrent relays for backup protection. To prevent operation of these relays on magnetizing inrush current when the transformer is switched on, the large proportion of currents at harmonic multiples of the line ·frequency contained in the magnetizing inrush current are filtered out and passed through the restraint winding so that the current unbalance required to trip is made much greater during the excitation transient than during normal operation.

4.5.2.2 *Medium-Voltage Protection.* The medium voltage (2.4 kV) substations shown in Fig 36 are designed primarily for the purpose of serving the medium- and large-size motors. Buses 2 and 3 fed by the 3750 kVA transformers are connected together by a normally closed tie circuit breaker which is relayed in combination with each main circuit breaker by means of a partial differential or totalizing relaying scheme (device 51). The current transformers are connected with the proper polarity so that the relay sees only the total current into its bus zone and does not see any current that circulates into a bus zone through either main and leaves through the tie. The relay backs up the feeder circuit breaker relaying connected to its respective bus or operates on bus faults to trip the tie and appropriate main circuit breakers simultaneously, thereby saving one step of relaying time over what is required when the tie and main circuit breakers are operated by separate relays. One possible disadvantage to this scheme occurs when a directional relay or main circuit breaker malfunctions for a transformer fault or when a bus feeder circuit breaker fails to properly clear a downstream fault. The next device in the system which can clear is the opposite primary feeder circuit breaker. If this occurs, a total loss of service to the substation will result. As a result, additional overcurrent relays (device 51) are sometimes added to the tie circuit breaker on systems where the possibility of this occurrence cannot be tolerated, although they are not shown on the system in Fig 36. These relays can be set so as not to extend any other relay operating time, while providing the necessary backup protection to afford proper circuit isolation for faults upstream from either main circuit breaker.

The source ground relaying for the double-ended 2.4 kV primary unit substation is similar to that described for the 13.8 kV transformer secondary. The single-ended 1500 kVA 2.4 kV primary unit substation illustrates a method for high-resistance grounding utilizing an isolation transformer in the neutral circuit. This scheme limits the magnitude of ground current to a safe level while permitting the use of a lower voltage rated resistor stack. The remainder of the 2.4 kV relaying shown in Fig 36 is, in one form or another, provided for protection of the motor loads.

The application of a combination motor and transformer as shown connected to the 13.8 kV bus 3 is referred to as the unit method. This is done in order to take advantage of the lower cost of the motor and the transformer at 2.4 kV, as compared to the motor alone at 13.8 kV. Motor internal fault protection is provided by instantaneous overcurrent relays, arranged to afford differential protection (device 87M), energized by zero-sequence (or doughnut type) current transformers located either at the motor terminals or, preferably, in the starter. This will also afford protection to the cable feeder. Three current transformers and three relays are applied in this form of differential protection. Thermal overload protection is provided by device 49 using an RTD as the temperature sensor. Surge protection is provided by the surge arrester and capacitor located at the motor terminals, while undervoltage and reverse-phase-rotation protection is provided by device 47 connected to the bus potential transformers. The sudden pressure relay (device 63) is used for detection of transformer internal faults. Branch circuit phase- and ground-fault protection is provided by devices 51/50 and 50GS, respectively.

The 500 hp induction motor served from the 2.4 kV bus 1 is provided with a nonfused class E contactor. The maximum fault duty on this 2.4 kV bus is well within the 50 000 kVA interrupting rating of the contactor, and therefore fuses are not required. Motor overload protection is furnished by the replica-type thermal relay (device 49) with the instantaneous overcurrent element (device 50) applied for phase-fault protection. Separate relaying for motor locked-rotor protection is normally not justified on motors of this size. Undervoltage and single-phasing protection is provided for this and the other motors connected to this bus by device 46V, a negative-sequence voltage relay connected to the bus potential transformers.

Due to the essential function of the motors applied on this bus, a high-resistance grounding scheme is utilized. A line-to-ground fault produces a maximum of 2 A as limited by the 1.72 Ω resistor applied in the neutral transformer secondary. A voltage is developed across relay device 59N which initiates an alarm signal to alert operating personnel.

The 1250 hp induction motor connected to 2.4 kV bus 2 is provided with a fused class E contactor for switching. The fuse provides protection for high-magnitude faults. Motor overload protection is furnished by a replica-type thermal relay (device 49). Locked-rotor and circuit protection for currents greater than heavy overloads is furnished by device 51. Protection against single phasing under load is provided by the current-balance relay (device 46). Instantaneous ground-fault protection is provided by device 50GS, which is connected to trip the motor contactor since the ground-fault current is safely limited to 800 A maximum. Undervoltage and reverse-phase rotation protection is provided by device 47. Surge-voltage protection is properly accomplished by the surge protection equipment connected to the bus since the motor feeder circuit length is short.

4.5.2.3 *Low-Voltage Protection.* Fig 36 illustrates several different types of 480 V unit substation operating modes. Buses 1, 2, and 3, for example, represent a typical low-voltage industrial spot network system which is often used where the size of the system and its importance to the plant operation require the ultimate in service continuity and voltage stability. Multiple sources operating in parallel and properly relayed provide these features. The circuit breakers are provided with either electromechanical or static trip devices as the overcurrent protection means. Ground-fault protection is also indicated and would be supplied either as an optional modification to the static trip device on the respective circuit breaker, or as a standard zero-sequence relaying scheme. A third approach to ground-fault tripping of the main circuit breakers would be transformer neutral relaying.

Since the trip devices of the three main circuit breakers supplying 480 V buses 1, 2, and 3 would normally be set identically to provide selectivity with the tie circuit breakers feeding the 3000 A bus and the other 480 V feeder circuit breakers for downstream faults, directional relays must be provided on these circuit breakers. This will permit selective operation between all 480 V feeder circuit breakers and the main circuit breaker during reverse current flow conditions for transformer or primary faults. Directional relays might also be applied to each of the service tie circuit breakers feeding the 3000 A bus duct so as to provide selective operation between these interrupters for transformer secondary bus faults.

To protect the 600 A frame size feeder circuit breakers from the high level of available fault current at secondary buses 1, 2, 3, and 5, current-limiting fuses are applied in combination with each circuit breaker. Since the tie circuit breaker at bus 5 is normally closed, the main circuit breakers are also provided with directional relays to ensure selective operation between mains for upstream faults.

The unit substation feeding 480 V bus 4 is a conventional radial arrangement and, except for the addition of ground-fault protection, the circuit breakers are equipped with standard trip devices. Bus 6 is fed from an ungrounded transformer and is provided with a ground-fault detection system with both a visible and an audible signal. The small low-current frame-size circuit breakers have standard trip devices only and do not require the assistance of current-limiting fuses as a result of the lower fault duty on the load side of the 1000 kVA transformer.

4.5.3 *Relaying for an Industrial Plant with Local Generation* [19], [24], [25]. When additional power is required in a plant that has been generating all its power, and a parallel-operated tie with a utility system is adopted, the entire fault-protection problem must be reviewed together with circuit breaker interrupting capacities and system component withstand capabilities. In Fig 37 the following assumptions are made.

(1) All circuit breakers in the industrial plant are capable of interrupting the increased short-circuit current.

(2) Each plant feeder circuit breaker is equipped with inverse-time or very-inverse-time overcurrent relays with instantaneous units.

(3) Each of the generators is protected by differential relays and also has external fault backup protection in the form of generator overcurrent relays with voltage-restraint or voltage-controlled overcurrent relays as well as negative-sequence current relays for protection against excessive internal heating for line-to-line faults.

(4) The utility company end of the tie line will be automatically reclosed through synchronizing relays following a tripout.

(5) The utility system neutral is solidly grounded and the neutrals of one or both plant generators will be grounded through resistors.

(6) The plant generators are of insufficient capacity to handle the plant load, and no power is to be fed back into the utility system under any condition.

Protection for the utility end of the tie line might consist of three distance relays or time overcurrent relays without instantaneous units. If the distance relays were used, they would be set to operate instantaneously for faults in the tie line up to 10 percent of the distance from the plant, and with time delay for faults beyond that point in order to allow one step of instantaneous relaying in the plant on heavy faults. If time overcurrent relays were used, they would be set to coordinate with the time delay and instantaneous relays at the plant. At the industrial plant end of the tie at circuit breaker 1 there should be a set of directional overcurrent relays for faults on the tie line, and reverse power relaying to detect and trip for energy flow to other loads on the utility system should the utility circuit breaker open.

The directional overcurrent relays are designed for optimum performance during fault conditions. The tap and time dial should be set to ensure operation within the short-circuit capability of the plant generation, but also to be selective to the extent possible with other fault-clearing devices on the utility system.

The reverse power or power directional relay is designed to provide maximum response for flow of energy into the utility system where coordination with the utility protective devices is not a requisite of proper performance. A sensitive tap setting can be used,

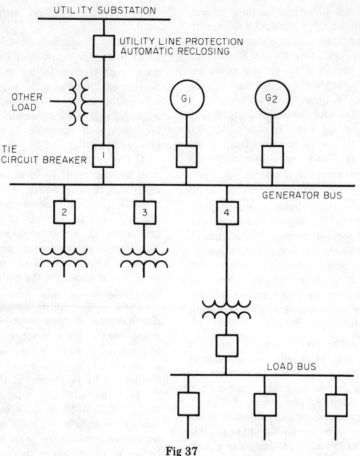

Fig 37
Industrial Plant System with Local Generation

although some time delay is required to prevent nuisance tripping that may occur from load swings during synchronizing.

Due to this time delay a reverse power relay trip of circuit breaker 1 alone is usually too slow to prevent generator overload in the event of loss of the utility power source. Further, the amount of power flowing out to the other utility loads may not at all times be sufficient to ensure relay pickup. A complete loss of the plant load can only be prevented by early detection of generator frequency decay to immediately trip not only circuit breaker 1, but also sufficient nonessential plant load so that the remaining load is within the generation capability. An underfrequency relay to initiate

the automatic load shedding action is considered essential protection for this system. For larger systems two or more underfrequency relays may be set to operate at successively lower frequencies. The nonessential loads could thereby be tripped off in steps depending on the load demand on the system.

The proposed relay protection for a tie line between a utility system and an industrial plant with local generation must be thoroughly discussed with the utility to ensure that the interests of each are fully protected. Automatic reclosing of the utility circuit breaker with little or no delay following a tripout is usually normal on overhead lines serving more than one customer. To protect against the possibility of

Table 13
Maximum Overcurrent Protection (in percent)

Transformer Rated Impedance	Primary Over 600 V		Secondary Over 600 V		600 V or Below
	Circuit Breaker Setting	Fuse Rating	Circuit Breaker Setting	Fuse Rating	Circuit Breaker Setting or Fuse Rating
No more than 6%	600	300	300	150	250
More than 6% but no more than 10%	400	200	250	125	250

the two systems being out of synchronism at the time of reclosure, the incoming line circuit breaker 1 can be transfer tripped when the utility circuit breaker trips. The synchro-check relaying at the utility end will receive a dead-line signal and allow the automatic reclosing cycle to be completed. Reconnection of the plant system with the utility supply can then be accomplished by normal synchronizing procedures.

Generator external-fault protective relays, usually of the voltage-restraint or voltage-controlled overcurrent type, and negative-sequence current relays provide primary protection in case of bus faults and backup protection for feeder or tie line faults. These generator relays will also operate as backup protection to the differential relays in the event of internal generator faults, provided there are other sources of power to feed fault current into the generator.

4.6 Protection Requirements. The primary purpose of a coordination study is to determine satisfactory ratings and settings for the electric system protective devices. The protective devices must be chosen so that pickup currents and operating times are short but sufficient to override system transient overloads such as inrush currents experienced when energizing transformers or starting motors. Further, the devices should be coordinated so that the circuit interrupter closest to the fault opens before other devices.

Determining the ratings and settings for protective devices requires familarity with the NEC requirements for the protection of cables, motors, and transformers, and with IEEE Std 462-1973, General Requirements for Distribution, Power, and Regulating Transformers (ANSI C57.12.00-1973) for transformer magnetizing inrush current and transformer thermal and magnetic stress damage limits.

4.6.1 Transformers. The NEC provides minimum standards of overcurrent protection for the setting of transformer protective devices. If there is no secondary protection, transformers rated for more than 600 V require either a primary circuit breaker or a primary fuse that will trip at no more than 300 percent or 150 percent of transformer full-load current, respectively. Provided the nature of the load on the transformers will permit it, better protection will be realized with a setting lower than the 300 percent NEC circuit breaker requirement. If there is a secondary circuit breaker or fuse, the protection requirements depend on the rated transformer impedance and secondary voltage as well as the type of secondary protection and the corresponding rating as summarized in Table 13.

Transformers rated 600 V and less, on the other hand, essentially require primary protection at 125 percent of full-load current when no secondary protection is present, and 250 percent of rated full-load current as the maximum primary protection if there is secondary protection set at no more than 125 percent of transformer rating. Certain exceptions to these requirements for smaller sized transformers detailed in Article 450-3 of the NEC are intended to permit the application of protective devices having the ratings normally available. The permissible fuse rating is generally lower than the circuit breaker setting due to the difference in the circuit opening characteristics in the overload region.

The transformer primary protection devices should have the ability to do the following.

(1) Carry the transformer magnetizing inrush current. In general the transformer inrush current is approximately 8 to 12 times the transformer full-load current for a maximum period of 0.1 s. This point should be plotted on the coordination graph and it should fall below the transformer primary protection device curve.

(2) Clear a bolted secondary short circuit before the transformer is damaged. In accord with IEEE Std 462-1973, standard transformers are designed to withstand the internal stresses caused by short circuits on the external terminals within the following limitations:

25 times base current for 2 s

20 times base current for 3 s

16.6 times base current for 4 s

14.3 or less times base current for 5 s

At levels of current in excess of about 400—600 percent of full load the transformer withstand characteristic can be conservatively approximated by a constant $I^2 t$ (heating) plot which is represented by a straight line of -2 slope extending to and terminating at the appropriate short-circuit withstand points described above. For proper protection it should fall above the transformer primary protective device curve when plotted on a coordination graph. Another consideration is a relative shift in the damage point which occurs in delta—wye transformers with the wye on the low side and the neutral point grounded. A secondary single-phase-to-ground fault of one unit value will produce a fault current of one unit in the delta winding on the primary side, but cause only 57.8 percent current which is seen by the protective device in the line to the delta winding. Therefore a second damage point, corresponding to that provided by IEEE Std 462-1973 and derated for a wye-wound solidly grounded neutral secondary, should be plotted at 57.8 percent of the normal point.

In selecting a primary protective device, the following items must be known and considered:

(1) Voltage rating of the system

(2) Rated load and inrush current of the transformer

(3) Short-circuit load of the supply system in kilovolt-amperes

(4) Type of load, whether steady, fluctuating, or subject to heavy motor, welding, furnace, or other starting surges

(5) Coordination with other protective devices

Relays, when used in combination with power circuit breakers for protection of a transformer primary circuit, should have a time—current curve shape similar to that of the first downstream device. Pickup of the time-delay element may typically be 150—200 percent of the transformer primary full-load current rating. The instantaneous pickup setting should be set at 150—160 percent of equivalent maximum secondary three-phase symmetrical short-circuit current to allow for the direct-current component of fault current during the first half-cycle. If there is more than one transformer connected to this feeder, the pickup of the time-delay element should not exceed 600 percent full-load current of the smallest transformer, assuming that the transformers have secondary protection and an impedance of 6 percent or less. When used in the transformer secondary circuit, the pickup of the time-delay element should also be between 150 and 200 percent full-load current of the transformer secondary rating. A typical circuit configuration is illustrated by the one-line diagram insert in Fig 42.

4.6.2 *Feeder Conductors.* Restrictions that apply are provided in the NEC. Protection of feeders or conductors rated 600 V or less shall be in accordance with their current-carrying capacity as given in NEC tables, except where the load includes motors. In this case it is permissible for the protective device to be set higher than the continuous capability of the conductor (to permit coordination on faults or starting the largest connected motor while the other loads are operating at full capacity), since running overload protection is provided by the collective action of the overload devices in the individual load circuits. Where protective devices rated 800 A or less are applied which do not have adjustable settings that correspond to the allowable current-carrying capacity of the conductor, the next higher device rating may be used.

Feeders rated more than 600 V are required to have short-circuit protection which may be provided by a fuse rated at no more than 300 percent of the conductor ampacity or by a circuit breaker set to trip at no more than 600 percent of the conductor ampacity. Although not required by the NEC, improved protection of these circuits is possible when running overload protection is also provided in accordance with the conductor ampacity.

The flow of short-circuit current in an electric system imposes mechanical and thermal stresses on cable as well as circuit breakers, fuses, and the other electric components. Consequently to avoid severe permanent damage to cable insulation during the interval of short-circuit current flow, feeder conductor damage characteristics should be coordinated with the short-circuit protective device. The feeder conductor damage curve, which is available from the cable manufacturer, should fall above the clearing-time curve of its protective device without overlapping.

4.6.3 Motors

4.6.3.1 *Large Alternating-Current Rotating Apparatus.* The protection of an alternating-current induction motor is a function of its type, size, speed, voltage rating, application, location, and type of service. In addition, a motor may be classified as being in essential or nonessential service, depending upon the effect of the motor being shut down on the operation of the process or plant. Although the discussion earlier in this chapter on the different types of protective devices indirectly touches on some of the problems associated with protecting motors, it is worthwhile to examine such an important subject from the standpoint of the machine itself.

Unscheduled motor shutdowns may be caused by

(1) Internal faults
(2) Subtained overloads and locked rotor
(3) Undervoltage
(4) Phase unbalance or reversal
(5) Voltage surges
(6) Reclosure and transfer switch operations

The ideal relay scheme for an induction motor must provide protection against all these hazards. In the following text the relaying approach to protect against each of these problems will be discussed in general terms. Later in the chapter, several specific applications will be discussed in detail. For a complete discussion on motor protection, see Chapter 9 of IEEE Std 242-1975.

(1) *Internal Faults.* Internal fault protection for induction motors can be obtained by either overcurrent relays or percentage differential relays as described in Section 4.3.5.1. When the supply source is grounded, separate and more sensitive ground-fault protection can be provided using the relaying schemes described in Sections 4.3.7.1 and 4.3.7.2. The preferred solution is to use the zero-sequence approach for ground-fault relaying. All three phase leads are passed through the single window-type current transformer. This eliminates false tripping due to unequal current-transformer saturation and allows the use of a sensitive ground-fault relay setting.

(2) *Sustained Overloads and Locked Rotor.* Conventional overcurrent relays do not provide suitable protection against sustained overloads because they will "overprotect" the motor if set to pickup for the normal overloads encountered. That is, the relay will not allow full use of the thermal capability of the motor and in many cases will not provide sufficient time delay to permit complete starting. This is shown in Fig 38, where for most conditions less than locked-rotor current there is too much margin between the motor thermal capability curve and the relay operating time characteristic. With a higher pickup and appropriate time-dial setting as shown the overcurrent relay will provide excellent locked-rotor and short-circuit protection of the motor.

Thermal relays, on the other hand, will give adequate protection for light and medium overloads, allowing loading of the motor close to its thermal capability. In general, however, thermal relays will not give adequate protection for heavy overloads, locked rotor, and short circuits. Therefore, in most cases both types of relays should be used to provide optimum protection for overloads, locked rotor, and short circuits and allow maximum

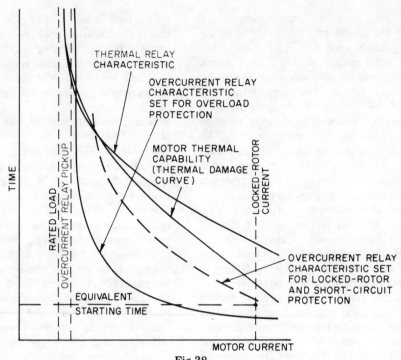

Fig 38
Motor and Protective Relay Characteristics

use of the motor capability. In this way the characteristics of the protection can be closely "shaped" to the motor thermal damage curve.

There are two common types of thermal relays available for motor protection as discussed in Sections 4.3.14 and 4.3.16. One operates in response to resistance temperature detectors embedded in the machine windings, and the other operates in response to motor current. The latter type normally has adjustable pickup and trip characteristics to compensate for the motor service factor as well as the ambient temperature differences between motor and relay.

Frequently medium-voltage motors are protected by a contactor with thermal overload relays applied in combination with current-limiting fuses, which are intended to open the circuit for high fault currents. In addition to matching the overload protection to the motor thermal capabilities for such applications, it is equally important to select the fuses so that

they protect the contactor by opening faster on currents in excess of the contactor interrupting rating. Likewise, the overload relays must prevent fuse blowing by tripping before the fuse clears on currents within the contactor capabilities.

(3) *Undervoltage.* Low voltages can prevent motors from coming up to normal operating speed or they can result in overload conditions. Although thermal overload relays will detect an overload resulting from undervoltage, large motors and medium-voltage motors should have separate undervoltage protection. An induction-type undervoltage relay is usually provided to prevent starting when the voltage is unacceptably low, and to prevent operation on momentary voltage dips.

Single-phasing under load or three-phase dissymmetry can cause overheating within the motor at a rate above that perceived by the thermal overload relays, even in the phase with the largest current. If the machine is only partially loaded, the overload relays may never detect that damage is occurring.

(4) *Phase Unbalance or Reversal.* When starting from rest, single phase (one line open) will prevent starting, while reverse-phase rotation can have immediate disastrous results on the motor or the driven equipment. In all cases where such conditions exist, a phase-failure and reverse-phase relay should be applied.

If not properly protected, three-phase motors are vulnerable to damage when loss of voltage occurs on one phase. There are numerous causes of such loss of voltage, and these can occur anywhere in the distribution system. The chief problems resulting from single-phasing of three-phase motors is overheating, which can cause reduction of life expectancy or complete failure.

The modern practice of applying three overload devices on three-phase motors assures detection of single-phasing in most cases. When operating at normal load, loss of voltage on one phase causes an abnormal current in the remaining phases which is sensed by the protective devices. However, under some conditions of light loading, three-phase motors can overheat when single-phased, without proper operation of the overload protective devices. Even when operating near rated horsepower under single-phase conditions, the motor can be damaged prior to response by conventional protective devices.

Negative-sequence voltage or current-balance relays should be considered for protection of motors above 1000 hp against this condition.

(5) *Voltage Surges* [26]–[31]. Voltage surges are transient overvoltages caused by switching or lightning strokes. They are characterized by a steep wave front. Surge protection equipment consists of a protective capacitor and arrester which should be connected as close to the motor terminals as possible. Further application criteria for treating this problem are discussed in Section 4.9 and IEEE Std 142-1972.

(6) *Reclosure and Transfer Switch Operations* [3]. Under normal operating conditions, the self-generated voltage of an alternating-current motor lags the bus voltage by a few electrical degrees in induction motors and by 25–35 electrical degrees in synchronous motors. The operation of a reclosure on the utility power supply or the transfer to an alternate source will cause the power to be interrupted for a fraction of a second or longer. When power is removed from a motor, the terminal voltage does not collapse suddenly, but decays in accordance with the open-circuit machine time constant (time for self-generated voltage to decay to 37 percent of rated bus voltage). The load inertia acts as a prime mover which attempts to keep the rotor turning. The frequency or phase relationship of the motor self-generated voltage no longer follows the bus voltage by a fixed torque angle, but starts to separate farther from it (out of phase in electrical degrees) as the motor decelerates.

If the motor is reconnected to the bus voltage with its self-generated voltage at a high level and severely out of phase, dangerous stresses, both mechanical and electrical, are placed on the motor. In addition to possible damage to the motor, excessive torque may also damage the motor coupling. Furthermore, the excessive current drawn by the motor may trip the overcurrent protective device.

A check should be made to determine that re-energization occurs at a point where the motor and load will not be subjected to excessive forces. Protection against this problem can be provided by certain types of frequency relays that operate as a function of the rate of change of frequency to remove the motor from the line. An alternate method is to prevent re-energization until the residual voltage has decayed to a safe value.

Automatic transfer switching means can be provided with accessory controls that disconnect motors prior to transfer and reconnect them after transfer and when the residual voltage has been substantially reduced. Another method is to provide in-phase monitors within the transfer controls that prevent transfer until the motor bus voltage and the source are nearly synchronized.

4.6.3.2 *Small Motors.* The specific requirements for the protection of small induction motors are specified in Article 430 of the NEC. Each motor branch circuit must be provided with a disconnect means, branch circuit protection, and a motor running overcurrent protective device. Two examples are shown in Fig 39.

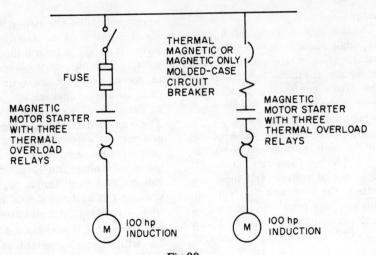

Fig 39
Motor Protection Acceptable to the NEC

The branch circuit disconnect and protective means are generally combined in one device such as a molded-case circuit breaker or a fused disconnect switch. The motor running overcurrent protection is provided by overload relays. The motor is energized and de-energized by a controller. This unit may be operated either manually or electrically (magnetic type). The overload relays open the motor controller to provide motor running overcurrent protection. It is not uncommon to have the motor controller included in the same enclosure as the motor branch circuit disconnect and overcurrent protection device. The complete unit is called a combination motor starter, providing motor branch circuit disconnect and overcurrent protection along with motor control and running overcurrent protection.

The motor branch circuit overcurrent device must allow the motor to start (without opening on motor inrush current), but it must open for short circuits.

A combination disconnect and overcurrent protective device must safely open the circuit under a short-circuit condition and disconnect the circuit with the motor running at full load or with locked rotor. The short-circuit overcurrent protective device must be capable of interrupting the circuit under the maximum available short circuit and in so doing protect the

branch circuit. The switch should be quick-make—quick-break, horsepower rated, and capable of being closed in on a fault of the magnitude available at its application point without damage. The switch must safely withstand the $I^2 t$ and peak let-through current by the fuses used in conjunction with it without realizing an immediate failure or change in operating characteristics which could lead to problems during normal operation some time later.

For running overcurrent protection it is necessary to select the proper thermal unit for the overload relay. All manufacturers' tables of thermal units are based on the operation of the motor and controller in the same ambient temperature of 40°C or less. To apply these devices properly the following must be determined: (1) motor full-load and locked-rotor current from motor nameplate, (2) motor service factor from nameplate, (3) ambient temperature for motor, (4) ambient temperature for controller, and (5) motor starting time with load connected. With this information, and following the manufacturer's recommendation an adjusted motor full-load current can be determined to select the proper overload relay thermal unit from manufacturers' tables. Then it must be verified that the trip characteristic will permit starting.

Undervoltage protection is inherent with the use of a magnetic controller and three-wire control, with the control voltage taken from the line or the primary side of the controller. Most magnetic motor controllers will drop out when the operating coil voltage drops to 65 percent of its rating. All units do not have the same drop-out characteristics, so the actual drop-out voltage should be determined by test. For many motors the three overload devices may not provide complete single-phase protection, which in such cases can be furnished as a special equipment modification.

4.7 Use and Interpretation of Coordination Curves

4.7.1 *Need and Value.* Determining the settings for the overcurrent protection on a power system can be a formidable task which is often said to require as much art as technical skill. Continuity of plant electric service requires that interrupting equipment operate selectively. This normally demands slower operation or longer opening delays (for a given current) of the interrupters successively closer to the power source during faults. The necessity for maximum safety to personnel and electric equipment, on the other hand, calls for the fastest possible isolation of faulted circuits.

The coordination curve plot provides a graphical means of optimizing the competing objectives of selectivity and protection by examining the problem of interrupter device coordination. This method of analysis is useful when designing the protection for a new power system, when analyzing protection and coordination conditions in an existing system, or as a valuable maintenance reference when checking the calibration of protective devices. The coordination curves provide a permanent record of the time—current operating relationship of the entire protection system.

Actual plotting of the curves on a single sheet of graph paper using a common current scale is essential because rarely do all the fault protective devices involved have time—current curves of the same shape, and it is difficult to visualize the relationship of the many different shapes of curves. A scale corresponding to the currents expected at the lowest voltage level works best. For example, fault-current protec-

tive devices on both sides of a 2400—480 V transformer should be plotted on the 480 V current scale. To plot 2400 V device time—current curves on the 480 V scale, first determine the desired time and current settings in the usual manner on the basis of current expected in the 2400 V circuit. Then plot time directly since that scale is unchanged, but multiply the 2400 V currents by 5 (ratio of 2400/480) to obtain equivalent current at 480 V before plotting on the 480 V scale.

Usually the coordination plot is made on log—log graph paper with current as the abscissa (horizonatal axis) and time as the ordinate (vertical axis). A choice of the most suitable current and time settings is made for the device to provide the best possible protection and safety to personnel and electric equipment and also to function selectively with other protective devices to disconnect the faulted equipment with as little disturbance as possible to the rest of the system.

4.7.2 *Device Performance.* The manufacturers of protective devices publish time—current operating characteristic curves and other performance data for all devices used in a protection system. The individual time—current characteristics of thermal overload relays, fuses, static and series trip devices, and relays are transposed onto a common plot for selecting coordinated settings. Typical relay curves are shown in Fig 23. Relay time—current curves normally begin at multiples of 1.5 times minimum closing current or pickup setting, since their performance cannot be accurately predicted below that value. However, curves showing the approximate expected time—current performance of lower values can usually be obtained from the manufacturer if required. The relay time—current curves define the operation of the relay alone and do not include any circuit breaker interrupting time.

4.7.2.1 *Time-Delay Relays in Series.* The time—current curves of direct-acting time-delay trip devices, fuses, and time-delay thermal devices include the necessary allowance for overtravel, manufacturing tolerances, etc. The time—current characteristics of relays are represented by families of single-line curves to which tolerance bands must be added. This tolerance band is based on the fact that the second relay in a chain continues to see fault current until the circuit breaker associated with the first relay has opened and the arc extinguished. This is nominally 5—8 cycles for the circuit breakers commonly used in industrial systems, although the actual opening time will be 3—5 cycles. After the first circuit breaker has opened the circuit and de-energized the second relay, the contacts of the second relay will continue to coast for 0.1 s (standard inverse-time relay) due to the inertia of the induction disk to which the movable contact is attached.

A minimum total time margin of 0.4 s at maximum fault current is sufficient to afford satisfactory selectivity between inverse-time relays. This margin allows for the 0.13 s circuit breaker opening time (8 cycles), 0.1 s overtravel, and a safety factor of 0.17 s to cover manufacturing variations and inaccuracies in positioning of the time dial or level when setting the relay. If the relays are more accurately set to a specific curve by instrument, this margin can be safely reduced to approximately 0.33 s. Faster circuit breaker operation and solid-state relays or applications with electromechanical relays which result in less disk overtravel will permit further reduction of this margin (to approximately 0.2—0.25 s).

When two induction relays in series are set to coordinate at the maximum available fault current, they will be selective on lower values of current if they have the same shape time—current curves, and the minimum operating current setting (pickup) of the slower relay is higher than that of the faster relay. If the pickup setting is lower, the operating-time curves of the two relays will cross each other at some low value of fault current, and the "slow" relay will beat the "fast" one for all currents below that value.

Plotting the time—current curves on a single graph with a common current for all the relays and other devices in series is illustrated in Fig 40 which reveals two conflicts which might otherwise have escaped notice. In this case the power supply was sufficient to maintain a 250 000 kVA short-circuit duty without appreciable decrement, and there was no fault current contribution from synchronous equipment connected to the 2.4 kV system. On this basis the maximum 2.4 kV system symmetrical fault current is 20 000 A and the maximum 13.8 kV system symmetrical fault current is 10 460 A (60 000 A on a 2.4 kV base). It was also assumed that the end relay in the chain (D) was to be set at a minimum of 0.5 s.

The three sets of 2.4 V relays were coordinated by selecting time—current settings that would make their operating times 0.4 s apart at the maximum current of 20 000 A. Then the single set of relays on the 13.8 kV system was coordinated with those on the 2.4 kV system at the same value of fault current (3480 A at 13.8 kV) because this is the current that the relays at A would see during a fault on the 2.4 kV side of the transformer. Coordination was accomplished by selecting time—current settings that would give 0.4 s delay between relays A and B for a 20 000 A fault on the 2.4 kV system.

Following the foregoing procedure could possibly give satisfactory results without actually plotting the curves if the basic rules for coordination relays were observed, that is, (1) relays with the same shape curves were used in series with each other and (2) relays farthest from the source of power had current settings below that of the relays ahead of them.

Unfortunately it is easy to overlook these basic rules, and because of this the effort required to plot the curves proves worthwhile.

As shown in Fig 40, time—current settings on relay D that would give either curve 1 or curve 2 would satisfy the requirement that it take a minimum of 0.5 s at 20 000 A. The curve 2 setting, however, is slower and less sensitive than that represented by curve 1 throughout most of its range. Furthermore, curve 2 crosses curve 3 which represents the desired setting for relay C, thereby requiring desensitization of C to make it selective with D. Thus the curve 2 setting on relay D would greatly reduce the short-circuit protection provided by either relay C or D.

Curves 4, 5, 6, and 7 of Fig 40 illustrate what would happen if relay B had a very-inverse-time characteristic instead of an inverse-time curve as the others do. Curve 4 meets the requirement that it be 0.4 s slower than curve 3 representing relay C, when both are operating at 20 000 A. Also, curve 6 satisfies the requirement that relay A be 0.4 s slower than B when A is operating at the equivalent of the 20 000 A 2.4 kV

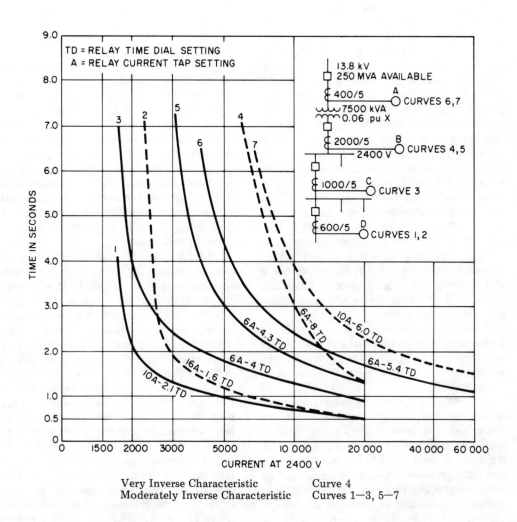

Fig 40
Selecting Time—Current Curves and Relay Tap Settings for an Industrial Plant Distribution System

NOTE: The curves are plotted on semilog coordinates for illustrative purposes only. Normally log-log coordinates are used. Also, in actual practice tap settings above 8A are seldom required.

system short-circuit current. If the curves had not been plotted, there would be reason to believe that the contemplated settings for relays A and B as represented by curves 6 and 4 would be satisfactory. Actually, the very-inverse-time characteristic of relay B causes its curve to cross that of relay A at a high level of fault current, so that the tripping sequence of the circuit breakers would be reversed. For this particular circuit it would not be too serious, since tripping either circuit breaker would shut down the whole circuit, but it would still nullify the effectiveness of the relays in giving indication as to where the trouble was. If the very-inverse-time relay were retained at B, relay A's setting would have to be desensitized and increased in time (curve 7) to be selective with B. This would result in greater damage during a short circuit, so an inverse-time relay should be substituted for the very-inverse-time relay at B, making it possible to set B to give performance as shown by curve 5. This would mean that A and B could both be more sensitive and faster, and consequently would give better protection to the system.

If the very-inverse-time relay were used at B, the backup protection that it could provide would be very poor due to the big "gap" in the pickup currents of relays B and C, as shown by curves 3 and 4.

When choosing between two combinations of current-tap and time-dial settings, either of which will give a desired operating time at maximum fault current, the combination with the lower current and higher time-dial setting is usually preferable, because the relay will be more sensitive and faster on low values of fault current. Suppose that an operating time of 0.5 s is desired with a relay connected to 1000/5 A current transformers in a circuit with an available symmetrical fault current of 20 000 A. Relays with 6 A tap and 2.1 time-dial setting, or 10 A tap and 1.7 time-dial setting will both give the desired time. But in case of a fault involving only 3000 A the relay with the 6 A setting would operate in 1.25 s compared with 2 s for the 10 A combination. If the current is still further reduced to 2000 A, the first relay will still operate in 2.1 s, but the second one will be very slow, since operation is uncertain when the current is only 1.0 times relay pickup.

A special problem arises when attempting to coordinate overcurrent devices on opposite sides of a delta—wye-connected transformer. For line-to-line faults occurring and measured on the wye side, the available fault current through the transformer will be reduced to approximately 87 percent ($\sqrt{3}/2 \times 100$) of the available three-phase fault current. The highest of the unbalanced phase currents on the delta side, however, will be 100 percent of the value experienced for a balanced three-phase fault, and the overcurrent device in this phase will operate faster relative to the protection on the wye side. The effect that this change in relative operating characteristics has on coordination for line-to-line faults can be examined graphically by shifting the normal plot of the delta-side protective device by the ratio of 0.87/1.0. This technique will be illustrated by an example in Section 4.8.

4.7.2.2 *Instantaneous Relays.* When two circuit breakers in series both have instantaneous overcurrent relays, their selectivity is dependent solely on their current settings. Therefore the relays must be set so that the one nearest the source will not trip when maximum asymmetrical fault current flows through the other circuit breaker. This requires sufficient impedance in the circuit between the two circuit breakers to cause faults beyond both circuit breakers to receive less current than faults near the source circuit breaker, so that the relays of the source circuit breaker can determine the fault location. If this differential is insufficient, selective operation is impossible with instantaneous overcurrent relays and the opening of both circuit breakers on through faults must be tolerated or the instantaneous relays of the circuit breaker farthest from the source must be made inoperative.

Usually the impedance of a transformer is sufficient to achieve selectivity between an instantaneous relay on a primary feeder and the instantaneous trip coil of a low-voltage secondary circuit breaker. Also the reactance of open transmission lines may be sufficient to provide the necessary differential in short-circuit-current magnitude to permit the use of instantaneous relays at both ends.

Generally instantaneous relays at opposite ends of in-plant cable systems cannot be co-

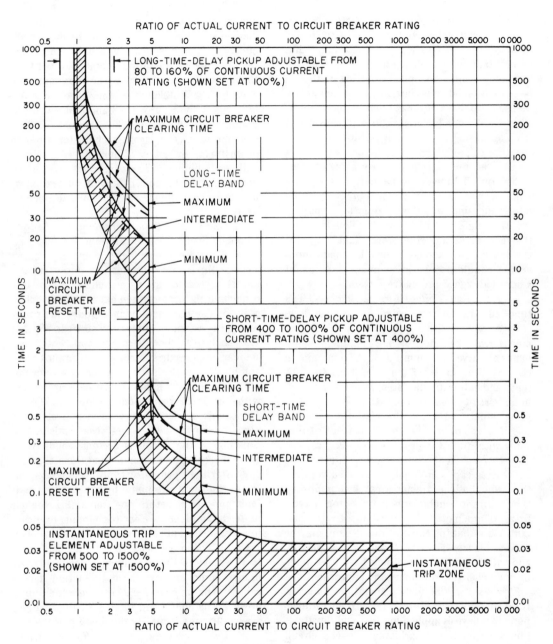

Fig 41
Adjustability Limits of Low-Voltage Power Circuit Breaker Trip Devices

ordinated because the circuit impedance is too low to provide the necessary current differential.

Applications involving both phase overcurrent and residually connected ground relays must be reviewed carefully to determine that steady-state and transient error currents are below the instantaneous pickup setting of the relay. Some solid-state instantaneous relays are designed to have inverse operating characteristics to allow a lower pickup setting without tripping during the transient portion of the error current.

Instantaneous attachments are generally furnished on all time-delay overcurrent relays on switchgear equipment so that they will be interchangeable, but they should be employed for tripping only when applicable. The fact that a relay setting study reveals that some of the instantaneous relays must be made inoperative should not be interpreted as a sign of a poorly designed protective system.

4.7.2.3 *Low-Voltage Circuit Breakers.* The time—current characteristics for typical low-voltage power circuit breakers with electromechanical trip devices are represented by bands of curves shown in Fig 41. The maximum clearing time includes the total time to both sense and interrupt the fault, whereas the minimum circuit breaker reset time is a current value which can cause the circuit breaker to trip if it is not reduced to 80 percent of that value within the time shown on the curve. The shapes of different manufacturers' direct-acting trip curves vary somewhat, but the curves have the same basic characteristics. The published time—current characteric curves for circuit breakers are plotted in multiples of rated current. Note that the top part of the band may be moved horizontally by changing the pickup, and the knee of the band may be shifted vertically by changing the time-band adjustment.

Fig 42 shows the relative operating characteristics of two circuit breakers with electromagnetic trip devices applied in series and illustrates how coordination is achieved between circuit breakers having different combinations of long-time, short-time, and instantaneous trip elements. Since all tolerances and operating times are included in the published characteristics for low-voltage circuit breakers, to establish

that selectivity exists requires only that the plotted curves do not intersect.

4.7.2.4 *Fuses.* Typical power fuse performance curves are shown in Fig 43. As with low-voltage circuit breakers, the total time for circuit opening is described by a band which represents the manufacturing tolerance of the fuse, the upper boundary being the maximum operating time. Total fuse performance, however, is defined by a broader band with a lower boundary representing the thermal damage characteristic. The clearing characteristic of a downstream interrupter must be coordinated with the thermal damage characteristic of an upstream fuse to prevent any deterioration of its rating or change in its normal opening delay.

4.7.2.5 *Ground-Fault Protection.* As discussed in Section 4.2.4, care must be exercised when using a conventional one-line diagram to analyze three-phase systems. An important application consideration associated with this problem is illustrated when examining the selectivity of a system employing a ground-fault protector in the main service as shown in Fig 44. The analytical flaw becomes apparent when standard time—current curves are employed in combination with the one-line diagram to evaluate coordination of the downstream protectors with the main ground-fault protector (GFP). The simple overlay of curves can indicate complete selectivity and suggest that satisfactory isolation of the circuit is accomplished. However, the change in circuit conditions that occurs after the opening of (only) one phase protector can cause a response by the main ground-fault protector before the remaining branch or feeder protectors clear the circuit, resulting in a loss of the entire electrical service.

The addition of sensitive ground-fault protective devices to each downstream feeder or branch could improve the coordination significantly. The improvement in operation must, however, be considered in view of the cost and availability of adequately rated equipment.

4.7.3 *Preparing for the Coordination Study* [2]. The following information will be required for a coordination study:

(1) A system one-line diagram showing the complete system details including all protective devices and associated equipment

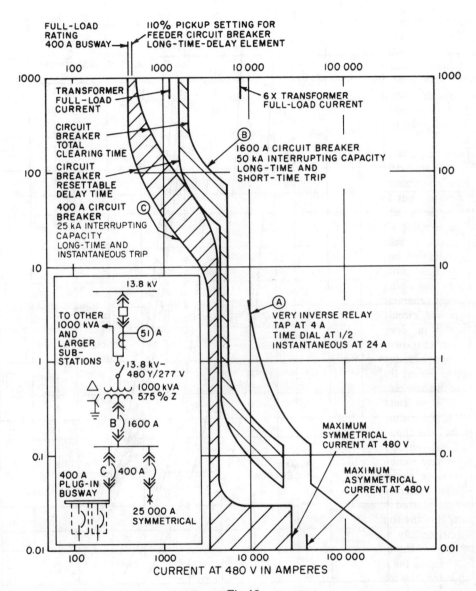

Fig 42
Selective Tripping Time—Current Characteristic Curves for
Low-Voltage Power Circuit Breakers on Secondary Unit Substation

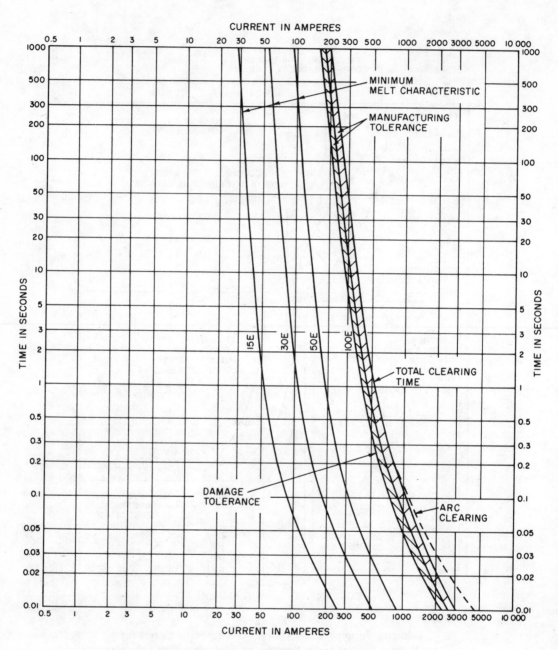

Fig 43
Typical Time—Current Characteristic Curves of Fuses

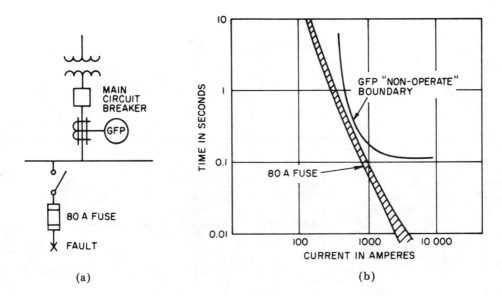

(a)

(b)

Fig 44
One-Line Diagram and Time—Current Coordination Curve
Misrepresenting Proper Fault Clearing

(2) Schematic diagrams showing protective device tripping functions

(3) A short-circuit analysis providing the maximum and minimum values of short-circuit current that are expected to flow through each protective device whose performance is to be studied under varying operating conditions

(4) Normal loads for each circuit and the anticipated maximum and minimum operating loads and special operating requirements

(5) Machine and equipment impedances and all other pertinent data necessary to establish protective device settings and to evaluate the performance of associated equipment such as current and potential transformer ratios and accuracies

(6) All special requirements of the power company intertie, including the time—current characteristic curve of the utility protection immediately upstream from the system

(7) Manufacturers' instruction bulletins, time—current characteristic curves, and interrupting ratings of all electric protective devices in the power system

(8) NEC or other governing code requirements as a reference

4.8 Specific Examples — Applying the Fundamentals. To illustrate some of the many factors that must be considered and the problems that arise when applying the information and principles provided in the previous sections of this chapter to an actual industrial power system, the completed coordination curves (Figs 45—54) for the system shown in Fig 36 will be discussed in detail. Strong emphasis has been placed on the prime objectives of equipment protection and selective interrupter performance. Relays that are unresponsive to system overcurrents and have no time—current characteristic are not shown on the graphical coordination plots. The selection of settings for these devices is beyond the scope of this text but can be readily determined by referring to the manufacturers' instruction material covering the relays in question.

The examples given are only intended to be illustrations. Each system encountered in practice must be analyzed in detail since effective protective device selection and coordination must apply to a specific situation and not a general case.

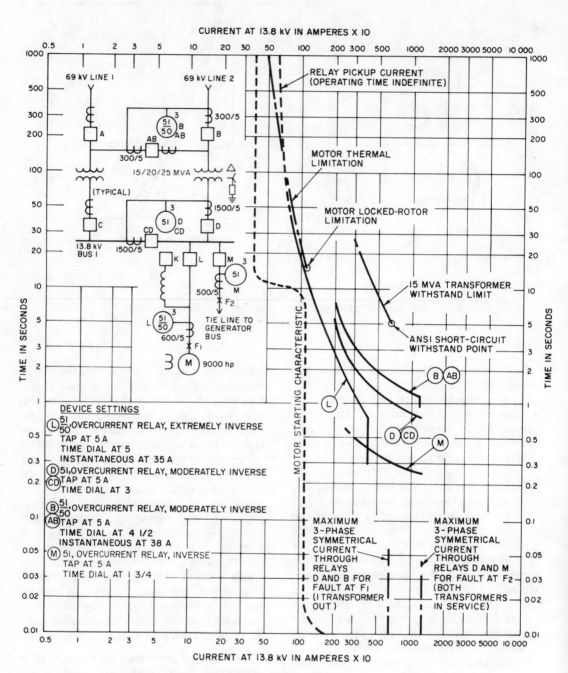

**Fig 45
Phase-Relay Time—Current Characteristic Curves for 13.8 kV
Feeders L and M and Incoming Line Circuits**

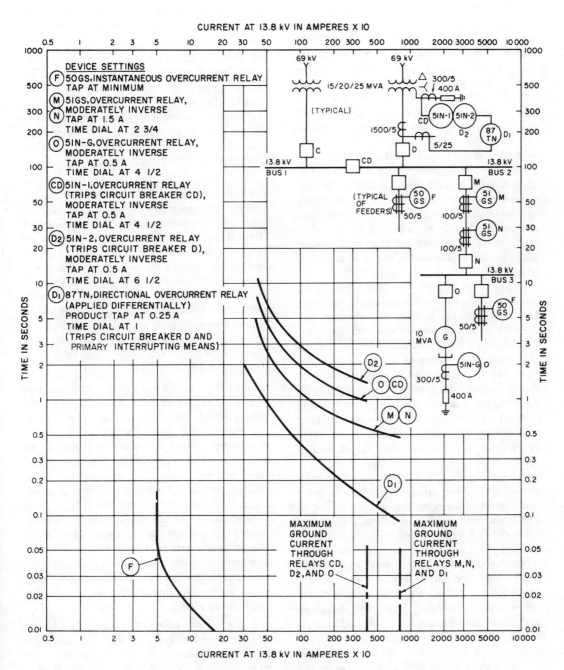

Fig 46
Ground-Relay Time—Current Characteristic Curves for 13.8 kV
Source and Feeder Circuits

4.8.1 *Setting and Coordination of the 13.8 kV System Relaying*

4.8.1.1 *Primary Feeders Supplying Transformer.* The overcurrent relays applied on the 13.8 kV circuits that energize load center distribution transformers provide the dual function of primary protection for phase and ground faults occurring on the 13.8 kV cable and transformer primary winding, and backup protection for faults normally cleared by the secondary devices. Since backup protection requires selective tripping with the secondary main circuit breaker, the primary protection is usually compromised to the extent necessary to obtain coordination. This compromise can be minimized by selecting a relay characteristic which follows the time—current characteristic of the secondary device as closely as possible.

As shown in Fig 52, the overcurrent relays (device 51/50) chosen for feeders E, G, and J, have an extremely inverse characteristic, and the settings provide a curve which ensures selective tripping with the secondary circuit breaker over most of the range of secondary fault current. The upper edge of the secondary main circuit breaker trip curve represents total clearing time, and a margin of 0.2 s between it and the relay curve at the maximum secondary fault current level is recommended. An intersection or crossover with the secondary circuit breaker occurs at a low region of fault current, as shown in Fig 52, for relay G, and this is regarded as an acceptable compromise in order that the transformer would be fully protected within its withstand limits. The transformer withstand limits are plotted on the curve sheet as three-phase fault limitations and also as the equivalent current (57.8 percent) appearing in the primary protective device for secondary line-to-ground faults when the secondary neutral is solidly grounded. The same degree of protection on secondary line-to-ground faults is not provided for the transformers supplied by feeders E and J, since the associated relaying is set higher to accommodate the other connected loads. Installation of separate protection ahead of these transformers having a trip characteristic no higher than that shown for relay G could correct the problem. Clearing could be accomplished either by a separate interrupter ahead of each transformer (not shown) or by transfer tripping of circuit breakers E and J.

The relay pickup or tap value setting is based on three considerations, (1) to enable the feeder and transformer to carry its rated capability plus any expected emergency overloading, (2) to provide for coordination with the transformer secondary circuit breaker, and (3) to provide protection for the transformer and cable within the limitations set forth in the NEC, Articles 450-3 and 240-100.

The adjusted trip characteristic for relay G demonstrates the relative performance of the primary and secondary protective devices for a secondary line-to-line fault. As compared to a three-phase fault, there is a slightly larger range of possible fault currents over which coordination between the primary relay and secondary circuit breaker is compromised. Had the secondary overcurrent device been a relay with the same characteristic as the primary relay, complete coordination could have been realized, provided a sufficient clearance was maintained between the curves at a current of approximately 25 000 A (28 700 × 0.87) at 480 V.

The instantaneous overcurrent element (device 50), employed in conjunction with the time element (device 51), is set to be nonresponsive to the maximum asymmetrical rms fault current which it will see for a transformer secondary three-phase fault. The symmetrical value of transformer let-through current is calculated and an asymmetry multiplying factor applied as determined from the X/R ratio of the impedance to the point of fault. In addition, a 10 percent safety margin is added to this calculated setting. When the relayed feeder is energizing more than one transformer, the magnetizing inrush of the transformer group may be the limiting factor for the instantaneous trip setting.

In Fig 50 the overcurrent relays (device 51/50) for feeders H and I are set to operate selectively with the totalizing relay at the maximum expected 2.4 kV fault current with about a 0.4 s delay between curves. The instantaneous element is set to pick up above the 2.4 kV system asymmetrical fault availability. The pickup setting and characteristics of the extremely inverse relay afford excellent protection of the

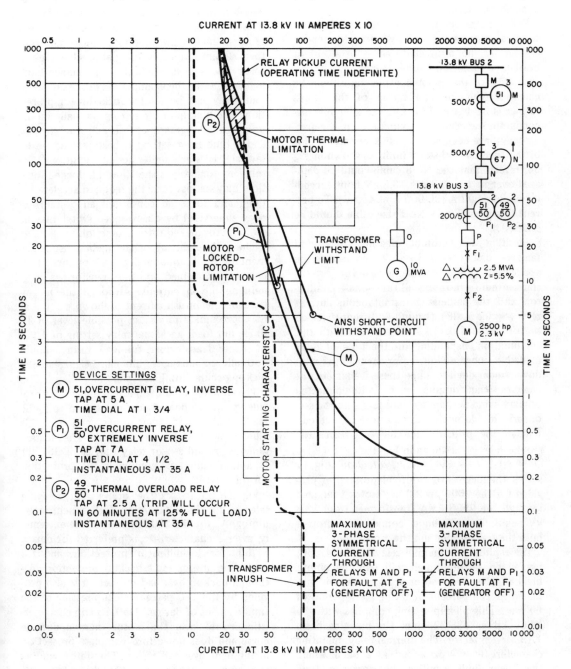

Fig 47
Phase-Relay Time—Current Characteristic Curves for Feeder Relay at 13.8 kV Bus 3

transformer by staying below the damage curve at all current levels.

For feeders serving relatively small transformers, such as the 750 kVA transformer energized by the bifurcated feeder from circuit breaker J, the short-time thermal withstand capability of the selected cable size should be checked. The full-load rating of this transformer is 31.4 A at 13.8 kV, and a No 8 three-conductor cable of approximately 45 A capacity may have been selected as adequate. A plot of the thermal withstand limit, as shown in Fig 54, reveals that the No 8 cable could be damaged over its length for a 13.8 kV fault exceeding about 3000 A (90 000 A at 480 V). To prevent this possibility a No 1 size cable should be selected.

Sensitive and prompt clearing of ground faults is possible on all 13.8 kV feeders with the application of the zero-sequence-type current transformer surrounding all three-phase conductors and the associated instantaneous current relay (device 50GS). Ground-fault sensitivity on the order of 4 to 10 A is achieved with this combination, depending on the type of relay used. No coordination requirement exists with downstream devices, since these feeder circuits energize transformers with delta-connected primary windings, and ground faults on the secondary side do not produce zero-sequence current in the primary side feeder circuit. The ground relaying is shown in Fig 46.

4.8.1.2 *13.8 kV Motor Protection.* Fig 47 shows the degree of overload protection provided for the 2500 hp 2.3 kV motor energized through the 2500 kVA transformer from 13.8 kV bus 3. The relaying as applied must protect both the transformer and the motor.

The replica-type thermal relay (device 49/50) has a tap setting which will trip the circuit breaker when the motor load current is sustained at 125 percent of rating for a period of 60 min. This pickup setting complies with the NEC, Article 430-32, since the machine has a 1.15 service factor. The thermal relay operating characteristic is represented as a band where the lower limit signifies the operating time when the overload occurs after a period of 100 percent load, and the upper limit signifies the operating time when the overload occurs following zero loading.

The setting for the time element of the phase overcurrent relays (device 51/50) is determined by the normal starting-time and starting-current requirement for the motor and its locked-rotor thermal limitation. If the permissible locked-rotor time is greater than the required accelerating time, as in the example shown, the overcurrent relay can be set for locked-rotor protection. The pickup or tap setting is usually on the order of 50 percent of locked-rotor current, and a time lever setting is best determined by several trial starts under actual conditions. For some motor designs the allowable locked-rotor time may be less than the required accelerating time, and for such conditions an overcurrent relay supervised by a zero-speed switch may be required for locked-rotor protection.

The pickup setting of the instantaneous element of the thermal and phase overcurrent relays is determined by the transformer magnetizing inrush current. Although the magnitude of the inrush current is shown plotted at its approximate minimum possible level of 10 times full load, an actual relay setting of 12 to 14 times transformer full-load rating should normally be adequate, but can be increased if pickup occurs during trial starts.

The device 51/50 overcurrent relays also provide primary phase-fault protection for the feeder cable and transformer, and for this reason two relays are applied. The instantaneous ground sensor relay (device 50GS), set at a minimum tap, completes the protection for this circuit.

Fig 45 illustrates the overcurrent relaying selected for the 9000 hp 13.2 kV synchronous motor. As for the 2500 hp motor, the extremely inverse characteristic is preferred for device 51/50, for locked-rotor protection, and for cable and motor fault backup protection. Although backup overload protection is also provided by the 160 percent pickup setting of the time element of device 51/50, its trip characteristic crosses over the motor thermal damage curve and does not afford complete protection in the light overload region. The stator winding temperature relay (device 49) whose operating characteristics are not customarily plotted on the time—current curves and which is set to trip rather than alarm, will protect the machine in the region where device 50/51 does not. The

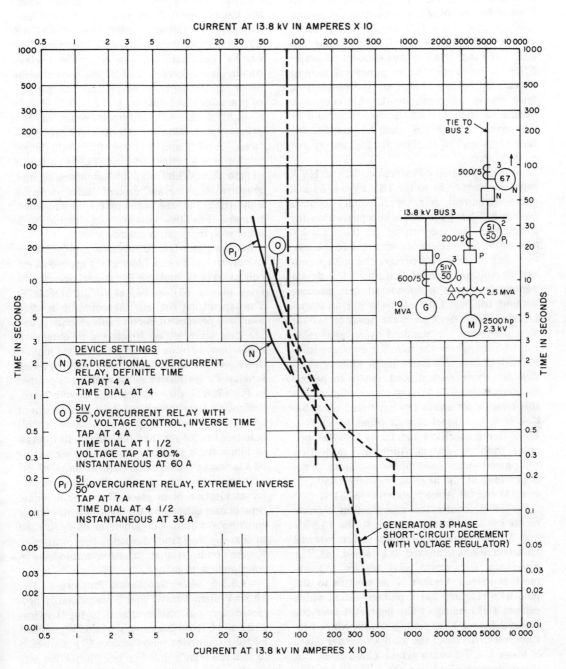

CURRENT AT 13.8 kV IN AMPERES X 10

DEVICE SETTINGS

(N) 67, DIRECTIONAL OVERCURRENT
RELAY, DEFINITE TIME
TAP AT 4 A
TIME DIAL AT 4

(O) $\frac{51V}{50}$, OVERCURRENT RELAY WITH
VOLTAGE CONTROL, INVERSE TIME
TAP AT 4 A
TIME DIAL AT 1 1/2
VOLTAGE TAP AT 80%
INSTANTANEOUS AT 60 A

(P₁) $\frac{51}{50}$, OVERCURRENT RELAY, EXTREMELY INVERSE
TAP AT 7 A
TIME DIAL AT 4 1/2
INSTANTANEOUS AT 35 A

GENERATOR 3 PHASE
SHORT-CIRCUIT DECREMENT
(WITH VOLTAGE REGULATOR)

Fig 48
Phase-Relay Time—Current Characteristic Curves for Generator Relay at 13.8 kV Bus 3

time dial setting falls within the limits of allowable locked-rotor time and required motor-starting time. Since the motor-starting inrush current is limited by the starting reactor, the setting for the instantaneous element of device 51/50 is based on the asymmetrical current which may be contributed by the motor to a fault on an adjacent circuit. This current is calculated from the subtransient reactance of the machine, and 1.6 and 1.1 multiplying factors are applied to allow for asymmetry and a safety margin.

4.8.1.3 *Generator Protection.* The 10 MVA generator connected to the 13.8 kV bus 3 has a phase-overcurrent relay with voltage control (device 51V) applied as backup protection for three-phase faults occurring on the 13.8 kV bus, or on feeder circuits connected to the bus, including the generator circuit. The voltage control or voltage restraint feature of the device permits moderate overloads of the machine without tripping. On systems where no operator may be present to make adjustments for such occurrences, a standard overcurrent relay can be used where the machine characteristics are such that current levels during a fault are sufficiently above normal load values to permit discrimination. The instantaneous element for this relay is set above the generator contribution including direct-current offset to back up the differential relays for faults into the machine from the system. Additional protection for phase-to-phase and phase-to-ground faults is provided by the negative-sequence relay (device 46) and the ground relay (device 51G).

Fig 48 illustrates the coordination requirements for the circuits connected to the 13.8 kV bus 3. The generator output under external fault conditions is plotted as a dashed line. The directional overcurrent relay (device 67), applied at circuit breaker N as backup to the pilot-wire relaying, has a pickup setting which permits full loading of the generator over the tie line. A time lever setting is used which provides selectivity with the 13.8 kV feeder relays of buses 1 and 2 to the extent allowable by the generator short-circuit current, which is too low in comparison to the system contribution to permit coordination in every case.

The plotted inverse characteristic for the voltage-controlled overcurrent relay (device

51V) at circuit breaker O is in effect only when the bus voltage is 80 percent of normal or less. This level of voltage can be expected for 13.8 kV feeder faults, and the relay operating time has to coordinate with the overcurrent relays on circuit breakers P_1 and N. The current pickup or tap setting is approximately 115 percent of generator rated output.

4.8.1.4 *13.8 kV Tie Circuit Protection.* The primary protection for the cable tie circuit between buses 2 and 3 is line differential, using pilot-wire type relays (device 87L) at each end of the line. This relay is instantaneous and sensitive to phase and ground faults occurring only within the area zoned by the current transformers. For this reason, coordination with other relaying is not required.

Backup phase-fault protection is provided by the overcurrent relay (device 51) applied at circuit breaker M, and the directional overcurrent relay (device 67) applied at circuit breaker N. The tap setting for relay M would be selected near 100 percent of circuit cable ampacity, and the time lever setting is selected to obtain coordination with the characteristic of the longest delay overcurrent relay it "overlooks," which is relay P_1 on feeder P. This setting is plotted in Fig 47. The time lever setting provides a coordinating time interval of 0.6 s at the current level at which relay P_1 goes instantaneous. This is somewhat longer than the usual 0.4 s interval to allow for a possibly required higher setting of the instantaneous trip element on relay P_1.

The selection of a directional overcurrent relay at location N in place of a nondirectional type is necessitated by the limited fault-current contribution from the generator as contrasted to that supplied from the utility source. If relay N were nondirectional, its setting would have to coordinate with relay P_1.

4.8.1.5 *Main Substation Protection.* The 13.8 kV main buses 1 and 2 are primarily protected by bus differential relaying (devices 87B1 and 87B2). Backup protection is provided by the overcurrent relays (device 51) connected so that the relay applied on one bus section sees the total contribution from both transformers for a fault on that bus section. The setting for the device 51 relay is plotted in Fig 45 and identified as relay D. Relay D must coordinate with the longest delayed feeder relay connected

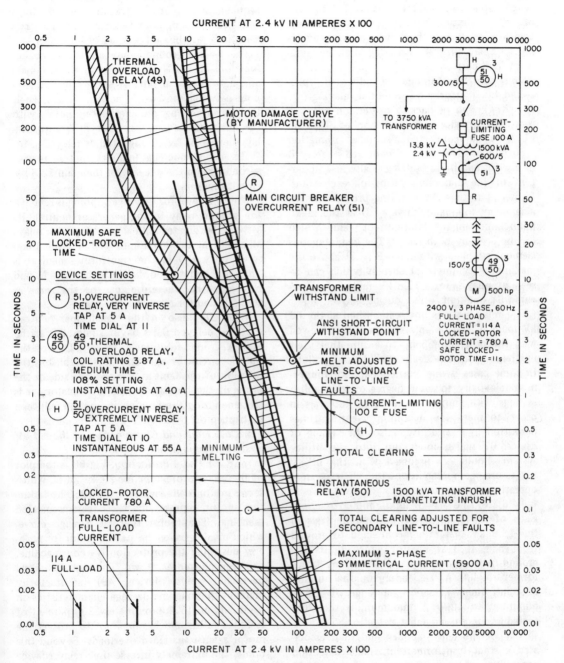

Fig 49
Phase-Relay Time—Current Characteristic Curves for 2.4 kV Bus 1 Coordination

to buses 1 or 2 and the tie line feeder relay M. Its pickup setting is approximately 140 percent of the maximum force-cooled rating of one transformer, and its time lever setting provides a 0.4 s delay interval with relay M at the maximum fault current level.

The main transformers are individually protected by transformer differential relaying (device 87T), and as backup protection overcurrent relays (device 51/50) are applied at the 69 kV level and connected also in a summation arrangement. This relay is identified as relay B in Fig 45, and its tap setting is also at 140 percent of the maximum rating of each transformer. The time lever setting provides a suitable delay interval with relay D at the maximum simultaneous fault-current value which can be seen by both relays. The instantaneous element supplied with relay B is set above the maximum asymmetrical current which can be seen by relay B for a 13.8 kV fault, which occurs with one transformer out of service.

The relay settings now established at the 69 kV main substation entrance must be reviewed with the power company to ensure that their upstream protective devices will be compatible. In some cases it may be necessary to compromise selectivity to some degree or to establish settings with shorter coordinating intervals in order to meet the maximum clearing times permitted by the utility. Also, the setting of device 67 which looks out into the utility system should be discussed with the utility to assure compatability with their system operating procedures.

4.8.1.6 *13.8 kV Ground-Fault Protection.* Each of the three wye-connected neutrals in the 13.8 kV system are connected to ground through a 19.9 Ω resistor which limits the ground-fault current available from any transformer to 400 A. Depending on the number of transformers in service, a range of 400 A minimum to 1200 A maximum is therefore available for ground relay detection. The sensitivity of the relays applied and their associated current transformers provide a detection capability less than 10 percent of the 400 A minimum available.

The ground overcurrent relay settings are plotted in Fig 46. All transformer feeders and the 13.8 kV motor feeder are protected with instantaneous current relays energized by zero-sequence-type current transformers of 50/5 ratio. In terms of primary current their pickup sensitivity will be on the order of 5 to 10 A.

The tie-line ground relays at circuit breaker locations M and N are necessarily time delayed and have identical settings since selective tripping between the two is not important. Their time lever setting provides coordination with the main transformer neutral differential relay (device 87TN), designated as relay D_1, for faults in the zone that includes the transformer secondary and the line side of the main 13.8 kV circuit breakers.

The next level for selective tripping is relay O, the ground relay in the generator neutral. Its setting coordinates with 0.4 s delay with relays M and N at the maximum 400 A level. Likewise, relay CD in the transformer neutral is only required to coordinate with relays M and N. Its setting, therefore, can be the same as that selected for relay O.

Relay D_2, also in the transformer neutral, must be delayed 0.4 s beyond relay CD at the 400 A ground-fault level. Relay CD trips the bus tie circuit breaker and thus establishes the location of the fault as being on one side or the other of the bus tie. Whichever transformer is still energizing the faulted bus section will then be tripped off by relay D_2.

4.8.2 *Setting and Coordination of the 2.4 kV System Relaying*

4.8.2.1 *Phase Protection.* Fig 50 is the plot of the phase protection for 2.4 kV bus 3 which serves motor loads including the 1250 hp induction motor, representing the largest connected machine. The motor thermal damage curve, which must serve as the starting point for properly designing the protection for any machine, has been plotted as shown. The motor thermal overload relay (device 49) satisfactorily matches the machine damage characteristics on overloads up to approximately 200 percent of full-load rating and has been set to protect the motor against sustained overloads. Beyond this point, the extremely inverse time relay (device 51) matches the motor-damage curve better than devices 49 and 50 and provides protection for currents up to locked rotor by cutting under the maximum safe locked-rotor time. The 2.4 kV fuse is present to protect the con-

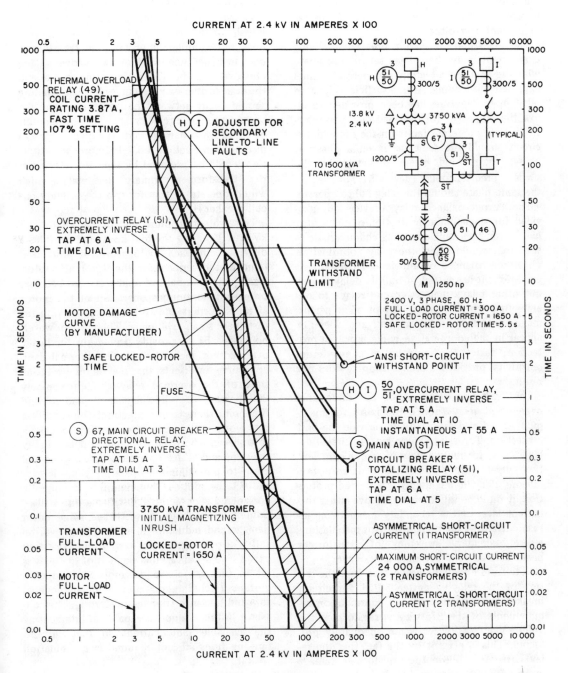

Fig 50
Phase-Relay Time—Current Characteristic Curves for 2.4 kV Bus 2 Coordination

tactor by interrupting heavy fault currents and is sized to withstand locked-rotor current at 10 percent overvoltage.

The main and tie circuit breaker totalizing overcurrent relays have been set to coordinate with the motor protection and permit normal expected bus loading. Also, a sufficient delay in relay operating time has been provided to allow the contactor (or overload relay) to selectively clear moderate faults on the same motor circuit should the fault occur on a phase or phases that would escape detection by the single overcurrent relay (device 51), or even on the same phase should the relay fail to operate.

The current-balance relay (device 46) provided for single-phase protection of the motor has no time—current operating characteristic as such that would affect the relay coordination on either balanced or unbalanced overcurrent condition. It has sufficient built-in delay to permit other relays such as ground relays to operate first when required. A plot of its performance therefore is not relevant and does not appear in the coordination curve. For best protection, device 46 should be set at maximum sensitivity provided nuisance tripping does not result.

The time-delay element of the directional overcurrent relay (device 67) is set for maximum speed and sensitivity to provide the best protection. The relay should have sufficient delay (0.1 s) for reverse flow of motor contribution current into a primary fault. If an instantaneous element was used for improved protection, it must be set to pick up above either this current level or the secondary magnetizing inrush current of the transformer for installations where secondary switching of transformer banks with a "dead" primary may occur.

Note from Fig 49, however, that since the setting of the relaying device 50/51 for circuit breaker H is dictated by the coordination requirements of the 3750 kVA transformer, the 1500 kVA transformer is not adequately protected. This is evident by the fact that the relay curve falls above the transformer damage curve. A 100E primary fuse has been applied to fill this protection void and appears to do so since its clearing time curve falls to the left of the transformer short-circuit withstand point. Also, it provides sufficient operating delay to

withstand expected transformer loading continuously as well as transformer magnetizing inrush for the required 0.1 s. However, the fuse does not completely protect the transformer on low-magnitude (arcing) faults due to the crossover of the fuse-interrupting curve and the transformer-damage curve. If such a failure should occur between the transformer and the secondary main circuit breaker R, some amount of transformer damage may be expected, as was discussed in Section 4.8.1.1. Improved protection would be provided by separate relaying at the transformer primary terminals, which would operate to transfer trip the primary circuit breaker H.

The protection illustrated in Fig 49 for 2.4 kV bus 1 serving the 500 hp motor provides a different approach to locked-rotor protection than that described for the 1250 hp motor. Again, the thermal overload relay has been set to permit continuous operation of the motor at rated current and to provide protection for sustained small overloads. There is, however, no device 51 to provide protection for heavy overloads or locked rotor. The thermal damage curve intersects the maximum operating time of the overload relay at approximately 300 percent of full-load current, and beyond this point there is no protection. The condition gets worse with increased current up to locked-rotor level. The overload relay will eventually operate, and if its operating time should fall close to the minimum operating time, it may protect the motor. This condition is normally considered economically justifiable on small or noncritical machines. Nevertheless, some motor damage and reduction in expected life will result if a prolonged high overcurrent condition exists. The instantaneous element of the relay is set to pick up above the motor locked-rotor current (including direct-current component) to avoid nuisance tripping on starting. The contactor in this example is capable of interrupting the available fault current so that current-limiting fuses are not required in combination with it.

Protection against motor damage from single-phase operation such as would occur following interruption by one transformer primary fuse is provided by the negative-sequence voltage relay (device 46V). The main circuit breaker overcur-

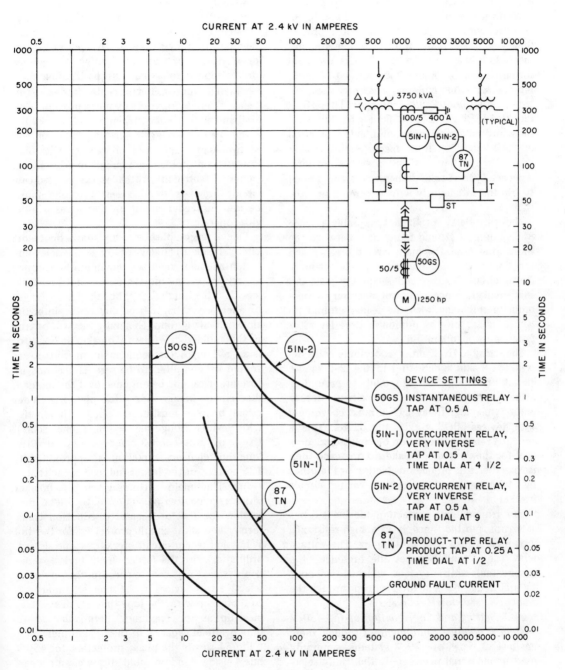

Fig 51
Ground-Relay Time—Current Characteristic Curves for 2.4 kV Buses 2 and 3

rent relay (device 51) has been set to provide transformer overload protection and also to coordinate with the downstream motor protection and the upstream transformer primary fuse. Coordination with the fuse is doubtful for currents above about 3000 A due to the difference in the fuse and relay characteristics.

4.8.2.2 *Ground-Fault Protection* [8], [32]. (See also IEEE Std 242-1975.) Fig 51 illustrates the coordination of the ground-fault protection on 2.4 kV bus 2. All the feeder circuit breaker ground relays (device 50GS) operate from zero-sequence-type current transformers and are set to trip instantaneously with maximum sensitivity.

The time-delay product type relay (device 87TN) detects ground faults only between the transformer and the main circuit breaker and functions to trip the appropriate primary feeder circuit breaker and secondary main circuit breaker. Since it is not necessary to coordinate this relay with the feeder ground relays it is set on the minimum time lever for fastest possible operation.

The relay 51N-1 must coordinate with relays 50GS and 87TN to trip the 2.4 kV tie circuit breaker ST for bus faults to ground or as backup to the 2.4 kV feeder circuit breaker ground relaying. Relay 51N-2 must coordinate with devices 50GS, 87TN, and 51N-1. This is the final relay to operate on bus faults and must wait for the tie circuit breaker to open and then isolate the fault on one bus section or the other by tripping the appropriate main circuit breaker.

The ground fault protection for 2.4 kV bus 1 is not plotted since it is a high-resistance grounded system with no tripping. The voltage relay (device 59N) senses the presence of a ground fault on the system, which is evidenced by a current flow and voltage drop through the resistor R, and operates an alarm.

4.8.3 *Setting and Coordination of the 480 V System Protective Equipment.* Although there are several types of 480 V systems commonly used in industrial plants, only the "fully selective — fully rated" system will be described since the other systems (such as partially selective systems) are similar but compromise either a degree of equipment protection or selectivity for reduced cost.

4.8.3.1 *Phase Overcurrent Protection*

(1) *480 V Radial System.* 480 V buses 4 and 6 each represent a small distribution system feeding motors or other loads such as lighting or heating. For optimum protection and coordination the following must be considered.

(a) The motor control center feeder circuit breaker series trip devices must provide overload and short-circuit-current protection for the feeder cables serving the lighting, heating, and motor loads, and also coordinate with any downstream molded-case circuit breaker for all values of fault current, that is, the molded-case circuit breakers should clear load-side faults before the feeder circuit breaker series trip devices operate.

This is accomplished using magnetic feeder circuit breaker trip devices (long-time delay and instantaneous) when the maximum let-through energy $I^2 t$ of the molded-case circuit breaker is less than the unlatching $I^2 t$ of the feeder circuit breaker series trip device during the interruption of fault current. Generally this will be true if the molded-case circuit breaker is 75 A or less, although the equipment manufacturer should be consulted to be sure. If accurate information cannot be obtained or if the molded-case circuit breaker $I^2 t$ is too high, the feeder circuit breakers must be equipped with selective trip devices (long-time — short-time delay). Overload protection is afforded by setting the long-time element to pick up at approximately 125 percent of the feeder full-load capacity.

(b) The main secondary circuit breaker series trip devices must provide overload protection for the load center distribution transformer and short-circuit protection for the 480 V bus and feeder circuit breakers, while also coordinating with the feeder circuit breaker series trip devices. This can be accomplished by using selective trip devices in the main secondary circuit breaker with the long-time element set to pick up at the maximum permissible short-time overload capacity of the transformer.

Fig 54 shows the phase protection for 480 V bus 4 using standard molded-case circuit breakers and selective trip devices.

(2) *480 V Network System.* 480 V buses 1, 2, and 3 comprise a typical spot-network system. For optimum protection and coordination the following factors must be considered.

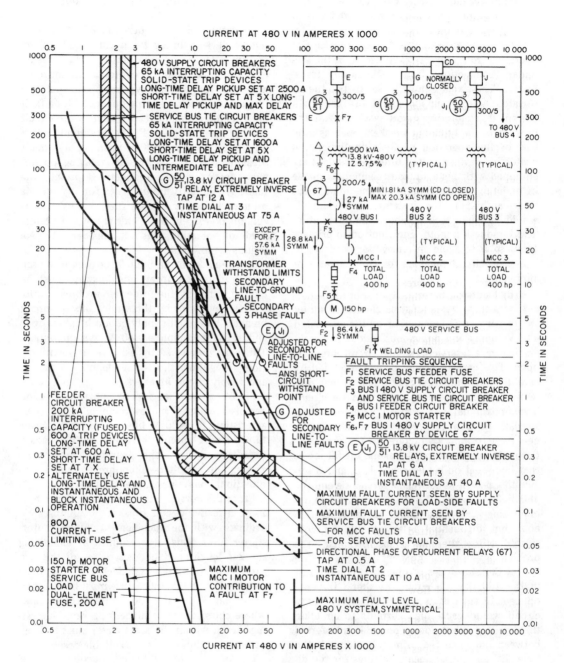

Fig 52
Phase-Protection Time—Current Characteristic Curves for 13.8 kV Feeder E, G, and J and
480 V Bus 1, 2, and 3 Network Coordination

(a) Power must not flow out of the network into the supply system as a result of faults in the supply circuits. This is accomplished by using directional overcurrent relays (device 67) operating from current transformers in the secondary connections from each supply transformer. These relays must be set to trip the associated 480 V supply circuit breaker when current flows into the supply system before the other two supply circuit breakers open as a result of the operation of their nondirectional protective devices. If the directional relays are equipped with instantaneous elements capable of directional discrimination (some do not have this capability), the instantaneous elements would be set to pick up above the momentary 480 V induction motor contribution to primary faults so as to avoid an unnecessary opening of the main circuit breakers for trouble on a remote circuit.

(b) Faults on buses 1, 2, or 3 or on feeders fed from these buses must be cleared either by the feeder circuit breakers or by the associated 480 V supply and service bus tie circuit breakers before all service bus tie circuit breakers open. This can sometimes be accomplished by careful selection and setting of the feeder protective devices and the supply and tie circuit breaker trip devices. In most cases, however, special relays will be required to obtain the necessary tripping characteristics for either the tie or the supply circuit breakers. Fig 52 shows that coordination between bus 1, 2, and 3 service ties as well as between a service tie and the 480 V feeder circuit breakers is difficult to obtain using directional relays alone. Instantaneous bus differential relays should be considered as the alternate means for solving the problem.

(c) Certain modes of operation could result in the 13.8 kV tie circuit breaker CD being opened. In such cases a fault on any primary circuit would cause high currents to circulate through this substation as power is transferred between the separated primary systems. If the time dial and instantaneous settings for the directional relays were selected so as to provide coordination between the 480 V main circuit breakers and any 13.8 kV feeder circuit breakers not serving this substation for remote primary faults, a loss of coordination between the

480 V main supply circuit breakers would result.

To ensure selective operation of all the protective devices during the normal mode of operation with the 13.8 kV bus tie circuit breaker CD closed, loss of coordination must be accepted for certain types of faults should this circuit breaker ever be opened. With a fault on a remote primary feeder such as the circuit fed by circuit breaker F, the directional relays on the bus 1 and 2 480 V main supply circuit breakers will trip before the relaying on primary feeder F would selectively isolate the fault. A similar situation arises for any remote primary feeder fault with the 13.8 kV tie circuit breaker open. Since this is not the normal mode of operation, it is not considered to be a serious compromise.

(d) Faults on the 3000 A busway or on service feeders fed from this busway must be cleared either by the feeder circuit breaker protective devices or by the service bus tie circuit breakers before all supply circuit breakers open and cause a complete 480 V system outage. This is accomplished by using selective trip devices (long-time—short-time delay) on the tie and supply circuit breakers.

Fig 52 shows that selective phase protection has been achieved (no overlapping of trip characteristics) for 480 V buses 1, 2, and 3 using fused feeder circuit breakers with standard series trip devices, and provided that the tie and supply circuit breakers are equipped with directional relays or differential relays.

From the curves it is not obvious that bus 1, 2, or 3 feeder circuit breaker 800 A fuses coordinate with the dual-element 200 A motor-starter fuses. This is determined by examining the I^2t let-through characteristics of the two fuses. To coordinate satisfactorily, the total clearing I^2t of the 200 A fuse must be less than the damage I^2t of the 800 A fuse. In some cases where coordination is difficult, several fuse ratings (even within the same class) must be evaluated to establish the most suitable design.

(3) *480 V Double-Ended Secondary Unit Substation with Normally Closed Secondary Tie.* 480 V bus 5 is essentially a variation of the 480 V spot network (buses 1, 2, and 3) previously discussed. The basic protection considerations also apply for this example. The fol-

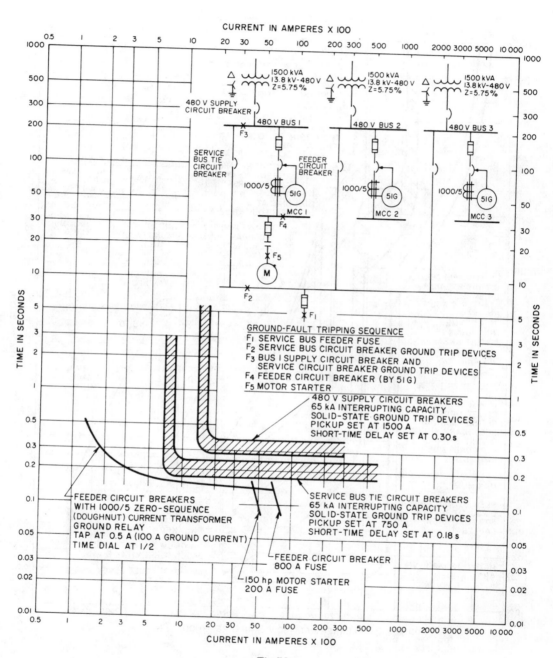

**Fig 53
Ground-Relay Time—Current Characteristic Curves for 480 V Bus 1, 2, and 3 Network**

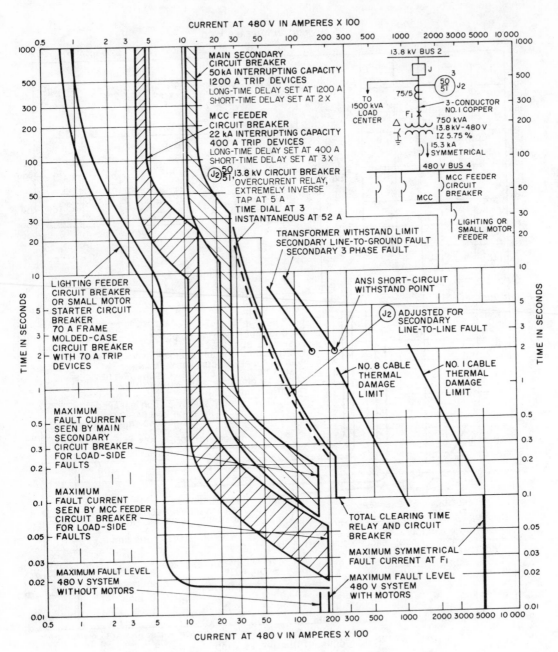

CURRENT AT 480 V IN AMPERES X 100

MAIN SECONDARY
CIRCUIT BREAKER
50 kA INTERRUPTING CAPACITY
1200 A TRIP DEVICES
LONG-TIME DELAY SET AT 1200 A
SHORT-TIME DELAY SET AT 2 X

MCC FEEDER
CIRCUIT BREAKER
22 kA INTERRUPTING CAPACITY
400 A TRIP DEVICES
LONG-TIME DELAY SET AT 400 A
SHORT-TIME DELAY SET AT 3 X

50 13.8 kV CIRCUIT BREAKER
51 OVERCURRENT RELAY,
EXTREMELY INVERSE
TAP AT 5 A
TIME DIAL AT 3
INSTANTANEOUS AT 52 A

TRANSFORMER WITHSTAND LIMIT
SECONDARY LINE-TO-GROUND FAULT
SECONDARY 3 PHASE FAULT

LIGHTING FEEDER
CIRCUIT BREAKER
OR SMALL MOTOR
STARTER CIRCUIT
BREAKER
70 A FRAME
MOLDED-CASE
CIRCUIT BREAKER
WITH 70 A TRIP
DEVICES

ANSI SHORT-CIRCUIT
WITHSTAND POINT

J2 ADJUSTED FOR
SECONDARY
LINE-TO-LINE FAULT

MAXIMUM
FAULT CURRENT
SEEN BY MAIN
SECONDARY
CIRCUIT BREAKER
FOR LOAD-SIDE
FAULTS

NO. 8 CABLE
THERMAL
DAMAGE
LIMIT

NO. I CABLE
THERMAL
DAMAGE
LIMIT

MAXIMUM
FAULT CURRENT
SEEN BY MCC FEEDER
CIRCUIT BREAKER
FOR LOAD-SIDE
FAULTS

TOTAL CLEARING TIME
RELAY AND CIRCUIT
BREAKER

MAXIMUM SYMMETRICAL
FAULT CURRENT AT F1

MAXIMUM FAULT LEVEL
480 V SYSTEM
WITHOUT MOTORS

MAXIMUM FAULT LEVEL
480 V SYSTEM
WITH MOTORS

TIME IN SECONDS

CURRENT AT 480 V IN AMPERES X 100

13.8 kV BUS 2

J
50 3
51 J2
75/5
TO
1500 kVA
LOAD
CENTER
F1
3-CONDUCTOR
NO. I COPPER
750 kVA
13.8 kV–480 V
IZ 5.75 %
15.3 kA
SYMMETRICAL
480 V BUS 4
MCC FEEDER
CIRCUIT
BREAKER
MCC
LIGHTING OR
SMALL MOTOR
FEEDER

Fig 54
Phase-Protection Time—Current Characteristic Curves for
13.8 kV Feeder J and 480 V Bus 4 Coordination

lowing factors must be considered for optimum protection and coordination.

(a) Directional relays must be applied to the main secondary circuit breakers as described in the discussion for 480 V buses 1, 2, and 3 to selectively isolate primary system faults.

(b) Faults on either bus section or on feeders fed from the bus must be cleared by the feeder circuit breaker protective devices or by the bus tie circuit breaker and the associated 480 V supply circuit breaker before the other 480 V supply circuit breaker operates and causes a complete 480 V system outage. This is accomplished by using long-time—short-time delay selective trip devices in the tie and supply circuit breakers.

Fig 52, although specifically representing buses 1, 2, and 3, illustrates the delay sequence which would be used for selecting coordinated device settings for bus 5 as well.

As in the case of 480 V buses 1, 2, and 3, the I^2t characteristics of the motor starter and feeder circuit breaker fuses must be evaluated to ensure coordination between fuses.

4.8.3.2 *Ground-Fault Protection.* The 480 V spot network in Fig 36 is solidly grounded. The coordination curves for the ground-fault protection for this portion of the system will, therefore, illustrate the principles that apply to obtaining selective performance of the circuit breakers during ground faults. With solid grounding the maximum line-to-ground fault short-circuit current is virtually equal to the three-phase value, but because of the fault and ground-return impedance much lower magnitude currents flow. Sensitive relaying must therefore be used to ensure that ground-fault currents which are too low to be tripped instantaneously or even detected by the standard phase series trip devices or static trip devices are safely cleared.

Common methods employed to provide this protection are as follows:

(1) Relays fed from residually connected current transformers in feeder circuits, tie circuit breakers, and main supply circuit breakers

(2) Large zero-sequence (doughnut) current transformers enclosing all phase conductors in feeder circuits rather than residually connected current transformers. The connections are otherwise the same as described in (1)

(3) Static ground trip devices in feeder, tie, and main supply circuit breakers

(4) Overcurrent relays connected to the output of a current transformer sensing the current in the distribution transformer neutral (not illustrated in this diagram)

Very special care is required when applying ground relays to four-wire secondary selective systems where the neutral circuit is not switched if coordination is to be achieved between main and bus tie circuit breakers (see IEEE Std 142-1972). The principles involved in selecting coordinated ground-relay settings are the same as those described for phase protective devices. Fig 53 shows selective ground relaying for bus 1, 2, and 3 spot-network systems using a definite-time ground relay fed from a large doughnut current transformer around the feeder cables to motor control centers, and static ground trip devices in the service bus tie and main 480 V supply circuit breakers.

In a solidly grounded system the zero-sequence current transformer must be of such a ratio that it will provide an output current low enough to prevent damage and sufficiently undistorted to accomplish accurate relaying during maximum ground-fault conditions. For the system illustrated a ratio of 1000/5 has been selected.

A time-delay relay is suggested for the feeder circuit breakers to allow motor-starter and feeder circuit breaker fuses to interrupt high-magnitude ground faults. Selective fault isolation for faults on bus 1, 2, or 3 requires bus differential relays, since coordination cannot be achieved between ground directional relays on the tie and service main circuit breakers.

4.9 Surge-Voltage Protection

4.9.1 *The Problem.* Except for outdoor overhead circuits, electric power distribution system conductors are not normally subject to direct lightning exposure. If lightning strikes the overhead power lines serving the plant, an overvoltage surge whose magnitude is limited by the sparkover potential of the overhead conductor system or the protective level of surge arresters will be imparted to the high-voltage conductors. This overvoltage surge will have the form of a steep front wave which will travel away from the stricken point in both directions along the power system conductors. As the surge travels along the conductors, losses cause the magnitude of the voltage surge to constantly diminish. If the voltage magnitude is sufficient to produce corona, the decay of the voltage surge will be fairly rapid until below the corona starting voltage. Beyond this point the decay will be more deliberate. Properly rated arresters at the plant terminal of the incoming line will reduce the overvoltage to a level which the terminal station apparatus can withstand.

In instances where the local industrial plant system is itself without lightning exposure except from the exposed high-voltage lines through step-down transformers effectively protected with high-side surge arresters, any surges induced by lightning are infrequent and quite moderate. Likewise, surges due to switching phenomena, although more common, are generally not severe. Only occasionally would line-to-ground potentials on the local system reach arrester sparkover values. The multiplicity of radiating cable circuits with their array of connected apparatus acts to greatly curb the slope and magnitude of the voltage surge which reaches any particular item of connected apparatus. Transformers and other equipment items connected as a single load at the end of circuits are particularly vulnerable. Experience has indicated that certain types of apparatus are susceptible to voltage surges for almost any circuit connection arrangement, and it is advisable to fully investigate the subject of damaging voltage surges.

The appearance of abnormal applied voltage stresses, either transient, short-time, or sustained steady state, contributes to premature insulation failure. This section considers electrical insulation protection. Electrical insulation deterioration to the point of failure is a result of an aggregate accumulation of insulation damage finally reaching the critical stage where a conducting path is rapidly driven through the insulation sheath and failure (short circuit) takes place. Large amounts of current are driven through the faulted channel, liberating large amounts of heat. The excessive temperature rise rapidly expands the zone of insulation damage, and complete machine destruction can occur rather quickly unless the input of electric power supply is interrupted. Even small insulation punctures such as might be discovered

after special nondestructive testing of apparatus require repair or rewinding.

An acceptable system of insulation protection will be influenced by a number of factors. Of prime importance is a knowledge of the insulation system withstand capability and endurance qualities. These properties are indicated by insulation-type designations and specified high-potential and surge-voltage test withstand capabilities. Another facet of the problem relates to the identification of likely sources of overvoltage exposure and the character, magnitude, duration, and repetition rate which are likely to be impressed on the apparatus and circuits. The appropriate application of surge protective devices will lessen the magnitude and duration, and is the most effective tool for achieving the desired insulation security. A working understanding of the behavior pattern of electric surge voltage propagation along electrical conductors is necessary to achieving the optimum solution.

The problem is complicated by the fact that insulation failure results not only because of excess-magnitude impressed overvoltages, but also because of the aggregate sum total duration of such overvoltages. No simple devices are available which can correctly integrate the cumulative effects of sequentially applied excessive overvoltage. The time factor must be estimated and then factored into the design and application of the protection system.

Lightning is a major source of transient overvoltages. Numerous industrial operations use open-wire overhead lines which are subject to direct exposure to lightning, allowing lightning surges to be introduced into the industrial distribution system.

Direct lightning strokes are rare to overhead outdoor plant wiring because of the shielding effect of adjacent structures. However, direct strokes to objects 25 to 50 ft away can induce substantial transient overvoltages into the overhead line so that surge arresters must be installed at the end of an overhead line which connects to the building wiring. Switches may also be installed to disconnect the overhead line when not in active use.

ANSI C2, National Electrical Safety Code (1976 edition), Part 2.2-1960, Safety Rules for the Installation and Maintenance of Electric

Table 14
Representative Certified Test Withstand Voltages

Voltage Class (kV)	60 Hz 1 min High-Potential Test (kV Crest)				1.2 by 50 μs Impulse Test (kV Crest, full wave)			
	Liquid-Filled Power Transformer	Liquid-Filled Distribution Transformer	Dry-Type Transformer	Switch-gear	Liquid-Filled Power Transformer	Liquid-Filled Distribution Transformer	Dry-Type Transformer	Switch-gear
1.2	14.4	14.4	5.66		45	30	10	
2.4	21.2	21.2	14.4		60	45	20	45
2.5								
4.16				26.9		60	25	60
4.8								
5.0	26.9	26.9	16.9		75			
7.2				51				75(95)*
8.32	36.8	36.8	26.9		95	75	35(65, 75)*	
8.7				51				
13.8							50(65, 95)*	95
14.4	48.1	48.1	43.9		110	95		110
15.0								
25.0	70.8	70.8			150	150		
34.5	99	99			200	200		

From ANSI C37.4a-1958 (R 1971); ANSI C37.6-1971; ANSI C37.41-1969 (R 1974); IEEE Std 20-1973 (ANSI C37.13-1973); and IEEE Std 462-1973 (ANSI C57.12.00-1973).

*Voltages in parentheses are often available as options.

Table 15
Rotating Machine 60 Hz 1 min High-Potential Test Voltages

| | Rated Motor Voltage (volts) | | | | | |
	480	2300	4000	4600	6600	13 200
Test Potential, kV Crest	3.11	7.92	12.73	14.43	20.1	38.8
Per unit of normal crest	7.94	4.21	3.9	3.83	3.72	3.59

From ANSI C50.10-1975 and ANSI C50.13-1975 for synchronous
motors, and ANSI C52.1-1973 (NEMA MG1-1972) for induction motors.

Supply and Communication Lines, requires that open conductors of different voltages installed on the same support must have the highest voltage on top and the lowest voltage below. When a higher voltage conductor breaks for any reason such as an automobile striking the supporting pole or a tree limb falling across the line, the higher voltage conductor may fall across the lower voltage conductor, impressing the higher voltage on the lower voltage circuit. Little can be done to protect against this condition other than to ensure that the protection equipment on the higher voltage line will disconnect the line as quickly as possible, generally as a result of a fault due to an insulation failure on the lower voltage line.

Transient overvoltages are generated in plant wiring by switching actions which change the circuit operation from one steady-state condition to another. Switches which tend to chop the normal alternating-current wave such as silicon-controlled rectifiers (SCR), vacuum switches, current-limiting fuses, and high-speed circuit breakers, force the current to zero which accelerates the collapse of the magnetic field around the conductor, generating a transient overvoltage.

A short circuit is a switching action which creates a by-pass around part of a circuit. The heat generated by the heavy flow of current across the short circuit melts or even vaporizes the conductor, creating a gap, and an arc is established. Heated air rising from the arc creates a draft causing the arc resistance to fluctuate rapidly, which produces transient overvoltages. An insulation failure results in an arc through the failure path with similar results.

The nonlinear inductance of an iron-core transformer and a capacitor in the same circuit may go into oscillation and produce a condition called ferroresonance, which will generate overvoltages. Other sources of overvoltages are described in [5], [29].

Transient overvoltages are propagated along the electric power conductors to create insulation distress far removed from the origin of the voltage surge. Furthermore, the voltage stress imposed on insulation far removed from the point of surge origin may exceed that appearing at the source point.

4.9.2 *Insulation Withstand Properties.* Insulation standards have been developed which recognize the need to meet a limited amount of temporary excess-voltage stress over and beyond the normal operating voltage. The ability of the equipment insulation system to withstand these stresses is certified by excess-voltage tests applied to the electrical products at the completion of manufacture. A number of different types of tests for rating equipment are in use. The most common standard factory tests are the 1 min 60 Hz high-potential test and the 1.2×50 BIL impulse wave test. The 1.2×50 designation means that a voltage pulse increases from zero to peak value in $1.2 \ \mu s$ and declines to one half peak value in $50 \ \mu s$. Typical values of test voltage in use, which are prescribed by ANSI standards, appear in Tables 14 and 15.

The security of major insulation (line to ground) and turn insulation (turn to turn) must be considered separately.

Circumstances will exist in which one of these insulation systems may be overstressed, while the other one is not subjected to any abnormal stress at all. The security of each must be independently examined and protected as necessary.

One of the confusing aspects of an insulation system capability and its protection is the progressive accumulation of deterioration within the dielectric which results from the complete history of voltage stress exposure. An item of equipment subjected to a 60 Hz high-potential test may withstand the voltage application for 50 s and then break down. The device failed the test. It may have withstood the voltage application for the entire 60 s period and passed the test, but it might have failed had the test voltage been continued for another 10 cycles.

With certain types of electric apparatus having a combination liquid and solid insulation system, the cumulative stress failure mechanism only occurs within a narrow band of stress voltages just below the breakdown voltage. Exposure to a lesser overvoltage may initially cause an incomplete failure of the solid insulation, but the subsequent penetration by the liquid material will partially "repair" the deteriorating region.

In all cases a large fraction of the insulation system's capability to withstand applied voltage can be destroyed simply by the process of testing it. For this reason overtesting with dynamic alternating-current voltage should be avoided, and nondestructive direct-current testing is preferred.

The design of the electric system, including the use of surge suppression devices (to assure adequate insulation security), should correctly interpret the effect of the inverse relationship between imposed voltage magnitude and the allowable duration. A 30 percent increase in the applied alternating-current voltage magnitude for most equipment will result in a tenfold reduction in insulation life. The high-magnitude surges require careful attention because of the very rapid loss of life. The system design engineer must, largely by judgment, set the margin of safety between the design controls of allowed overvoltage exposure and the certified withstand capability of the insulation system, based on his knowledge of the probable character and repetition rate of troublesome surge voltage transients.

The problems relating to the achievement of insulation security for turn-to-turn insulation in multiturn coils are many and complex. The normal 60 Hz voltage developed in a single turn will range from perhaps a small fraction of 1 V in a contactor magnet coil to 20 V in a medium-

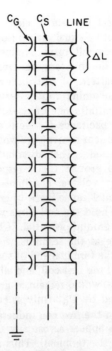

Fig 55
LC Network in a Multiturn Winding

size induction motor to several hundred volts in a large transformer. If it were necessary to only insulate for this normal operating voltage developed in a single turn, the problem would be simple. However, the voltage stress which appears across a single turn-to-turn insulation element when high rate-of-rise voltage surges occur, may be much greater than the single-turn operating voltage.

This aggravated voltage stress is most pronounced at turn insulation adjacent to the coil terminals and is intensified by the increased shunt capacitance between winding sections and ground, such as exists inherently in motor windings as a result of each coil in the construction being surrounded by grounded stator core iron.

The controlling parameters (Fig 55) are the elemental values of coil series inductance ΔL, the elemental values of capacitance C_S shunting the above elemental coil segments, and the elemental capacitance to ground C_G.

Under normal 60 Hz excitation, the voltage distribution is controlled almost entirely by the series-connected chain of elemental coil inductances, creating equal division of the impressed voltage across all turns. A steep-front voltage applied to the line terminal creates an entirely different pattern of distributed voltage. Consider the incoming voltage wave to be a step voltage of infinite rate of rise. If only the turn shunting capacitance C_S were present, the incoming transient overvoltage would be uniformly distributed. The distributed capacitive coupling to ground C_G is responsible for the nonuniform voltage distribution. Note that each elemental capacitance to ground C_G tends directly to hold the coil turns with which it is coupled at ground potential. If C_S were zero, a step voltage at the terminal would create full voltage at the terminal end of the first coil. The inner end of the first coil, and all coils deeper in the winding, would remain at ground potential as controlled by C_G. Only as current flow begins through the first coil inductance ΔL, could any voltage appear across any C_G (except the one unit at the terminal). Thus the initial voltage distribution would display the full surge voltage across the terminal coil with zero voltage across all coils deeper in the winding.

Following the initial steep-front voltage application, current flow will build up in the ΔL which will redistribute the voltage more uniformly. In the process the internal $L-C$ oscillations will create a voltage drop across some inner coil (or coils) substantially greater than the uniform-distribution value, but seldom as great as the initial voltage stress across the terminal coil.

In an actual motor winding up to 80 percent of an applied surge voltage with a $1/10$ μs front can appear across the terminal coil.

Unfortunately the internal electric network by which most apparatus can be represented is not commonly found in equipment specifications or industry standards. In the case of alternating-current rotating machines an unofficial application guide rule for turn-insulation security is to limit the voltage rate of rise at the machine terminals to a value no greater than that represented by a uniform rise from zero to the 60 Hz high-potential-test crest voltage in 10 μs.

4.9.3 *Surge-Voltage Control — Magnitude.* The methods most commonly employed to curb the magnitude of surge voltages include surge arresters, gaps, shielding, and other devices, along with superior design practices in the power system.

Surge arresters have been developed for clipping the peak of an excessive voltage surge diverting the excess surge current to ground, and resealing in the presence of normal-frequency voltage. An internal gap unit which controls the sparkover voltage that is required to establish a conducting path to ground is provided. Designs have been perfected to produce consistent sparkover value and negligible time delay in establishing the conducting path. Rated sparkover voltages for standard arresters are given in Table 16.

The arrester contains a valve section in the circuit to ground which interrupts the power follow current at the next power-frequency current zero. The valve unit must allow a large current flow to ground (20 000 A and sometimes more) without excessive voltage drop. The discharge-current voltage drop values of standard arresters are also shown in Table 16.

When selecting the arrester voltage rating for a specific application, care must be taken that the recovery voltage across the arrester following surge-current discharge will not exceed the arrester voltage rating (its voltage seal-off limit). Reconduction of power follow current would result in arrester destruction within a few half-cycles. For this reason ungrounded or impedance-grounded industrial power systems require line-to-line voltage rated arresters. Otherwise arrester failure can occur if an overvoltage appears on one phase while another phase is faulted to ground.

Plain open gaps, called "rod gaps," are sometimes connected between line and ground to provide a degree of control of surge-voltage magnitude. The absence of special design features at the gap lead to considerable variations in sparkover voltage and considerable voltage turn up in the very-short-time region. The gap circuit contains no valve section and, hence, is not self-extinguishing. Gap sparkover will create a forced interruption on the line. Such protective gaps may prove economically useful when no critical expensive equipment is in jeopardy and the voltage surge incidence rate is low. Low-voltage gap protectors are available which are self-extinguishing at 50 V and less for use

Table 16
Representative Performance Data for Surge Arresters

Arrester Rating (kV)	Station Type			Intermediate Type			Rotating Machine		
	Sparkover† (kV)	Discharge Voltage (kV)* at 5kA	at 10kA	Sparkover† (kV)	Discharge Voltage (kV)* at 5kA	at 10kA	Sparkover‡ (kV)	Discharge Voltage (kV)* at 1.5kA	at 3 kA
3	12	8.5	9	12	10	10.8	12	8	8.8
4.5							16	12	13
6	24	17	19	24	19.6	21.6	20	16	17.5
7.5							25	20	22
9	35	24	26	35	29	32	30	24	26
12	45	32	35	45	36.5	40.5	39	32	35
15	55	40	44	55	46	51	48	40	44
16.5							52	44	48
18							57	48	53
19.5							62	52	57.5
21	72	55	60	72	63	70	66	56	61.5
22.5							71	60	66

From ANSI C62.2-1969.

* Average voltage drop with an 8 by 20 μs discharge current wave at the stated peak current.
† Average sparkover on an impulse wave, based on a voltage rise to sparkover of 100 kV/μs per 12 kV of arrester rating.
‡ Average sparkover on an impulse wave, based on 10 μs to sparkover.

on telephone lines, local communication, and signaling circuits.

The principle of shielding can be applied effectively to reduce the magnitude of lightning-derived surge voltages [30], [31], [33]—[35]. The design pattern involves the installation of strategically located overhead grounded conductors as a sort of umbrella over the power conductors to be benefited. The presence of the grounded overhead conductors lessens the space voltage gradients beneath them which, in turn, lessens the induced voltage magnitude imparted to the power conductors by a nearby lightning stroke. The overhead shield wires will intercept a direct stroke [33, chap 10]. If the maximum discharge current is not too high, the stroke current will be diverted to ground without a flashover to the power conductors. At some level of discharge current the voltage gradients will become high enough to flash over to the power conductor. Even then the voltage on the power conductor will be no greater than if the initial stroke had contacted the power conductor. Since apparatus insulation destruction is the result of the aggregate accumulation of imposed transient overvoltage stress, any reduction in magnitude of the impressed surges is of direct benefit.

Shielding is used on outdoor substations containing critical apparatus, such as transformers and rotating machines, in which a large percentage of the transient overvoltage surges are of lightning origin, communicated to the station by the power supply conductors. Proper attention to effective grounding in the station is essential. Lightning masts and shielding conductors over the bare-conductor portions of the station must be used [17], [33], [34]. Overhead shielding conductors, grounded at every pole or tower, are recommended for the last 2000 ft of the line conductors approaching the station. A set of line-type surge arresters at the remote end of the 2000 ft line section will intercept severe surges and dissipate a large portion of the transient energy to ground. Only the attenuated voltage surge (perhaps one half or less of the original value) continues along the 2000 ft line section to the station. The lessened duty on the station surge protective devices results in a corresponding reduction in the surge-voltage magnitude arriving at the terminals of vulnerable apparatus.

A rather wide variety of additional protective devices are available for surge-voltage reduction under special or unusual conditions. Distribution transformers with lower basic impulse levels of insulation are available for installations which are sheltered from severe overvoltage transients of lightning origin. Dry-type transformers are also intended for such sheltered application. These require careful attention to protective devices to prevent equipment failures.

Rotating machines are particularly vulnerable to damage to the insulation system, and special rated lightning arresters must be used to match motor rated voltages. Surge protective devices must be as close to the motor windings as possible for optimum protection [30], [31], [35] (see also IEEE Std 142-1972).

Apparatus employing solid-state devices may have surge-voltage withstand levels well below other conventional apparatus. In some cases special arresters may be available to provide sufficiently close overvoltage protection for this type of equipment. In other situations an acceptable solution may involve the use of less costly semiconductors to protect expensive units in a sacrificial manner. The protective semiconductors may be selected to display a voltage withstand of lower value than the expensive units. Another needed property is that the protective unit, when it fails, retains a conducting path for at least the time required for the line protective interrupter to operate.

4.9.4 *Surge-Voltage Control — Rate of Rise.* A rotating alternating-current machine is vulnerable to the destruction of turn-to-turn insulation. The conventional method for protecting motors is shown in Fig 56. A capacitor is installed line to ground on each motor phase conductor. The rate of rise of the surge voltage across the motor winding terminals will be limited by the charging rate of the capacitor.

Typically the capacitor would limit the rate of rise of voltage to 1000 V/μs, requiring 10μs to reach the crest value of the applicable 60 Hz 10 000 V high-potential test, and the slope reduction for acceptable turn insulation protection would be obtained. Capacitor units for this type of service are designed to minimize internal inductance.

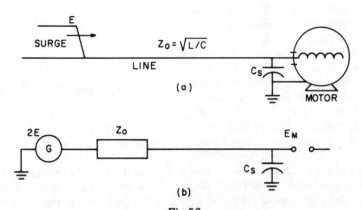

Fig 56
Application of Shunt-Connected Surge-Protective Capacitors for Wavefront Control
(a) Physical. (b) Equivalent

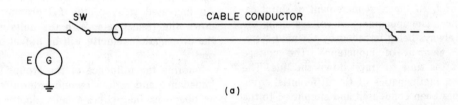

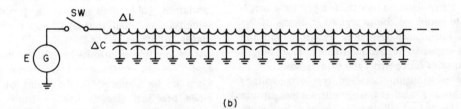

Fig 57
Distributed-Constant Transmission Circuit
(a) Physical. (b) Equivalent

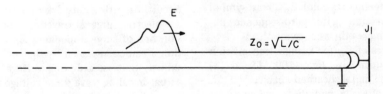

Fig 58
Surge-Voltage Wave in Transmit along a Line of Surge Impedance Z_0

4.9.5 *Surge-Voltage Propagation.* The electric power circuits transmit undesired surge voltages equally as well as normal-frequency voltages and can do so efficiently, even for frequencies into the megahertz range. The usual concepts of line impedance, expressed as resistance and reactance in ohms, used in power-frequency computations do not apply for the solution of short-time transitory voltages. The problem begins with a distributed-constant electric transmission line, either an open-wire line or a solid-insulated cable (Fig 57). Such a transmission line consists of a continued succession of small incremental series inductances with evenly distributed increments of shunt capacitance. When the switch SW is closed, the voltage E becomes connected to the line terminal. The first increment of capacitance is charged to a voltage E. Current begins to flow through the first increment of L to the next increment of shunt capacitance. The appearance of voltage along the line is always being impeded by the next incremental element of inductance. The voltage wave takes time to travel down the line. The complex mathematics of the differential equations has been condensed and simplified to the following practical concepts.

The electrical behavior of a distributed-constant transmission line can, for practical surge-voltage problems, be expressed in terms of the series' inductance per unit length L and the shunt capacitance per identical unit of length C. Consider L in henrys and C in farads.

The relationship between the surge-voltage magnitude E and the surge-current magnitude I is $E/I = \sqrt{L/C} = Z_0$, and is depicted in Fig 58.

With the prescribed units, the quantity Z_0 has the dimensions of ohms and relates with E expressed in volts and I in amperes. The quantity $\sqrt{L/C}$ is called the surge impedance and assigned the reference symbol Z_0. This symbol has no connection with zero-sequence impedance, which uses the same symbol.

Typical values of Z_0 for a single transmission line are (1) open-wire line, 400 Ω and (2) solid-dielectric metal-sheathed cable, 40 Ω. The applied voltage E and the surge current I are in phase. The current flow duplicates the wave shape of the impressed voltage and is in time phase with it. At this point it appears exactly as a resistor of ohmic value $\sqrt{L/C}$, but

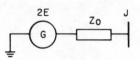

Fig 59
Equivalent Circuit Representing the Arrival of Surge at Junction Point J

the behavior differs from that of a resistor. In a true resistor the I^2R line loss energy is converted to heat. In the distributed-constant line the electric energy is stored in the inductance and capacitance as the $LI^2/2$ and $CE^2/2$ of an electric surge existing on a finite length of the transmission line.

The transit of the surge along the line is propagated at a rate controlled by the quantity LC. An increased value of the LC product slows down the transit rate. With the units chosen, the propagation velocity will be in feet per second.

Ignoring the influence of the ground circuit impedance and various second-order effects on an open-wire line with air dielectric, the propagation velocity is approximately that of the speed of light, 1000 ft/μs, or 1 ft/ns. A solid-insulation cable will display a propagation velocity about half that of the open-wire line, that is, 500 ft/μs or 1/2 ft/ns.

Were the line of infinite length, the bundle of energy would continue to travel forever, never again to be observed at the point of origin. Since practical circuits have a finite length, problems develop as the surge reaches the end of the line. The equivalent circuit, shown in Fig 59, allows evaluation of the effect of continuing along a line of different surge impedance which terminates at an open circuit, at a short circuit, or with a network of lumped constants. An electric surge E traveling along a transmission line of surge impedance Z_0 toward a junction J can be replaced by the equivalent circuit shown, that is, a driving voltage of twice the actual traveling-wave surge voltage magnitude in series with a resistor of ohmic value Z_0.

If at the junction bus J every circuit that exists in the real system, whether a lumped impedance or distributed-constant line, is connected line to ground, the resulting network

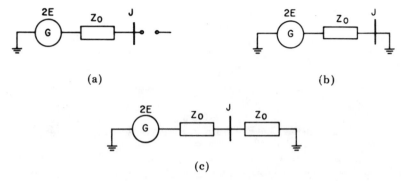

Fig 60
Equivalent Circuits Representing a Line Terminated in Different Ways
(a) Open Circuit at Junction Point J. (b) Grounded Circuit at Junction Point J
(c) Line Continuation of Equal Surge Impedance

correctly satisfies the voltage—current relationships which will prevail at the junction bus J. Every distributed-constant line connected to bus J is represented by a line-to-ground-connected resistor of ohmic value Z_0 for each respective line.

An examination of some familiar line-termination cases will aid in developing a conviction that the equivalent circuit is indeed a valid one.

(1) An open-ended line at bus J [Fig 60(a)]. The equivalent circuit yields the following, which we know to be correct:

$$\text{bus J voltage} = 2E$$

$$\text{line terminal current} = 0$$

(2) A short-circuited line at bus J [Fig 60(b)]. Therefore the equivalent circuit yields the following familiar relationships:

$$\text{bus J voltage} = 0$$

$$\text{line terminal current} = 2E/Z_0$$

(3) A line joining another line of equal surge impedance at bus J [Fig 60(c)]. Again, the equivalent circuit correctly yields

$$\text{bus J voltage} = 2E\,(Z_0)/2Z_0$$

$$\text{line terminal current} = 2E/2Z_0$$

A verification of these solutions appears in Note 1.

The construction of an equivalent circuit for more complex combinations is accomplished using the techniques described. The simplest equivalent circuit used to accommodate distributed-constant lines is valid only until a returning reflected wave arrives at the junction under study. In many cases the entire critical voltage excursion at the bus under study will have passed before the first reflected wave returns.

To account for the effect of a returning reflected voltage wave, the correct equivalent circuit for surge arrival of that reflected wave at bus B_1 must be created as if it were an independent surge voltage initially approaching bus B_1. The computed voltage which this reflected wave contributes to bus B_1 is then added with proper polarity to that still being contributed to bus B_1 by the initial surge. When more than a few wave reflections must be accepted, "lattice diagram" techniques must be used to ensure correct results [36, chap 9, p 215].

The energy of a traveling wave can be dissipated completely if the traveling wave is directed to a junction whose equivalent circuit displays a real resistance termination of ohmic value equal to the Z_0 value of the transmission line. The wave energy is disposed of as heat in the terminating resistor. Although such terminating devices are seldom applied, they are sometimes called transient snubbers.

4.9.6 *Amplification Considerations.* A surge-voltage wave traveling along a distributed-constant line upon encountering a junction terminated in a higher surge impedance Z_0 will in-

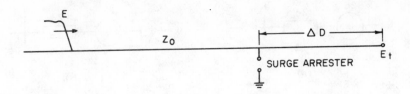

Fig 61
Transmission Line Extended Beyond a Surge-Voltage Arrester

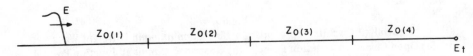

Fig 62
Voltage Amplification by a Series Chain of Line Sections
of Progressively Higher Surge Impedance

crease in voltage to as much as double if the junction is terminated in an open circuit. A surge arrester installed a finite distance ahead of such a junction (Fig 61) can result in a voltage at the junction well above the voltage at the arrester. The junction voltage rise will depend on (1) the steepness of the surge voltage wave, (2) the propagation velocity along the line, (3) the distance of the line extension ΔD, and (4) the magnitude of surge impedance connected to the terminal junction. The rise in terminal voltage will be aggravated by (1) a steeper front wave, (2) greater ΔD, (3) slower line propagation velocity, and (4) greater magnitude of terminating surge impedance. The terminal voltage will not exceed twice the traveling-wave value with any possible value of parameters, and the mechanism associated with this amplification is discussed in Note 2.

Many application charts are available which display the maximum allowable ΔD for specific application conditions. With voltage wave fronts no steeper than 0.5 μs, a separation spacing ΔD of 25 ft is quite generally allowable. The protection system design should locate the protective device as close to the terminals of the critical protected apparatus as is reasonable. Surge-voltage waves may have steeper fronts than the standard reference wave.

A traveling surge voltage, encountering in succession junctions with higher surge impedance,

may have its voltage magnitude elevated to a value in excess of twice the magnitude of the initial voltage (Fig 62). Assume the surge impedance of line sections 1–4 to be 10, 20, 40, and 70 Ω, respectively. Next assume each line section to be long enough to contain the complete wave front, distributed along its length. At the junction between sections 1 and 2 the refracted wave, which continues, will have a magnitude of 1.33E. This wave encountering the junction between sections 2 and 3 would create a refracted wave of 1.78E. In like fashion, this voltage wave, in turn, encountering the junction of sections 3 and 4, would be increased to 2.27E. This voltage wave, upon reaching the open-end terminal at section 4 would be doubled to 4.54E.

Typically the change in surge impedance might result from a different cable construction, which may be additionally modified by a different number of cables run in multiple. The example might well have been represented by four 500 kcmil conductors in parallel in section 1, two in parallel in secton 2, one alone in section 3, and a section of bus duct in section 4. The presence of an open-wire line (400 Ω Z_0) extension from a cable feeder (40 Ω Z_0) could develop, at an open-end terminal, a voltage of 3.64 times the surge voltage traveling in the cable.

In most instances a surge voltage approaching

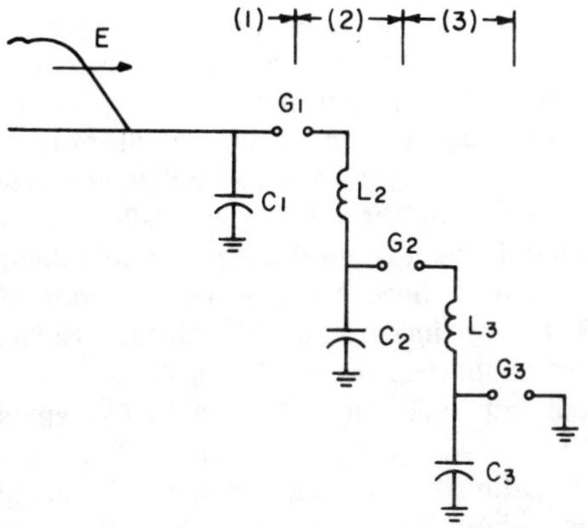

Insulation Breakdown

G_1	10 kV
G_2	15 kV
G_3	25 kV
C_1	1.0 μF
C_2	0.05 μF
C_3	0.002 μF
L_2	100 μH
L_3	10 μH

Fig 63
Vulnerability of a Chain of Insulation Systems in Series

a junction bus will encounter a surge impedance of lower value, resulting in a stepdown rather than a stepup in voltage magnitude. Where step-up conditions exist, supplementary protective devices may be required.

The use of several insulation systems (of different character) operating in series may, under a transient surge condition, be broken down at far less voltage that what had been expected by a cascade chain type failure within an equipment insulation system, which actually amplifies the magnitude of the incident wave. Fig 63 illustrates the mechanics involved. Three insulation systems 1, 2, and 3 are in series, having individual breakdown strengths of 10, 15, and 25 kV. Dependence was placed mainly on insulation systems 2 and 3 to assure a withstand capability of 35 kV.

Suppose that a surge voltage E has charged capacitor C_1. The circuit geometry is such that when the voltage on C_1 reaches 12 kV, the voltage across gap 1 reaches 10 kV, the insulation fails, and a sparkover occurs. Conduction across gap 1 causes the 12 kV to be impressed across the L_2–C_2 circuit. The response will be oscillatory with a period of about 15 μs per cycle. The voltage across C_2 will crest in about 7 μs with perhaps 70 percent overshoot. The voltage across C_1 will drop about 5 percent during this excursion. The resulting crest voltage across C_2 will be 19.4 kV, more than enough to break down gap 2. Consider that the insulation system 2 fails when the voltage level reaches 17 kV

on C_2. Conduction across gap 2 energizes the L_3–C_3 circuit, initiating an oscillatory response very similar to the L_2–C_2 circuit, except that the time period for a first half-cycle swing to crest voltage on capacitor C_3 will take only about 0.5 μs. A 70 percent overshoot would be reasonable with a 10 percent drop in voltage across capacitor C_2. The resulting first-crest voltage on capacitor C_3 would be expected to be 26 kV, sufficient to spark over gap 3. The series insulation system which was expected to withstand 35 kV failed with an applied voltage of 12 kV.

One solution is to introduce voltage-dividing impedances across the several individual insulation systems to force a proportional distribution of voltage across the individual systems under applied voltage conditions of direct-current, power-frequency, and high-frequency transient waves. Another solution would use an independent protective device across each insulation system. If the unbalanced-voltage distribution occurs only in the presence of a fast-front transient voltage wave, a wave-sloping capacitor network in the line connection of the same character used for turn-insulation protection of motor windings could be used (Note 3).

Surge voltages as they arrive at utilization apparatus usually display a substantial delay time in the buildup to crest value of perhaps 1–5 μs. Standard protection procedures recognize a wave front of 0.5–1.5 μs to crest. (In certain cases a chopped wave test is included which involves a fast-front voltage change of substantial magnitude.) Since standard test procedures are based on a 1.2 μs wave front, it follows that most of the available application data will be based on a surge having this characteristic.

There are practical circumstances which generate surge-voltage waves having extremely fast fronts. In the case of vehicular collisions with power poles which cause high- and low-voltage conductors to be brought together, the result can be a severe steep-front voltage surge to the low-voltage system. A lightning stroke to an overhead ground wire may create high voltages on the ground wire and tower structures which may flash back to a power conductor over the line insulator. The result is a step-like steep-front voltage transient on the power line conductor. The voltage rise on the ground mat of a high-capacity substation may be high enough to spark back to low-voltage conductors, resulting in a severe steep-front voltage transient on the low-voltage conductors. Switching interrupters, especially when switching capacitor banks, may make several unsuccessful attempts to interrupt after current zero. When a restrike occurs at the switching contacts, the voltage previously stored on the capacitor is abruptly impressed on the line conductors, creating a very-steep-front surge voltage on the line. Multiple restrikes can cause the voltage stored on the capacitor to become elevated to a value much greater than the normal operating voltage [11, chap 5, p 286].

Some conditions tend to reduce the slope of the wave front. A voltage surge in traveling through an isolating transformer, as distinguished from an autotransformer, will emerge with a less steep slope. A voltage surge traveling along a system containing a multitude of stub tap circuits will display a more gradual rise in voltage as a result of the initial diversion of charge into the stub circuits. Where installation circumstances seem favorable for the creation of large-magnitude extremely steep-front surge voltages with no inherent wave-front desloping before reaching critical apparatus, a surge-capacitor protective device should be installed. Most application tables showing allowable lead lengths are based on wave fronts of 0.5 μs or longer.

4.9.7 *Application Principles.* The design pro-

cedures involved in securing adequate insulation overvoltage protection are a combination of good basic design practices and appropriate overvoltage protective devices at proper locations. The objective is to limit the overvoltages imposed on the equipment insulation so that they do not exceed the certified withstand voltage limits, both in magnitude and rate of rise. The long-time insulation withstand capabilities are established by the 1 min 60 Hz high-potential test. (See Tables 14 and 15 for representative values of crest voltages.) The limits on short-time overvoltage withstand capabilities are established by the impulse tests which the insulation system is required to withstand (Table 14).

The magnitude and duration of modest overvoltages will be affected by the basic system design, especially system grounding and fault detection and protection. Generally the severity of such voltage exposure cannot be reduced by overvoltage protective equipment. The occurrence of the high-magnitude short-duration surge voltages will be substantially reduced by the good basic design. Those overvoltages which are inevitable can be controlled by overvoltage protective equipment. The overvoltage boundary limits set by available surge arresters are the sparkover voltages discussed earlier and listed in Table 16 [30], [31], [see also IEEE Std 28-1974, Surge Arresters for Alternating-Current Power Circuits (ANSI C62.1-1975), and ANSI C62.2-1969, Guide for Application of Valve Type Lightning Arresters for Alternating-Current Systems, including revised Appendix A-1972].

Rotating machine insulation systems are inherently complex. The presence of massive amounts of capacitive coupling to grounded core iron at each coil intensifies the problem of achieving turn-insulation security. Special protective capacitor units are designed with low internal inductance (Table 17) which control the rate of rise of incident overvoltages to protect the turn-to-turn insulation. Surge arresters of special design complete the rotating-machine insulation protection to ground by limiting the magnitude of the incident voltage wave.

Since the capacitive coupling to ground and the turn-insulation conditions are generally not as critical in transformers, adequate overvoltage

Table 17
Capacitance of Surge-Protective Capacitors
per Line Terminal Connected Line to Ground

	Rated Motor Voltage		
	650 V and Less	2400—6900 V	11 500 V and Higher
Capacitance, μF	1.0	0.5	0.25

protection is normally afforded by the use of station or intermediate-type surge arresters alone. These units are also suitable for protecting switchgear. The withstand levels of the particular equipment involved must always be considered.

The installation of surge arresters must secure a close direct shunt connection across the insulation to be protected. The additional impedance introduced in the connecting leads to the arrester will raise the magnitude of the surge voltage permitted by the protective device. With steep-front waves the inductance of connecting leads is more damaging than the resistance. Installation of the arresters upstream from the protected equipment will introduce another objectionable effect. If the apparatus connected to the line has a higher surge impedance than the line itself, which is a likely condition, this junction transition will cause an increase in the voltage at the terminal up to twice the voltage set at the arrester. The arrester location becomes more important as the operating system is faced with steep wave-front voltages (Section 4.9.6).

In establishing an appropriate safety margin between the apparatus withstand voltage curve and the allowed applied voltage stress, the fact that the insulation failure point is a function of the total duration of voltage stress must be considered. The factory insulation proof test need only be withstood once. The operating level set by the overvoltage protective system may be repeated many times. Since a 30 percent change in the applied voltage magnitude will create a tenfold change in the time of failure, a design objective aimed at limiting the permitted applied magnitude of voltage stress to approximately four fifths of the test-certified withstandard voltage provides a reasonable margin

of safety. A safe criterion for controlling the surge-voltage rate of rise where it is considered critical, should be a wave having a maximum crest equal to the 60 Hz high-potential-test (Section 4.9.2) peak voltage and a uniform rise of 10 μs.

4.10 Testing [37], [38]. In order to secure the full benefit which a well-designed protective installation is capable of providing it must be properly installed and tested. The tests are exacting and often complex, and must be performed very carefully to avoid endangering persons and equipment. Where possible, these services should be procured from specialists. Otherwise the following guide may be of assistance. Checking a new installation to ascertain that it has been correctly installed, connected, and adjusted must be more comprehensive than the testing which should be done periodically to determine deterioration, loss of adjustment, or tampering.

4.10.1 *Installation Checking.* Installation checking includes the following:

(1) General survey and diagramming
(2) Preliminary checking of equipment
(3) Calibration and setting of installed equipment
(4) Implementation of safeguards to prevent permanent magnetization of current transformers
(5) Final checking of equipment going into service

4.10.1.1 *General Survey and Diagramming*
(1) Study the intended function of each device and the manner in which all the devices are coordinated.
(2) Check the wiring diagrams to ensure that each device is so connected that it will perform its intended function. If no diagrams have been

provided, make them or obtain them, for it will be difficult to do a safe intelligent job of testing without them. Preserve the diagrams for future reference, and update them when changes or additions are made.

(3) Compare the diagrams with the actual connections, and when differences are found, determine whether the error is in the diagram or in the wiring and correct it.

4.10.1.2 *Preliminary Checking of Equipment*

(1) Inspect equipment for damage or maladjustment caused by shipment or installation.

(2) Verify that all protective relays, auxiliary relays, trip coils, trip circuit seal-in and target coils, fuses, and instrument transformers are the proper types and range.

(3) Remove wedges, ties, and blocks installed by the manufacturer to prevent damage during shipment.

(4) Make electrical continuity checks of all current, potential, and control circuits, referring constantly to the diagrams.

(5) Remove short-circuiting links from current transformers after checking that secondary circuits are complete.

(6) Make ratio and polarity checks of current and potential transformers where these somewhat time-consuming tests are considered necessary.

(7) Make tests of the insulation of all relays, wiring, instrument transformer secondaries, and instruments.

(8) Make provision for future testing of the equipment in conveniently small units. The installation of a relatively small number of test switches, test terminals, and test links before the equipment goes into service may make it possible to test the various elements of the protective installation one by one after they are in service, with a minimum of disturbance to production. The omission of such devices may require major parts of the plant to be shut down for testing. Where current transformers or wiring are shared between the relays of a single position and differential or other relays which are common to groups of positions, see that safe means are provided for separating each position's relays, current transformers, and wiring from the group being tested.

4.10.1.3 *Calibration and Setting of Installed Equipment*

(1) *Direct-Trip Circuit Breakers.* Low-voltage air circuit breakers are often tripped directly by the current through them without the interposition of current transformers and relays. They are usually set at the factory; and to check them in the field requires a test source capable of supplying trip currents. If they cannot be tested, at least verify that the marked instantaneous and time-delay settings are as required for coordination with other circuit breakers and fuses.

(2) *Relay-Operated Circuit Breakers.* The relays must be checked in accordance with the manufacturer's instructions and with the general guide below in which initial and maintenance checks are compared. However, the actual performance of the relays in service depends on the behavior of the instrument transformers which supply them with current and potential; and these in turn are influenced by the magnitude of their secondary burdens. It is therefore advisable to plan the testing in such a way as to obtain information about the performance of the relays, wiring, and transformer together as a unit, as well as separately.

Fig 64 shows one phase of a typical current-transformer circuit, indicating four different positions at which test current may be applied. The first three positions cause current to flow either toward the relay only, toward the current transformer only, or toward both in parallel. The test from position 2 or 3 toward the current transformer only is a secondary impedance or excitation test, and should include at least three points on the current-transformer saturation curve, with one at or slightly above the knee. One three-phase set of current transformers may differ widely in impedance from another set, yet each may be satisfactory for its own function if the values are consistent within the set.

(a) *Position 1. Test current applied at the individual relay location.* At this point it is possible to make three measurements, each of which is useful under certain conditions. In order to show position 1 in all three of its variations, an auxiliary current transformer has been added. These auxiliary current transformers are often used in multiple differentials and other

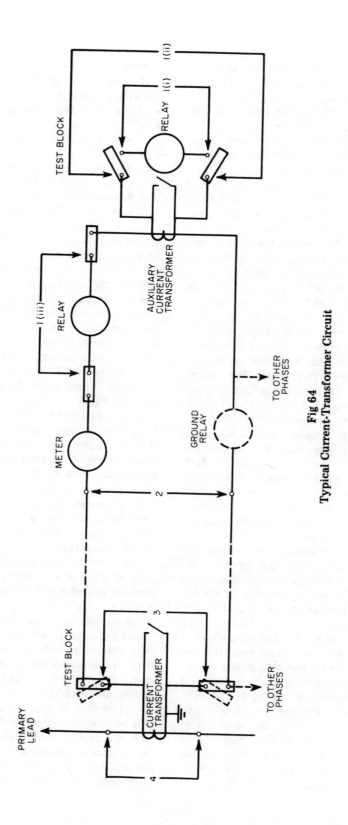

Fig 64
Typical Current-Transformer Circuit

complex schemes but are rarely employed in simple circuits like that shown in Fig 64. In testing they are treated the same as any other current transformers.

(i) The relay is disconnected, checked, and calibrated separately as an instrument.

Relays must be checked with current and an accurate timing device when they are placed in service to make sure that they have not been damaged in shipment and that they operate in the desired time shown by the coordination curves. If a relay does not give the desired operating time for a given current on the predetermined time dial setting, the desired time can usually be obtained by adjustment of the time dial. Other adjustments should not be attempted unless the adjuster is quite familiar with relay design and performance or has specific manufacturer's instructions.

(ii) With the relay disconnected and the main current-transformer primary effectively open, the test current is applied to the remainder of the secondary circuit. The current drawn should be low until the voltage is raised to the point where a main or auxiliary current transformer begins to saturate. This test checks for defects in the secondary circuit, except in the disconnected relay, but including current-transformer excitation current, open and short circuits, cross connections to other phases, etc.

If this test discloses appreciable differences in the test voltage required to produce a given value of current in the various phases, find out why. The cause may be an open or short circuit in the secondary wiring, a defective circuit transformer, or a legitimate unbalance of secondary burden, caused, for example, by single-phase metering or by the omission of a relay on one phase. It is not unusual to find that in the current-transformer common return lead of the three phases the burden, including relays, is much greater than in the phase leads. If it appears to be excessive, the tests should be extended at least to position 2.

(iii) The test current is applied to the relay current terminals, with the secondary wiring to the current transformers and other equipment normally connected. The relay is subjected to the same tests as in position 1(i), including timing, any difference in the results obtained being of course due to the fraction of current used in the excitation of current transformers.

This test is of particular value, not only because it provides a measure of the extent to which current-transformer performance affects relay pickup and timing, but also because it is the basis of much maintenance testing. If the values obtained in this position during later maintenance tests are substantially the same as those in the installation tests, the entire layout may be assumed unchanged. Only if the tests in positions 1(iii) and 1(ii) show unexplained unbalances, definitely noticeable saturation, or questionable residual burdens, are more extended tests necessary.

(b) *Position 2. Test current applied at switchboard terminals of current-transformer leads.* The test current is applied to an entire phase group of relays, meters, auxiliary current transformers, etc. Since the main current transformers remain in shunt with the burden, their effect on relay performance is included. This is a convenient and fairly effective means of determining whether special relay calibrations are required.

In testing ground relays from position 2, both phase- and ground-relay burdens will be included, which is the condition that will exist in actual operation to clear a ground fault. The phase relays will sometimes be called upon to operate on phase-to-phase faults (test current applied between two current-transformer phase leads) and sometimes on three-phase faults (test current applied between one current-transformer phase lead at a time and the neutral current level, with the neutral burden jumpered out). If there is any significant difference in readings, data should be recorded for both connections.

The connections at position 2 may be opened to test the current transformers without burden other than their leads to the panel. This is particularly advantageous in metalclad installations, where positions 3 and 4 are inaccessible or difficult to reach.

(c) *Position 3. Test current applied at current-transformer secondary terminals.* The test current is applied across the secondary terminals of the current transformer, or across the secondary leads in the proximity of the current transformer, with all meters, relays, and other burden normally connected and the primary open. The testing is the same as position 2, and the results are the same except that the secondary leads are included with the burden in the same manner as in normal service and that all devices can be readily identified with their respective current transformers. If leads were not positively identified, this is important. The current transformer can be tested alone from this position.

(d) *Position 4. Primary current check.* This is the best method of checking the performance of current transformers and relays together, since all burdens are included, together with their normal effect on the saturation characteristics of the current transformer. Unfortunately it is usually difficult to make the necessary high-current connections to the primary circuit and the equipment required for the high-current test source is unwieldy, so the primary current check is limited to special applications. When a primary current check is made, both ratio and polarity of the current transformers should be determined.

4.10.1.4 *Implementation of Safeguards to Prevent Permanent Magnetization of Current Transformers.* If during any of the above procedures the test current in the secondary winding of a current transformer is abruptly interrupted, the current transformer core iron may become permanently magnetized by residual flux to an extent determined by the current-transformer turns ratio, the hysteresis characteristics of the core steel, and the magnitude of test current interrupted in the current-transformer winding. Unremoved, this residual magnetism will significantly affect the accuracy of the current transformers when they are placed back into service and may cause the connected relays to nuisance trip or otherwise operate unpredictably.

This misbehavior can be avoided by using a continuously variable current supply for any tests involving current transformers in the connected circuits and instructing the operator to gradually reduce the test current from the test value to zero before opening the circuit to the test power source. As an alternative, the tests can be performed with all current-transformer secondaries short-circuited, and the test procedures modified accordingly.

4.10.1.5 *Final Checking of Equipment Going into Service.* Once the usual high-potential and phasing checks have been completed, the equipment is energized at normal potential for final check. Instrument-transformer cases have been grounded with conductors of adequate size, and the secondary wiring has been grounded either solidly or, if necessary, through well-designed spark gaps.

With proper grounds in place, suitable test switches, jacks, links, etc, installed, and with adequately insulated test leads, many users manipulate connections and make tests with the equipment energized. However, one has to make certain that all the necessary auxiliary testing devices are present, and that every step of the testing procedure has been planned and closely examined in advance to guard against unforeseen hazards.

(1) *Current-Transformer Secondary-Circuit Checks.* All changes of connection, insertion, and removal of meters, etc, must be made in such a manner that the secondary circuits of energized current transformers are not opened, even momentarily. An energized current transformer with an open secondary acts as a stepup transformer with a ratio equal to the turns ratio, and thus dangerously high voltages are generated. All current-transformer secondary circuits must therefore be provided with a test block which requires the current-transformer secondary terminals to be short-circuited before the secondary circuit can be disconnected.

(a) *Null checks.* If current is found where there should be none, a defect is indicated. However, a null check is inconclusive and must be supplemented by an additional check which would detect a false null caused by an open or short circuit.

For differential circuits in which the operating coil normally has no current, check that there is none. This, in conjunction with check (b) or (c), verifies that the current-transformer ratios are correctly balanced and the polarities and phases correctly related.

A zero or negligible current reading in the neutral or common return lead of a three-phase set of current transformers under balanced-load conditions indicates ratio balance and like polarity of the three secondaries.

(b) *Inspection of active relay circuits.* Use an ammeter, voltmeter, wattmeter, or phase-angle meter to check for the proper values and polarities of voltage and current in the various relay circuits. Check the contact positions of directional element and voltage relays, and compare with those expected in view of load conditions.

(c) *Relay operation checks with diverted load currents.* Whenever any relay in service is tested in a manner that may cause it to operate, the consequences of circuit breaker tripping must be considered. If a circuit breaker operation is not permissible, the trip circuit of the relay being checked must be opened.

(i) *Differential relays.* After determining that current is in the individual current-transformer circuits but none in the operating coil circuit, a current may be caused to flow in the operating coil by temporarily short-circuiting and disconnecting all but one of the current transformers.

The current from the remaining current transformer will check not only the operating coil circuit, but also the current transformer and its leads. This should be done with each current transformer circuit in turn if they have not been verified by other tests.

(ii) *Neutral or residual current relays.* Short-circuit and disconnect the current transformer leads of all but one phase. The remaining phase will supply current to the relay.

(2) *Potential-Transformer Secondary Circuit Checks.* Measure the potentials applied to all relay potential coils. If any are inadequate, investigate. Look for blown fuses, short circuits, excessive burdens, and improperly adjusted potential devices.

Check the potential transformer and phase to which each relay potential terminal is connected by removing one potential fuse at a time (at the potential-transformer secondary terminals) and noting the effect on the voltage applied to the relay.

(a) *Ground-fault potential relays and elements.* Voltage relays, or relay elements, that are connected in one corner of a potential-transformer secondary broken-delta connection have no voltage across them in the absence of a ground fault. Such a fault can be simulated as follows.

(i) De-energize the equipment so that it is safe to work on.

(ii) Disconnect the phase lead from one potential-transformer primary terminal and fasten this lead where it can safely be re-energized.

(iii) Short-circuit the vacated primary terminal to its neutral terminal.

(iv) Re-energize the equipment, read voltages, and observe relay operation.

(v) Return connection to normal and re-energize the equipment.

(b) *Ground-fault directional relays.* These have potential windings energized from one corner of a broken-delta-connected set of potential transformers, the same as the preceding type, and their current windings energized from the common (neutral or residual) connection of a set of current transformers. Check them with diverted load currents as follows.

(i) Determine the direction of power flow.

(ii) Alter one phase of the potential transformer primary as described in (a) (ii) and (iii).

(iii) Short-circuit and disconnect the current-transformer leads of this same phase. This should cause the ground-fault directional relay to indicate a direction of power flow which is the reverse of that actually existing in the line, that is, if the power flows toward the bus, the relay contacts should close to trip.

(iv) Restore the current-transformer leads and remove their short circuit. Then short-circuit and remove the current-transformer leads of the other phases. This should cause the relay to indicate a direction of power flow which is the same as that actually in the line.

(v) Restore all connections to normal.

(3) *Staged System Tests at Normal or Reduced System Voltage.* This method causes no difficulties with automatic throw-over devices and is the best method of testing them. Nearly all other system tests require setting up staged faults, which are the last resort in testing. The faults are applied to the system at carefully chosen times and places, under controlled and backup-protected conditions, and the action of relays and other equipment is recorded and analyzed. Such tests are seldom used, and can only be justified (a) if the scheme is intricate, new, or unfamiliar, (b) if the wiring is complicated or inaccessible, (c) if the relay response characteristics are believed to be so

critical that the use of normal or diverted load currents would introduce intolerable phase-angle errors, or (d) if the scheme has shown otherwise unexplainable misbehavior.

A staged fault test should be approved only when no other method of testing will suffice. The plan should be scrutinized from every conceivable safety consideration, and all parties who could possibly be affected should be notified.

4.10.2 *Maintenance and Periodic Testing.* Protective equipment should be checked periodically to assure that its operation and coordination have not been impaired by exposure to dusty, smoky, oily, or corrosive atmospheres, by mechanical vibration or shock, by excessive temperature, by tampering, or in any other manner. The consequences of neglecting proper testing may at first seem slight; but they are cumulative and such neglect can lead to an increasing loss of protection.

4.10.2.1 *Maintenance Benefits.* If a plant is interconnected with a utility, the utility will own or control the protective equipment adjacent to one or both sides of the tie point and will make its own initial and periodic tests necessary to protect equipment and the service to other customers. The acceptance of responsibility for the proper maintenance of other equipment may be covered by contracts and correspondence between the utility and the industrial power user; but frequently it is left to custom and practice. However, it is best that the role of the utility and the plant be clearly established in writing so that each party clearly recognizes its responsibilities to the other.

A regular maintenance program is required to minimize the number of unscheduled plant shutdowns. In some plants an hour of down time can be more costly than the expense of a thorough maintenance program. An effective maintenance program involves inspecting, testing, and maintenance of electric protective devices and all associated equipment at regular intervals.

4.10.2.2 *Recommended Maintenance Frequency and Procedures.* The time interval between tests of protective equipment is a variable, depending upon such factors as cleanliness of atmosphere and surroundings, average

operating temperatures, freedom from vibration and shock, quality of operating personnel, etc. The best interval in any given type of installation must be determined by experience, but will lie somewhere between a minimum of six months and a maximum of three years (with supplementary operating checks at intermediate periods). If no deterioration is found on several successive checks, the testing is probably more frequent than necessary, while if equipment is found in an inoperative condition or out of adjustment, the testing has been delayed too long.

Regularly scheduled testing should be supplemented with special tests made at any time that there is reason for suspecting that the protective equipment may have been damaged. An intelligent and loyal operating force can assist in noting and reporting signs of abnormal performance. Where such assistance is reliable, periodic tests may be scheduled less frequently. Standard test values must be on file, and although all tests may not be recorded, details of any defects found must be recorded. The records must cover not only relays, but all protective equipment for special as well as periodic tests.

In planning a maintenance program for a plant shutdown, the following preparatory work should be done.

(1) Select the date and time when equipment may be de-energized. In operations where summer vacations or other annual shutdowns occur, these are the best times for maintenance work. Maintenance programs may also be carried out on night shifts and over weekends.

(2) Update the system single-line diagram.

(3) Update the short-circuit analysis and protective-device coordination study, using the latest values of short-circuit duty available from the power company's system.

(4) Since maintenance work is generally done on premium time, and shutdown periods must be held to a minimum, the necessary materials, tools, and test sets must be assembled in advance and prepared for use. These include relay test set, circuit breaker test set, any necessary lighting and power during the shutdown, and cleaning aids consisting of clean, white, lint-free cloths, solvents, vacuum cleaners (with insulated attachments), and dry compressed air. Equipment required for personnel safety

will include an energized-phase detector, an energized-phase-detector tester, three-phase ground straps, rubber floor mats, safety glasses and shoes, and rubber gloves. Personnal must be made safety conscious by being required to wear nonconducting hard hats, keep sleeves rolled down, and remove watches, rings, metals, and other conducting articles.

4.10.3 *Scope of Testing.* Testing of protective equipment is generally considered as relay testing, but it goes far beyond mere verification of the relay calibration. The greater number of irregularities found do not involve the relays as instruments. Testing must be planned to check, as far as possible, the entire system from instrument transformers to circuit breaker operation.

During shutdown the maintenance program should include but not necessarily be limited to the following work.

(1) Remove, inspect, and electrically test the protective relays. Testing consists of passing current through the relay coil and recording the time required for contact closing to verify proper calibration and operation.

(2) Remove from service, inspect, and electrically test the electromechanical-trip low-voltage circuit breakers to ascertain that their tripping characteristics comply with the manufacturer's time—current characteristic curves or the coordination study. Testing consists of passing current through each pole and recording the current and time necessary to trip the circuit breaker.

A trip test kit, which applies a current signal to the trip device simulating the output of the circuit breaker current transformers during overload or fault conditions, should be used to test static-trip low-voltage circuit breakers to verify trip calibration. At least one test in each phase using the same procedure as recommended for electromechanical-trip circuit breakers should be performed to verify actual operation.

(3) Inspect and electrically test high- and medium-voltage circuit breakers if provisions are available. Check them mechanically, and lubricate at specified points. If slow closing is possible, determine if the arcing contacts make before the main contacts do and open after the main contacts open.

(4) Check medium-voltage fuses for correct application, size, continuity, and signs of overheating.

(5) Check the key interlocking system between the medium-voltage and low-voltage components to determine if it is operating correctly to provide the safety features required.

(6) Open medium- and high-voltage cubicles and visually inspect for general conditions and for loose bus and cable connections, cable terminations, bus insulator cracks, switchgear and conduit grounding connections, control wiring, etc.

(7) Check the grounding system for all switchgear and unit substations.

A specific procedure must be established for the maintenance work during the shutdown. First electrically test relays and circuit breakers with the unit substations energized, then de-energize the switchgear and unit substations and perform the necessary inspection and maintenance. With this procedure the test set can be connected directly to the substation. Make connections with heavy-duty alligator clips or bolt-type clamps to the load side studs of a cubicle that contains a spare circuit breaker.

4.10.3.1 *Instrument Transformers and Wiring.* Inspect equipment visually for obvious defects such as broken studs, loosened nuts, damaged insulation, etc.

The indication of normal potential at the relay by lamp or voltmeter is considered adequate verification for potential transformers and circuits. If the combined relay and current-transformer check made at installation is repeated and substantially the same results are obtained, this is sufficient proof that there is no short circuit in the current transformer or its leads. A check under load with a low-burden ammeter in series or shunt with the relay will establish proof of continuity. More elaborate checking as outlined in Section 4.10.1 is required if there has been any change in the equipment or wiring, or if a change in setting materially alters the current or potential-transformer burdens. If there is evidence of improper performance, the equipment must be completely de-energized, the protective ground connections removed, and the insulation of the current, potential, and control wiring tested.

4.10.3.2 *Relays.* Protective relays may be temporarily removed for inspection and test.

The relay is very accurate and dependable when operated in the proper environment and under the right conditions and should require little maintenance. Whenever removed from its case, it should be protected from dust, moisture, and excessive shock. Before testing any relay, its mode of operation and its relationship to the system should be known. Since the metallic enclosure forms a path for stray eddy currents which can affect relay performance, overcurrent relays should be tested while installed in a relay case. Manufacturers' instructions for specific relays provide useful information concerning connections, adjustments, repairs, timing data, and so on.

(1) Visually inspect the relay by removing the cover and checking for deposits of dust, dirt, or other foreign matter. Disable the circuit breaker trip circuit by removing the relay from service. Since the majority of relays operate from the secondaries of current transformers, removing the relay must be done very carefully because if the secondary circuit is opened, a very high voltage will result.

(2) Remove all foreign material such as dust or filings in the relay using a small hand air syringe or dry compressed air of 15 lb/in^2 or lower.

(3) Make sure all connections are tight.

(4) Remove any rust or filings from disk or magnet poles with a magnet cleaner or brush.

(5) Hold relay up to the light to make sure the disk has proper clearance between magnet poles.

(6) Inspect the relay for the presence of moisture.

(7) Check the condition of bearings and pivots by rotating the disk manually to close the contacts and observe that operation is smooth. Special oil is required for jewel-bearing lubrication.

(8) Clean pitted or burned contacts with contact burnisher or fine file.

(9) Check for damaged coils, resistors, or wiring, and for damaged or maladjusted indicator targets or holding devices.

The following list compares initial and periodic tests for some of the more common types of relays.

(1) *Differential Relays*

(a) *Initial.* Verify the published differential characteristics by checking at minimum operating current and at several other values.

(b) *Periodic.* Check minimum operating current with current in each restraint coil. Check operation and time at a higher value with one restraint coil.

(2) *Undervoltage Relays*

(a) *Initial.* Check and adjust pickup and dropout values. Record time, at least on drop from normal to zero voltage.

(b) *Periodic.* Check dropout voltage and time.

(3) *Overvoltage Relays*

(a) *Initial.* Check and adjust pickup and dropout values. Check time characteristics, if specified, and operational functions.

(b) *Periodic.* Check pickup. Check operation according to desired functions.

(4) *Directional Elements*

(a) *Initial.* On energizing the current winding or the potential winding alone, there should be no more than a slow drift toward closing. Using high current and low voltage, determine the minimum power in volt-amperes required for effective closing. Set test-feed for high (60 A or more) current and no voltage and apply momentarily; the contacts should not close. In applying time settings to a directional overcurrent relay, arrange the test to include the time of both the directional and the overcurrent elements.

NOTE: On current-polarized directional units and on current-product type directional relays, the current coils can usually be connected in series for ease in testing and calibrating.

(b) *Periodic.* Test for action of current or voltage alone as above. Check overall time, including the associated element. Check directional element time separately only if the selectivity margins are so close as to require it.

(5) *Instantaneous Overcurrent Relays and Elements.* In checking the pickup current of these relays, raise the current gradually in order to avoid transient effects. Also, some plunger-type relays exhibit a marked increase of burden impedance when the gap of the magnetic circuit closes in operating, so that it may be difficult to obtain certain values of current on test. This effect can be made less troublesome by the use of relatively high values of test voltage, in-

cluding a resistance in series with the test source. If a relay heats excessively during current adjustment, make a preliminary adjustment of the current with the relay bypassed.

(a) *Initial.* Check pickup current and adjust if required. Dropout is important in a few applications. If a plunger type relay has trip coils or other protective devices in series with the relay coil, check from position 2, 3, or 4 to be sure that the pickup relay does not make the total current-transformer burden too great for the operation of these other devices.

(b) *Periodic.* Check minimum pickup current and functional reponse.

(6) *Induction Time-Overcurrent Relays and Elements.* These are particularly susceptible to variation in both pickup current and timing if the waveform of the current used for testing is distorted. Such distortion can arise from nonlinear magnetic effects either in the relay itself or in a test transformer. It can be minimized by using a relatively high-voltage test source (several times the voltage required across the relay winding itself), absorbing the excess in a rheostat connected in series with the relay. Some relays also change characteristics with heating, which may arise from ambient temperatures, load currents or test currents.

(a) *Initial.* Check the pickup current for the tap setting selected. Verify the timing characteristic by checking two points on the relay operating curve, one at a minimum value of four times pickup and the second at a larger current. Make all these tests both with the relay isolated and with the current transformers and other burden included, and if there is any significant difference, record both sets of data for future use. If the relay does not operate according to its published characteristics, adjust following the manufacturer's instructions and repeat the test.

(b) *Periodic.* Check pickup current and timing, preferably in a combined test.

A very important part of the periodic maintenance is keeping adequate records for future use in detecting changes. Fig 65 illustrates a typical inspection and test form for a time-delay overcurrent relay with instantaneous attachment. It also serves as a record for model, type, and manufacturer with "as found" and "as left" calibration settings.

4.10.3.3 *Low-Voltage Circuit Breakers.* Manufacturers' literature should be consulted prior to inspection and testing. Visual and mechanical inspection should include the following.

(1) Test the interlock feature which prevents plug-in or drawout while the circuit breaker is closed to assure positive action.

(2) Trip the circuit breaker and draw it out from its enclosure.

(3) Lubricate the tracks, jack screws, or other mechanism present in the cubicle for drawing the circuit breaker in and out using the lubricant recommended by the manufacturer.

(4) Cover the opening to the cubicle if a spare circuit breaker is not available.

(5) Clean the circuit breaker.

(6) Examine the primary fingers on drawout contacts on the circuit breaker for evidence of overheating or embrittlement cracks. Check springs for tension. Clean and service contact surfaces as required. Apply a thin film of contact lubricant.

(7) Remove arc chutes; examine for cracks, dirt, splashed copper, broken or cracked snuffers. Clean.

(8) Inspect and clean main and arcing contacts in accordance with the manufacturer's instructions. Note that commercial products are available to resilver these contacts.

(9) Open and close circuit breaker several times to determine that operation is smooth and there is no binding.

(10) If provisions for slow closing are available, manually slow-close contacts and observe that all three phases make at the same time and that arcing contacts (if fitted) make first on closing and break last on opening. Make necessary adjustments.

(11) With the circuit breaker contacts open, insert a piece of paper backed with carbon paper between the fixed and moving contacts. Close the circuit breaker, then manually trip it. Remove paper and examine the "contact print" to make sure that both moving and fixed contacts are mating properly. Clean the contacts. Check proper operation of counter and flag.

(12) Tighten all screwed and bolted connections except pivot joints.

PLANT ENGINEERING SURVEY
OVERCURRENT RELAY INSPECTION AND TEST RESULTS

Plant Location _____ Date _____

Relay Location _____

Relay Description _____

Relay Type _____

Relay Manufacturer _____

Current-Transformer Ratio ___ Accuracy ___ Potential-Transformer Ratio ___ Accuracy ___

Burden: _____   Burden: _____

TEST RESULTS

Settings			Relay Tests					
Tap (amperes)	Time Dial	Instan-taneous	Pickup (amperes)	Tap Value × 4 (amperes)	Time (seconds)	Instan-taneous Pickup (amperes)	Target (dc amperes)	
								Specified
								As Found
								As Left
			For Future Tests					Date

Seal-in holds at (dc amperes) _____

Comments:

Fig 65
Typical Relay Inspection and Test Form

(13) Manually operate the armatures from all trip devices including undervoltage and shunt trip devices, strike the trip bar, and trip the circuit breaker.

(14) Lubricate mechanical joints and the racking device using lubricant recommended. Keep lubricants away from electrical contact surfaces.

(15) Examine all overload trip devices to verify that they are of the same type, have the same rating, and have proper settings.

Some of the common causes of malfunction of circuit breakers are as follows.

(1) Improper main contact pressure resulting in local overheating and burning of contact surfaces. In extreme cases, contacts may actually be welded together through this local overheating. Deterioration of insulation may result. The dimensional change or wear of parts may vary the wipe or overtravel of contacts resulting in improper contact pressure.

(2) Improper fit of circuit breaker primary fingers or drawout contacts on the bus stabs. Local overheating can destroy the tension in the primary finger springs.

(3) Cracked or dirty arc chutes can cause a phase-to-phase or phase-to-ground flashover when the circuit breaker is operated under load.

(4) Stiff or frozen mechanical joints in either the circuit breaker mechanism or trip device mechanism may cause delayed operation or failure to operate.

(5) Vibration may shift the trip bar so that the armature of the overload device fails to strike properly to release the latch to trip the circuit breaker.

(6) Worn parts or loss of oil in the overload time delay dashpot.

(7) Contaminated oil in the overload device dashpot freezing the armature, resulting in failure to trip.

(8) Improper calibration or setting of the overload trip device.

Improper delay in opening automatically under overload or fault is most dangerous to safety. Normal operational procedures and careful maintenance inspections will reveal most of the other conditions that may remove protection from the circuit. Unless overload tests are run, there is no way that improper delay can be detected until the device operates improperly.

Testing the direct-acting trip devices of low-voltage circuit breakers is becoming a widely accepted practice. The following procedures are used for electrical testing of circuit breakers.

(1) The connection of the circuit breaker drawout contacts to the stabs of the test set must be tight. This may present problems especially with the larger size circuit breakers. Hand-operated hydraulic platform trucks and chain hoist suspension systems have been used with some success to raise the circuit breaker to the proper height.

(2) The circuit breaker is tested one phase at a time.

(3) Trip devices must be allowed to fully reset between tests.

(4) Each of the trip features on each pole must be individually tested — long time, short time, and instantaneous — for proper action, operating current, and time to trip. Test current and trip times must be compared to the coordination study or manufacturers' time—current characteristic curves, and the trip mechanism should be adjusted if necessary to the proper setting.

(5) Several tests are recommended at each setting.

Test current must not be passed through the fuses when fused circuit breakers are tested. Testing must be arranged so that neither the test set nor a circuit breaker is overheated during the tests. Records must be kept of all tests and results. Fig 66 shows a typical inspection and test form for a circuit breaker with or without fuses.

Complete electrical test and maintenance for circuit breakers should be performed every two years and the circuit breakers should be exercised by tripping and closing several times every year. After the relays and circuit breakers have been tested electrically, the switchgear and substations must be de-energized, inspected, and maintained as necessary.

Records must be kept on the general condition of the apparatus for comparison at subsequent inspection periods. Fig 67 illustrates an inspection checklist for a unit substation that has a primary selective design. It is divided into a high-voltage section, a transformer section, and a low-voltage section. A similar check-

PLANT ENGINEERING SURVEY
LOW-VOLTAGE POWER CIRCUIT BREAKER INSPECTION AND TEST RESULTS

Plant Location _____ Date _____

Circuit Breaker Location _____

Circuit Description _____

Circuit Breaker Manufacturer _____ Type _____

Manufacturer's Serial Number or Shop Order Number _____

Trip Device Manufacturer _____ Type _____

Trip Coil Rating _____ Long-Time Delay Range _____

Short-Time Delay Range _____

Instantaneous Range _____

Cable Size _____

Fuse Manufacturer and Catalog Number _____

TEST RESULTS

Trip Unit		As Found Settings per Phase			As Left Settings per Phase		
		A	B	C	A	B	C
Long-Time Delay (amperes)							
Short-Time Delay (amperes)							
Instantaneous (amperes)							
Test Results	Long-Time Delay Current (amperes)						
	Time (seconds)						
	Short-Time Delay Current (amperes)						
	Time (seconds)						
	Instantaneous Current (amperes)						

Comments:

Fig 66
Typical Low-Voltage Power Circuit Breaker Inspection and Test Form

SUBSTATION # . . .

	Yes	No

A. *High-Voltage Section*
  1. Concerning the two-position, three-pole gang-operated air interrupter switches for each feeder,
     a. Do the switches operate freely and with the proper action if "quick-make"—"quick-break"? □ □
     b. Do the three blades make contact at the same time? □ □
     c. Are the arcing contacts intact? □ □
     d. Are there insulating barriers between the poles? □ □
  2. Has the circuit phase rotation been checked for the two feeders? □ □
  3. What are the nameplate data for the three current-limiting fuses on the transformer primary?

  4. Does the key interlocking system work correctly? □ □
  5. Are the stress cones and terminations made correctly? □ □
  6. Are the phase buses insulated completely? □ □
  7. Are bus connections tight? □ □
  8. Have the insulators been checked for cracks, cleanliness, and tracking? □ □
  9. Have the system ground connections been made to the ground bus? □ □
  10. Have the steel plates been provided to prevent access to the high-voltage compartment? □ □
  11. Has a mimic bus been painted on the switchgear? □ □
  12. Has any discoloration of current-carrying metal parts occurred? □ □

B. *Transformer Section*
  1. What are the transformer nameplate data?

  2. What are the insulation resistances of the primary and secondary windings?
     a. Primary to ground _____ megohms
     b. Secondary to ground _____ megohms
     c. Primary to secondary _____ megohms
  3. What are the resistances of the delta-connected primary windings?
     a. Phase A to phase B _____ ohms
     b. Phase A to phase C _____ ohms
     c. Phase B to phase C _____ ohms
  4. Has the transformer been checked for leaking conditions? □ □
  5. Have the provisions been made for four 2½% taps in the high-voltage windings (two above and two below rated primary voltage)? □ □
     Tap changer is set at position _____
  6. Have the connections at the transformer bushings for both the high side and low side been checked for tightness and good contact surfaces? □ □
  7. Are the connections to the transformer primary and secondary bushings made through flexible connectors? □ □
  8. Have the transformer bushings been checked for cracks? □ □
  9. Has the neutral of the transformer secondary been brought out through a bushing and has it been connected to the neutral bus? □ □

Fig 67
Typical Unit Substation Inspection Checklist

		Yes	No
B.	*Transformer Section* (Cont'd)		
10.	Has the neutral bus been connected to the grounding pad at the secondary end of the transformer by means of a removable bus-bar link?	☐	☐
11.	Have the transformer accessory devices been checked?	☐	☐
12.	Has the phase designation been maintained in connections from the transformer secondary bushing to the low-voltage bus? (Front to back, top to bottom, and left to right shall be phase A, phase B, and phase C, respectively)	☐	☐
13.	Has the gas absorber been installed completely?	☐	☐
14.	Have provisions been made to install auxiliary fan cooling with all necessary control devices wired?	☐	☐
C.	*Low-Voltage Section*		
1.	Have buses been provided for between the transformer secondary bushings and the transformer secondary main circuit breaker?	☐	☐
2.	Are the three-phase main bus bars continuous and rated for _____ amperes?	☐	☐
3.	Is the neutral bus continuous and rated for _____ amperes?	☐	☐
4.	Is the ground bus		
	a. Continuous?	☐	☐
	b. Connected to the system ground?	☐	☐
	c. Solidly bolted to the steel framework?	☐	☐
5.	Is the bus phase designation from front to back, top to bottom and left to right phase A, phase B, and phase C, respectively, when viewed from the front of the switchgear?	☐	☐
6.	Has the circuit phase rotation been checked in the bus-tie cubicle?	☐	☐
7.	Are all bus connections tight?	☐	☐
8.	Has any discoloration of current-carrying metal parts occurred?	☐	☐
9.	Is it possible to draw a circuit breaker in or out while it is in the closed position?	☐	☐
10.	If aluminum cable is used,		
	a. Have the outer strands of the conductor been scored or nicked causing oxidation at the point of connection?	☐	☐
	b. Are the cable terminator lugs designed for use on aluminum cable?	☐	☐
11.	Have primary current-limiting fuses been provided for potential transformers?	☐	☐
12.	Do the ammeter and voltmeter read correctly?	☐	☐
13.	Are the ammeter and voltmeter test blocks associated with the main secondary circuit breaker wired correctly?	☐	☐
14.	Are there transit barriers or protective boots over primary studs for each space cubicle?	☐	☐
15.	Are the cable connections made directly to the buses only for metering and/or control?	☐	☐
16.	Are all feeder conduits grounded within the substation?	☐	☐
17.	Are feeder cables adequately supported?	☐	☐
18.	Have feeder nameplates been installed on the switchgear?	☐	☐
19.	In the fire pump feeder circuit,		
	a. Is the circuit breaker overcurrent protective device set so as not to open the circuit under stalled rotor current or other motor starting conditions of the fire pump motor under maximum plant load?	☐	☐
	b. Does the circuit breaker overcurrent protective device (fuses are not recommended) have overcurrent setting for short-circuit protection only (instantaneous trip without long- and short-time delay)?	☐	☐

Fig 67 (Cont'd)

list should be tailored to other items of equipment.

4.10.3.4 *Control Wiring and Operation.* Periodic testing of protective equipment must ensure that the operation of any tripping relay will result in the circuit breaker being tripped. After all terminals and exposed portions of the trip circuit wiring and the condition and adjustment of any circuit breaker auxiliary switches in the trip circuit have been visually checked, the relay trip contacts should be manually closed to simulate an actual trip operation. Where there are too many relays to trip the circuit breaker from each, the trip circuit can normally be tested through at least one relay which is connected with the others to a common point on the opening control circuit wire. A record should be kept of each relay from which the circuit breaker was tripped, so that all relays may in turn be covered during successive tests.

4.10.3.5 *Completing the Job.* Before re-energizing the protective system, a visual inspection by at least two qualified persons should be made to ascertain that all temporary ground connections are removed, and that all tools, rags, and other cleaning aids have been removed from the interior of switchgear and unit substations.

4.10.4 *Summary — Requirements for Protective Equipment Testing*

(1) Adequate test sources.

(2) Adequate test switches or similar means of isolating equpment which will permit testing equipment both individually and in functional groups.

(3) Control and metering equipment for the applied currents and voltages. Some types of relays require special equipment, which is specified in the manufacturers' instruction books.

(4) A suitable means for timing trip circuit closures and openings and for timing circuit breaker closing and opening times.

(5) Complete records of all connections, electrical constants, settings, test values, operating performance, and failures or weaknesses found on test.

(6) Trained personnel.

(7) Written test procedures which are comprehensive and easy to follow.

4.11 Standards References. The following standards publications were used as references in preparing this chapter.

ANSI C2 (1976 ed), National Electrical Safety Code

ANSI C37.2-1970, Manual and Automatic Station Control, Supervisory, and Associated Telemetering Equipments

ANSI C37.4a-1958 (R 1971), Supplement to ANSI C37.4-1953 (R 1971), Definitions and Rating Structure, AC High-Voltage Circuit Breakers Rated on a Total Current Basis

ANSI C37.6-1971, Schedules of Preferred Ratings for AC High-Voltage Circuit Breakers Rated on a Total Current Basis

ANSI C37.40-1969 (R 1974), Service Conditions and Definitions for Distribution Cutouts and Fuse Links, Secondary Fuses, Distribution Enclosed Single-Pole Air Switches, Power Fuses, Fuse Disconnecting Switches, and Accessories

ANSI C37.41-1969 (R 1974), Design Tests for Distribution Cutouts and Fuse Links, Secondary Fuses, Distribution Enclosed Single-Pole Air Switches, Power Fuses, Fuse Disconnecting Switches, and Accessories

ANSI C37.46-1969 (R 1974), Specifications for Power Fuses and Fuse Disconnecting Switches

ANSI C37.47-1969 (R 1974), Specifications for Distribution Fuse Disconnecting Switches, Fuse Supports, and Current-Limiting Fuses

ANSI C50.10-1975, General Requirements for Synchronous Machines

ANSI C50.13-1975, Requirements for Cylindrical Rotor Synchronous Generators

ANSI C57.13-1968, Requirements for Instrument Transformers

ANSI C62.2-1969, Guide for Application of Valve Type Lightning Arresters for Alternating-Current Systems, including revised Appendix A-1972

ANSI C97.1-1972, Low-Voltage Cartridge Fuses 600 Volts or Less

IEEE Std 20-1973, Low-Voltage AC Power Circuit Breakers Used in Enclosures (ANSI C37.13-1973)

IEEE Std 28-1974, Surge Arresters for Alternating-Current Power Circuits (ANSI C62.1-1975)

IEEE Std 100-1972, Dictionary of Electrical and Electronics Terms (ANSI C42.00-1972)

IEEE Std 142-1972, Grounding of Industrial and Commercial Power Systems (ANSI C114.1-1973)

IEEE Std 242-1975, Protection and Coordination of Industrial and Commercial Power Systems

IEEE Std 273-1967, Protective Relay Applications to Power Transformers (ANSI C37.91-1972)

IEEE Std 313-1971, Relays and Relay Systems Associated with Electric Power Apparatus (ANSI C37.90-1971)

IEEE Std 462-1973, General Requirements for Distribution, Power, and Regulating Transformers (ANSI C57.12.00-1973)

NEMA FU 1-1972, Low-Voltage Cartridge Fuses

NEMA MG 1-1972, Motors and Generators (ANSI C52.1-1973)

NFPA No 70, National Electrical Code (1975), (ANSI C1-1975)

NFPA No 70B, Electrical Equipment Maintenance (1975), (ANSI C132.1-1975)

UL 198.1-1973, Class H Fuses

UL 198.2-1974, High-Interrupting-Capacity Current-Limiting Type Fuses

UL 198.3-1974, High-Interrupting-Capacity Class K Fuses

UL 198.4-1974, Class R Fuses

UL 198.5-1973, Plug Fuses

UL 198.6-1975, Fuses for Supplementary Overcurrent Protection

4.12 References and Bibliography

4.12.1 References

[1] ANANIAN, L. G., and COLVIN, F. L. The Industrial Electrical Engineer's Responsibilities and How They Reflect on Management. *IEEE Transactions on Industry and General Applications*, vol IGA-7, Mar/Apr 1971, pp 169-177.

[2] McFADDEN, R. H. Power-System Analysis: What It Can Do for Industrial Plants. *IEEE Transactions on Industry and General Applications*, vol IGA-7, Mar/Apr 1971, pp 181-188.

[3] CASTENSCHIOLD, R. Solutions to Industrial and Commercial Needs Using Multiple Utility Services and Emergency Generator Sets. *IEEE Transactions on Industry Applications*, vol IA-10, Mar/Apr 1974, pp 205-208.

[4] BRIGHTMAN, F. P. Selecting AC Overcurrent Protective Device Settings for Industrial Plants. *AIEE Transactions (Applications and Industry)*, pt II, vol 71, Sept 1952, pp 203-211.

[5] KAUFMANN, R. H. Nature and Causes of Overvoltages in Industrial Power Systems. *Iron and Steel Engineer*, Feb 1952.

[6] HARDER, E. L., KLEMMER, E. H., SONNEMANN, W. K., and WENTZ, E. C. Linear Couplers for Bus Protection. *AIEE Transactions*, vol 61, May 1942, pp 241-248.

[7] SHIELDS, F. J. The Problem of Arcing Faults in Low-Voltage Power Distribution Systems. *IEEE Transactions on Industry and General Applications*, vol IGA-3, Jan/Feb 1967, pp 15-25.

[8] CONRAD, R. R., and DALASTA, D. A New Ground Fault Protective System for Electrical Distribution Circuits. *IEEE Transactions on Industry and General Applications*, vol IGA-3, May/Jun 1967, pp 217-227.

[9] MASON, C. R. *The Art and Science of Protective Relaying*. New York: Wiley, 1956.

[10] Application and Protection of Pilot-Wire Circuits for Protective Relaying. AIEE S-117, 1960.

[11] BEEMAN, D. L., Ed. *Industrial Power Systems Handbook*. New York: McGraw-Hill, 1955.

[12] DALZIEL, C. F., and STEINBACK, E. W. Underfrequency Protection of Power Systems for System Relief: Load Shedding — System Splitting. *AIEE Transactions (Power Apparatus and Systems)*, pt III, vol 78, Dec 1959, pp 1227-1238.

[13] JACOBS, P. C., JR. Current-Limiting Fuses: Their Characteristics and Applications. *AIEE Transactions (Power Apparatus and Systems)*, pt III, vol 75, Oct 1956, pp 988-993.

[14] LARNER, R. A., and GRUESEN, K. R. Fuse Protection of High-Voltage Power Transformers. *AIEE Transactions (Power Apparatus and Systems)*, pt III, vol 78, Oct 1959, pp 864-878.

[15] CAMERON, F. L. The Coordination of High-Voltage Current-Limiting Fuses. *Proceedings, 1964 1st Annual Conference on Industrial and Commercial Power Systems*, IEEE T-163, pp 100-108.

[16] FITZGERALD, E. M., and STEWART, V. N. High-Capacity Current-Limiting Fuses Today. *AIEE Transactions (Power Apparatus and Systems)*, pt III, vol 78, Oct 1959, pp 937-947.

[17] IEEE COMMITTEE REPORT. Coordination of Lightning Arresters and Current-Limiting Fuses. *IEEE Transactions on Power Apparatus and Systems*, vol PAS-91, May/Jun 1972, pp 1075-1078.

[18] BRIGHTMAN, F. P. More About Setting Industrial Relays. *AIEE Transactions (Power Apparatus and Systems)*, pt III, vol 73, Apr 1954, pp 397-406.

[19] BRIGHTMAN, F. P. Short-Circuit Protection for Industrial Plant Generators. General Electric Company, Bull GER-779, 1953.

[20] HIGGINS, T. D., and PEACH, N. First Principles of System Coordination and Protection — A Proposed First Chapter for a Publication on Preferred Practice. *Conference Record, 1966 First Annual Meeting of the IEEE Industry and General Applications Group*, IEEE 34C-36, pp 245-252.

[21] AIEE COMMITTEE REPORT. Bibliography of Industrial System Coordination and Protection Literature. *IEEE Transactions on Applications and Industry*, vol 82, Mar 1963, pp 1-2.

[22] LATHROP, C. M., and SCHLECKSER, C. E. Protective Relaying on Industrial Power Systems. *AIEE Transactions*, pt II, vol 70, 1951, pp 1341-1345.

[23] WEDDENDORF, W. A. Evidence of Need for Improved Coordination and Protection of Industrial Power Systems. *IEEE Transactions on Industry and General Applications*, vol IGA-1, Nov/Dec 1965, pp 393-396.

[24] BARNES, H. C., MURRAY, C. S., and VERRALL, V. E. Relay Protection Practices in Steam Power Stations. *AIEE Transactions (Power Apparatus and Systems)*, pt III, vol 77, Feb 1959, pp 1360-1367.

[25] FAWCETT, D. V. The Tie Between a Utility and an Industrial when the Industrial Has Generation. *AIEE Transactions (Industry and Applications)*, pt II, vol 77, Jul 1958, pp 136-143.

[26] AIEE COMMITTEE REPORT. Switching Surges Due to Deenergization of Capacitive Circuits. *AIEE Transactions (Power Apparatus and Systems)*, pt III, vol 76, Aug 1957, pp 562-564.

[27] HENDRICKSON, P. E., JOHNSON, I. B., and SCHULTZ, N. R., Abnormal Voltage Conditions Produced by Open Conductors on Three-Phase Circuits Using Shunt Capacitors. *AIEE Transactions (Power Apparatus and Systems)*, pt III, vol 72, Dec 1953, pp 1183-1193.

[28] AIEE COMMITTEE REPORT. Power System Overvoltages Produced by Faults and Switching Operations. *AIEE Transactions*, vol 67, 1948, pp 912-922.

[29] KAUFMANN, R. H. Overvoltages in Industrial Systems — How to Reduce Them by Neutral Grounding. *Industrial Engineering News*, May/Jun 1951.

[30] DRAKE, C. W., JR. Lightning Protection for Cement Plants, Part I — Surge Voltages on the Power System. *IEEE Transactions on Industry and General Applications*, vol IGA-4, Jan/Feb 1968, pp 57-61.

[31] SARRIS, A. E. Lightning Protection for Cement Plants, Part II — Surge Voltages in the Cement Plant. *IEEE Transactions on Industry and General Applications*, vol IGA-4, Jan/Feb 1968, pp 62-67.

[32] DE WITT, J. S., Ed. *Ground Fault Protection Handbook*. Salem, MA: Electrical Protection, Inc

[33] LEWIS, W. W. *The Protection of Transmission Systems Against Lightning*. New York: Wiley, 1950.

[34] BEWLEY, L. V. *Traveling Waves on Transmission Systems*, 2nd ed. New York: Wiley, 1951.

[35] SHANKLE, D. F., EDWARDS, R. F., and MOSES, G. L. Surge Protection for Pipeline Motors. *IEEE Transactions on Industry and General Applications*, vol IGA-4, Mar/Apr 1968, pp 171-176.

[36] GREENWOOD, A. N. *Electrical Transients in Power Systems*. New York: Wiley, 1971.

[37] AIEE COMMITTEE REPORT. A Survey of Relay Test Methods. *AIEE Transactions (Power Apparatus and Systems)*, pt III, vol 75, Jun 1956, pp 254-260.

[38] BOURBONNAIS, T. L., II. The Coordination and Testing of Protective Relays in Industrial Plants. *AIEE Transactions (Power Apparatus and Systems)*, pt III, vol 78, Apr 1959, pp 1-10.

[39] RUDENBERG, R. *Electrical Shock Waves in Power Systems*. Cambridge, MA: Harvard University Press, 1968.

4.12.2 *Bibliography*

[40] ABETTI, P. A. Survey and Classification of Published Data on the Surge Performance of Transformers and Rotating Machines. *AIEE Transactions (Power Apparatus and Systems)*, pt III, vol 77, Feb 1959, pp 1403-1414.

[41] AIEE COMMITTEE REPORT. A Review of Backup Relaying Practices. *AIEE Transactions (Power Apparatus and Systems)*, pt III, vol 72, Apr 1953, pp 137-142.

[42] AIEE COMMITTEE REPORT. Bibliography of Relay Literature 1959—1960. *AIEE Transactions (Power Apparatus and Systems)*, pt III, vol 81, Jun 1962, pp 109-112.

[43] AIEE COMMITTEE REPORT. Voltage Gradients Through the Ground Under Fault Conditions. *AIEE Transactions (Power Apparatus and Systems)*, pt III, vol 77, Oct 1958, pp 669-692.

[44] *Applied Protective Relaying*. Newark, NJ: Westinghouse Electric Corporation, Relay Instrument Division, B-7235-D, 1974.

[45] BRIGHTMAN, F. P., McGEE, R. R., and REIFSCHNEIDER, P. J. Protecting AC Motors with Low-Voltage Air Circuit Breaker Series Trips. *AIEE Transactions (Applications and Industry)*, pt II, vol 76, Jul 1957, pp 114-119.

[46] CONCORDIA, C., and PETERSON, H. A. Arcing Faults in Power Systems. *AIEE Transactions*, vol 60, 1941, pp 340-346.

[47] *Electrical Transmission and Distribution Reference Book*. East Pittsburgh, PA: Westinghouse Electric Corporation, 1964.

[48] *Electric Utility Engineering Reference Book*; vol 3, *Distribution Systems*. Trafford, PA: Westinghouse Electric Corporation, 1965.

[49] Engineering Dependable Protection for an Electrical Distribution System, Part III — Component Protection for Electrical Systems. McGraw-Edison Company, Bussman Manufacturing Division, St Louis, MO, 1971.

[50] FAWCETT, D. V. How to Select Overcurrent Relay Characteristics. *IEEE Transactions on Applications and Industry*, vol 82, May 1963, pp 94-104.

[51] FINK, D. G., and CARROLL, J. M. *Standard Handbook for Electrical Engineers*. New York: McGraw-Hill, 1968.

[52] GILBERT, M. M., and BELL, R. N. Directional Relays Provide Differential-Type Protection on Large Industrial Plant Power System. *AIEE Transactions (Applications and Industry)*, pt II, vol 74, Sept 1955, pp 220-227.

[53] HOFFMANN, D. C. and REIFSCHNEIDER, P. J. The System Application of Low-Voltage Power Circuit Breakers with Solid-State Trip Devices. *Conference Record, 1967 IEEE Industrial and Commercial Power Systems and Electric Space Heating Joint Technical Conference*, IEEE 34C-55, pp 29-44.

[54] IEEE COMMITTEE REPORT. Protection Fundamentals for Low-Voltage Electrical Distribution Systems in Commercial Buildings. IEEE JH 2112-1, 1974.

[55] *Industrial Power Systems Data Book*. Schenectady, NY: General Electric Company, 1961.

[56] JOHNSON, I. B., SCHULTZ, A. J., SCHULTZ, N. R., and SHORES, R. B. Some Fundamentals on Capacitance Switching. *AIEE Transactions (Power Apparatus and Systems)*, pt III, vol 74, Aug 1955, pp 727-736.

[57] KELLY, A. R. Allowing for Decrement and Fault Voltage in Industrial Relaying. *IEEE Transactions on Industry and General Applications*, vol IGA-1, Mar/Apr 1965, pp 130-139.

[58] KELLY. A. R. Relay Response to Motor Residual Voltage During Automatic Transfers. *AIEE Transactions (Applications and Industry)*, pt II, vol 74, Sept 1955, pp 245-252.

[59] KIMBARK, E. W. *Power System Stability*, vol 2. New York: Wiley, 1950.

[60] LIAO, T. W., and LEE, T. H. Surge Suppressors for the Protection of Solid-State Devices. *IEEE Transactions on Industry and General Applications*, vol IGA-2, Jan/Feb 1966, pp 44-52.

[61] Lightning Protection of Metal-Clad Switchgear Connected to Overhead Lines. *General Electric Review*, Mar 1949.

[62] MONTSINGER, V. M. Breakdown Curve for Solid Insulation. *Electrical Engineering*, vol 54, Dec 1935, pp 1300-1301.

[63] *Protective Relays — Application Guide.* Stafford, England: QEC Measurements Ltd, 1975.

[64] SCHULTZ, A. J., VAN WORMER, F. C., and LEE, A. R. Surge Performance of Aerial Cable, Part I — Surge Testing of the Aerial Cable and Analysis of the Test Oscillograms. *AIEE Transactions (Power Apparatus and Systems)*, pt III, vol 76, Dec 1957, pp 923-930.

[65] SHOTT, H. S., and PETERSON, H. A. Criteria for Neutral Stability of Wye-Grounded-Primary Broken-Delta-Secondary Transformer Circuits. *AIEE Transactions*, vol 60, Nov 1941, pp 997-1002.

[66] SKEATES, W. F., TITUS, C. H., and WILSON, W. R. Severe Rates of Rise of Recovery Voltage Associated with Transmission Line Short Circuits. *AIEE Transactions (Power Apparatus and Systems)*, pt III, vol 76, Feb 1958, pp 1256-1266.

[67] STACEY, E. M., and SELCHAU-HANSEN, P. V. SCR Drives — AC Line Disturbance, Isolation and Short-Circuit Protection. *IEEE Transactions on Industry Applications*, vol IA-10, Jan/Feb 1974, pp 88-105.

[68] VAN WORMER, F. C., SCHULTZ, A. J., and LEE, A. R. Surge Performance of Aerial Cable, Part II — Mathematical Analysis of the Cable Circuits and Synthesis of the Test Oscillograms. *AIEE Transactions (Power Apparatus and Systems)*, pt III, vol 76, Dec 1957, pp 930-942.

[69] WARRINGTON, A. R. van C. *Protective Relays: Their Theory and Practice*, vol 1. New York: Wiley, 1968.

[70] WORLD, R. W., ROSE, C. L., and SKUDERNA, J. E. Relaying Tapped Substations for Faults on High-Voltage Transmission Lines. *AIEE Transactions (Power Apparatus and Systems)*, pt III, vol 77, Apr 1958, pp 73-78.

Note 1. Verifying the Validity of the Simplified Equivalent Circuits in Dealing with the
Three Stated Line Terminations [39, chap 6]

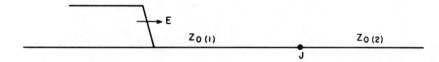

Fig N1.1
Surge Voltage Traveling on a Line of Surge Impedance $Z_{0(1)}$
Approaching a Junction where the Surge Impedance Changes to $Z_{0(2)}$

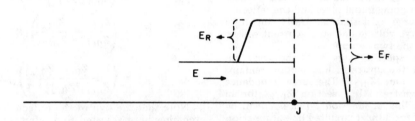

Fig N1.2
Junction Mismatch Accounted for by a Foreward-Going Refracted
Surge Voltage E_F and a Backward-Going Reflected Surge Voltage E_R

Table N1.1

Case	$Z_{0(2)}$	Initial Surge		Refracted Surge		Reflected Surge	
		Voltage	Current	Voltage	Current	Voltage	Current
A	∞	E	$E/Z_{0(1)}$	$2E$	0	E	$-E/Z_{0(1)}$
B	0	E	$E/Z_{0(1)}$	0	0	$-E$	$E/Z_{0(1)}$
C	Z_1	E	$E/Z_{0(1)}$	E	$E/Z_{0(1)}$	0	0

A surge voltage E is considered to be traveling to the right along the circuit having a surge impedance of $Z_{0(1)}$ (Fig N1.1). Associated with this voltage wave is a current wave of magnitude $E/Z_{0(1)}$ traveling in the same direction. When this surge encounters a change in surge impedance at junction J, the impedance mismatch will cause changes in the resulting traveling surges of voltage and current. The effect of the impedance mismatch is resolved in terms of a forward-going refracted surge voltage E_F which will be associated with a forward-going surge current $E_F/Z_{0(2)}$, and a reflected surge voltage E_R associated with a current surge of $E_R/Z_{0(1)}$. The current created by a positive backward-going surge voltage will be a nega-

tive surge current in the forward reference direction, $-E_R/Z_{0(1)}$.

The relationships between these components as the surge passes junction J are (Fig N1.2)

$$E_F = \left(\frac{2Z_{0(2)}}{Z_{0(1)} + Z_{0(2)}} \right)(E)$$

and associated current $I_F = E_F/Z_{0(2)}$, and

$$E_R = \left(\frac{Z_{0(2)} - Z_{0(1)}}{Z_{0(1)} + Z_{0(2)}} \right)(E)$$

and associated current $I_R = E_R/Z_{0(1)}$.

Table N1.2

Case	Bus Voltage	Line Current Upstream of J	Line Current Downstream from J
A	$2E$	0	0
B	0	$2E/Z_{0(1)}$	0
C	E	$E/Z_{0(1)}$	$E/Z_{0(1)}$

Note that when the reference direction of I_R is considered to be in the outgoing direction, the current expression becomes

$$I_R = -E_R/Z_{0(1)}$$

On the line section ahead of junction J where both the outgoing initial surge and the reflected returning surge exist simultaneously, the total value of surge voltage and surge current will be the sum of the two components.

For three specific cases A—C the surge impedance of the upstream line section remains at a fixed value $Z_{0(1)}$. The surge impedance of the downstream line section $Z_{0(2)}$ for an open-ended line at junction J (case A) will be $Z_{0(2)} = \infty$, for a short-circuited line at junction J (case B) it will be $Z_{0(2)} = 0$, and for a continuing line of the same surge impedance (case C) it will be $Z_{0(2)} = Z_{0(1)}$.

The component surge magnitudes for each of the cases are given in Table N1.1.

A first attempt at inserting $Z_{0(2)} = \infty$ leads to an indeterminate. This is avoided if the expressions for E_F and I_F are slightly revised by dividing both their numerators and denominators by $Z_{0(2)}$:

$$E_F = \left(\frac{2Z_{0(2)}}{Z_{0(1)} + Z_{0(2)}}\right)(E)$$

$$= \left(\frac{2}{Z_{0(1)}/Z_{0(2)} + 1}\right)(E)$$

$$E_R = \left(\frac{Z_{0(2)} - Z_{0(1)}}{Z_{0(1)} + Z_{0(2)}}\right)(E)$$

$$= \left(\frac{1 - Z_{0(1)}/Z_{0(2)}}{Z_{0(1)}/Z_{0(2)} + 1}\right)(E)$$

Note that the sign of the reflected surge current has been adjusted to a reference direction which "outwards" is positive.

Combining the component values of voltage and current listed in Table N1.1 yields the values of bus J voltage and the line currents upstream and downstream of the junction bus shown in Table N1.2.

The simplified procedures presented in the text yield these results in a more direct manner with a better understanding of the behavior pattern involved.

Note 2. The Degree of Voltage Amplification Produced by a Stub Line Extension Continuing Beyond the Arrester [33, chap 10]

The situation being discussed is portrayed in Fig N2.1. A transient voltage surge is traveling along a transmission channel toward a utilization terminal. A surge-voltage arrester is installed on the transmission channel at a distance d ahead of the terminal junction. As the wave front passes the point of arrester installation, no action by the arrester occurs until the voltage at the arrester reaches the arrester sparkover value. Fig N2.2 illustrates the voltage distribution pattern which will exist along the line if d extends to the right of the wave tip when the arrester sparks over. (To simplify the analysis of the action under study, consider that the stub line section is terminated in an open circuit.) It is quite evident that the entire wave beyond the arrester will be doubled at the open-end terminal.

If the stub end is somewhat shorter, as pictured in Fig N2.3, the front tip of the wave will have encountered the line end and been doubled before the arrester acts. That portion of the incident wave front which resides on the line between the arrester and the end terminal will also be doubled, unavoidably, upon reaching the open end. Hence the top terminal voltage will be twice the voltage level which has already passed the arrester location.

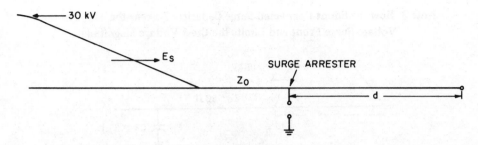

**Fig N2.1
30 kV Voltage Surge Traveling Toward a Line Termination Containing an Arrester Installation**

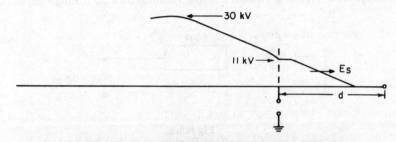

**Fig N2.2
Entire Surge Front Allowed by the Arrester Contained
on the Stub Section ahead of the Termination**

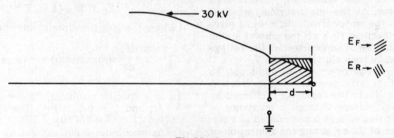

**Fig N2.3
Forward Section of the Traveling Surge Already
Reflected back from the Open End Before Arrester Conduction**

If the stub line section is still shorter, such that the tip of the backward reflected voltage wave has just arrived at the arrester location when the arrester sparks over, the voltage at the open terminal will still go to twice the arrester sparkover voltage value. The first half of the wave front has already been doubled as it reaches the open end. The second half of the rising front is a forward-traveling wave already on the line that cannot be modified by any subsequent arrester action. Hence it too will be doubled as it arrives at the open line terminal.

Thus, in general the portion of the traveling wave front which can be contained on a line section of 2d length will be subject to full, reflection amplification at the terminal. Line stubs of shorter lengths will proportionately reduce the amount of overshoot in voltage magnitude at the terminal.

With this basic understanding one can interpret the performance factors for other rates of rise of voltage wave fronts and for other values of propagation velocity. The adjustments needed to accommodate line terminals other than an "open circuit" are self-evident.

Note 3. How the Shunt-Connected Surge Capacitor Lessens the Slope of the Voltage Surge Front and Limits the Crest Voltage Magnitude

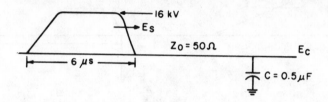

Fig N3.1
Surge Voltage Wave Traveling Toward a Motor Terminal on a 50 Ω Surge-Impedance Line

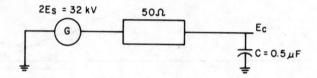

Fig N3.2
Accurate Lumped-Constant Equivalent Circuit for Analysis

Fig N3.1 illustrates a voltage surge traveling along a branch cable circuit $Z_0 = 50~\Omega$ to a 4160 V motor. At the line terminal, which is connected to the motor terminals, a set of surge protective capacitors (0.5 μF per phase) are installed. By the use of surge arresters the voltage crest has already been clipped to 16 kV.

The electrical equivalent circuit applicable to the connection arrangement of Fig N3.1 is shown in Fig N3.2. The capacitor is charged by a 32 kV surge voltage through a resistance of 50 Ω. The driving voltage is considered as a rectangular wave of 32 kV acting for a duration of 5 μs.

The motor major insulation security will be concerned primarily with the magnitude of the terminal voltage E_C, while the turn insulation security will be concerned primarily with the rate of rise of that voltage, dE_C/dt.

The fundamental current—voltage relationships associated with a capacitor, starting from a de-energized condition, are

$$E_C = \frac{Q}{C} = \frac{\int i\, dt}{C}$$

$$\frac{dE_C}{dt} = \frac{dQ}{dt} = \frac{I}{C}$$

When the capacitor is charged from a step-voltage source through a series resistor, as in Fig N3.2, the capacitor voltage builds up in ac-

cordance with

$$E_C = 2E_S\,(1 - e^{-t/t'})$$

where $t' = RC$, the circuit time constant, and

$$\frac{dE_C}{dt}\,(\text{maximum}) = \frac{I}{C}\,(\text{maximum})$$

In this specific problem, $2E_S = 32$ kV, $Z_0 = 50~\Omega$, and $C = 0.5(10)^{-6}$. The RC product is $50(0.5)(10)^{-6} = 25(10)^{-6}$ s.

The maximum input current to the capacitor I_C occurs at $t = 0$ when the surge voltage first arrives at the capacitor and $E_C = 0$. We then have

$$I_C = \frac{2E_S}{Z_0} = \frac{32\,000}{50} = 640~\text{A}$$

$$\frac{dE_C}{dt}\,(\text{maximum}) = \frac{I_C}{C} = \frac{640}{0.5(10)^{-6}}$$

$$= 1280(10)^6~\text{V/s} = 1280~\text{V/}\mu\text{s}$$

For a surge-voltage duration of 5 μs the quantity $e^{-t/t'}$ is equal to

$$e^{-0.2} = 0.8187$$

$$(1 - e^{-0.2}) = 0.1813$$

Thus at the end of the 5 μs interval the capacitor voltage is

$$E_C = 2E_S (0.1813) = 32(0.1813) = 5.81 \text{ kV}$$

In comparison with the 13 kV crest value of the motor high-potential test, the voltage level developed across the capacitor (5.81 kV) is well below that level, and also well below the sparkover value of the special 4.16 kV surge arrester. The rate of rise of voltage at the motor terminal (maximum value) meets the criterion of at least 10 μs to reach the crest level of the nameplate voltage.

In conclusion the following should be noted.

(1) Had the circuit construction involved spaced conductors or open-wire lines, the surge impedance would have been substantially greater, making the surge current values lower, which in turn would account for lower values of capacitor voltage and lower values of the rate of rise of the capacitor voltage.

(2) Had the surge voltage been alternating, each subsequent half-cycle of surge current flow would create cancellation effects in the capacitor voltage created by the previous half-cycle.

(3) A greater duration of unidirectional voltage surge would account for a greater voltage across the capacitor, limited by the level $2E_S$ or the arrester sparkover voltage, whichever is lower.

(4) The presence of series inductance in the capacitor circuit acts to deteriorate the wave-sloping action of the surge capacitor, and even inductances of as little as a few microhenrys can greatly impair performance [17, chap 2].

5. Fault Calculations

5.1 Introduction. Even the best designed electric systems occasionally experience short circuits resulting in abnormally high currents. Overcurrent protective devices such as circuit breakers and fuses must isolate the faults at a given location safely with a minimum of damage to circuits and equipment and with a minimum amount of shutdown of plant operation. Other parts of the system such as cables, bus duct, and disconnect switches must be capable of withstanding the mechanical and thermal stresses resulting from maximum flow of fault current through them. The magnitudes of fault currents are usually estimated by calculation. Equipment is selected using the calculation results.

The current flow during a fault at any point in a system is limited by the impedance of circuits and equipment from the source or sources to the point of fault and is not directly related to the load on the system. However, additions to the system which increase its capacity to handle a growing load, while not affecting the normal load at some existing parts of the system, may drastically increase the fault currents. Whether an existing system is expanded or a new system is installed, available fault currents should be determined for proper application of overcurrent protective devices.

The maximum fault currents should be calculated and, in most cases, also the minimum sustained values, which are needed to check the sensitivity requirements of the current-responsive protective devices.

This chapter has three purposes; first, to present some fundamental considerations of fault calculations; second, to illustrate some commonly used methods of making fault calculations with typical examples; and third, to furnish typical data which can be used in making fault calculations.

The size and complexity of many modern industrial systems may make long-hand fault calculations impractically time consuming. Computers are generally used for major fault studies. Whether or not computers are available, a knowledge of the nature of fault currents and calculating procedures is essential to conduct such studies.

5.2 Sources of Fault Current. Power frequency currents which flow during a fault come from rotating electric machinery. Power capacitors can also produce extremely high transient fault or switching currents, but usually of short duration and of natural frequency much higher than power frequency. These are discussed in Chapter 7. Rotating machinery in industrial-plant fault calculations may be analyzed in four categories:

(1) Synchronous generators
(2) Synchronous motors and condensers
(3) Induction machines
(4) Electric utility systems

The current from each rotating machinery source is limited by the impedance of the machine and the impedance between the machine and the fault. The impedance of a rotating machine is not a simple value, but is complex and variable with time.

5.2.1 *Synchronous Generators.* If a short circuit is applied to the terminals of a synchronous generator, the short-circuit current starts out at a high value and decays to a steady-state value after some time has elapsed from the inception of the short circuit. Since a synchronous generator continues to be driven by its prime mover and to have its field externally excited, the steady-state value of fault current will persist unless interrupted by some switching means. To represent this characteristic, one can use an equivalent circuit consisting of a constant driving voltage in series with an impedance which varies with time. This varying impedance consists primarily of reactance.

For purposes of fault-current calculations, industry standards have established three specific names for values of this variable reactance, called subtransient reactance, transient reactance, and synchronous reactance.

X_d'' = Subtransient reactance; determines current during first cycle after fault occurs. In about 0.1 s reactance increases to

X_d' = Transient reactance; assumed to determine current after several cycles at 60 Hz. In about 1/2 to 2 s reactance increases to

X_d = Synchronous reactance; this is the value that determines the current flow after a steady-state condition is reached

As most fault-protective devices, such as circuit breakers and fuses, operate well before steady-state conditions are reached, generator synchronous reactance is seldom used in calculating fault currents for application of these devices.

5.2.2 *Synchronous Motors and Condensers.* Synchronous motors supply current to a fault in much the same manner as synchronous generators. When a fault causes system voltage to drop, the synchronous motor receives less power from the system for rotating its load. At the same time the internal voltage causes current to flow to the system fault. The inertia of the motor and its load acts as a prime mover and with field excitation maintained, the motor acts as a generator to supply fault current. This fault current diminishes as the magnetic field in the machine decays.

The generator equivalent circuit is used for synchronous motors. Again, a constant driving voltage and the same three reactances X_d'', X_d', and X_d are used to establish values of current at three points in time.

Synchronous condensers are treated in the same manner as synchronous motors.

5.2.3 *Induction Machines.* A squirrel-cage induction motor will contribute fault current to a circuit fault. This is generated by inertia driving the motor in the presence of a field flux produced by induction from the stator rather than from a direct-current field winding. Since this flux decays on loss of source voltage caused by a fault at the motor terminals, the current contribution of an induction motor to a terminal fault reduces and disappears completely after a few cycles. As field excitation is not maintained, there is no steady-state value of fault current as for synchronous machines.

Again, the same equivalent circuit is used, but the values of transient and synchronous reactance approach infinity. As a consequence, induction motors are assigned only a subtransient value of reactance X_d''. This value is about equal to the locked-rotor reactance.

For fault calculations an induction generator can be treated the same as an induction motor. Wound-rotor induction motors normally operating with their rotor rings short-circuited will contribute fault current in the same manner as a squirrel-cage induction motor. Occasionally large wound-rotor motors operated with some external resistance maintained in their rotor circuits may have sufficiently low short-circuit time constants that their fault contribution is not significant and may be neglected. A specific investigation should be made to determine whether to neglect the contribution from a wound-rotor motor.

5.2.4 *Electric Utility Systems.* The remote generators of the electric utility system are a source of short-circuit current, often delivered through a supply transformer. The generator equivalent circuit can be used to represent the utility system. The utility generators are usually remote from the industrial plant. The current contributed to a fault in the remote plant appears to be merely a small increase in load current to the very large central station generators, and this current contribution tends to remain constant. The electric utility system is therefore usually represented at the plant by a single-valued equivalent impedance referred to the point of connection.

5.3 Fundamentals of Fault-Current Calculations.

Ohm's law, $I = E/Z$, is the basic relationship used in determining fault current, where I is the fault current, E is the driving voltage of the source, and Z is the impedance from source to fault including the impedance of the source.

Most industrial systems have multiple sources supplying current to a short circuit since each motor can contribute. One step in fault calculation is the simplification of the multiple-source system to the condition where the basic relationship applies.

5.3.1 *Purpose of Calculations.* System and equipment complexity and the lack of accurate parameters make precise calculations of short-circuit currents exceedingly difficult, but extreme precision is unnecessary. The calculations described provide estimates with reasonable accuracy of the maximum and minimum limits of short-circuit currents. These satisfy the usual reasons for making calculations.

The maximum calculated short-circuit current values are used for selecting devices of adequate interrupting rating, to check the ability of components of the system to withstand mechanical and thermal stresses, and to determine the time—current coordination of protective relays. The minimum values are used to establish the required sensitivity of protective relays. Minimum short-circuit values are sometimes estimated as fractions of the maximum values. If so, it is only necessary to calculate the maximum values of fault current.

5.3.2 *Type of Fault.* In an industrial system the three-phase fault condition is frequently the only one considered, since this type of fault generally results in maximum current.

Line-to-line fault currents are approximately 87 percent of three-phase fault currents. Line-to-ground fault currents can range in utility systems from a few percent to possibly 125 percent of the three-phase value, but in industrial systems line-to-ground fault currents of more than the three-phase value are rare.

Assuming a three-phase fault condition also simplifies calculations. The system including the fault remains symmetrical about the neutral point, whether or not the neutral point is grounded and regardless of wye or delta transformer connections. The balanced three-phase current can be calculated using a single-phase circuit which has only line-to-neutral voltage and impedance.

In calculating the maximum current it is assumed that the fault is a zero-impedance (bolted) fault with no current-limiting effect due to the fault itself. It should be recognized, however, that actual faults often involve arcing, which reduces the fault current magnitude. In low-voltage systems the minimum values of fault current are sometimes calculated from known effects of arcing. Analytical studies indicate that the arcing-fault currents, in per unit of bolted-fault values, may be typically as low as

(1) 0.89 at 480 V and 0.12 at 208 V for three-phase arcing

(2) 0.74 at 480 V and 0.02 at 208 V for line-to-line single-phase arcing

(3) 0.38 at 277 V and 0.01 at 120 V for line-to-neutral single-phase arcing

5.3.3 *Basic Equivalent Circuit.* The basic equation finds the current of a simple circuit having one voltage source and one impedance. In the basic equation the voltage E represents a single overall system driving voltage which replaces the array of individual unequal generated voltages acting within separate rotating machines. This voltage is equal to the prefault voltage at the point of fault connection. The impedance Z is a network reduction of the impedances representing all significant power system elements.

This equivalent circuit of the power system is a valid circuit transformation in accordance with Thevenin's theorem. It permits a determination of short-circuit current corresponding to the values of system impedances used.

Ordinarily the prefault voltage is taken as the system nominal voltage at the point of fault because this is close to the maximum operating voltage under fully loaded system conditions, and therefore the short-circuit currents will approach maximum.

The single-phase representation of a three-phase balanced system uses per-phase impedances and the line-to-neutral system driving voltage. Line-to-neutral voltage is line-to-line voltage divided by $\sqrt{3}$. Calculations may use impedances in ohms and voltages in volts, or both in per unit. Per-unit calculations simplify short-circuit studies for industrial systems which involve voltages of several levels. When using the per-unit system, if voltage bases are equal to system nominal voltages, the driving voltage is equal to 1.0 per unit.

The major elements of impedance must always be included in a short-circuit calculation. These are impedances of transformers, busways, cables, conductors, and rotating machines. There are other circuit impedances such as those associated with circuit breakers, current transformers, bus structures, and connections which are usually small enough to be neglected in short-circuit calculations, because the accuracy of the calculation is not generally affected. Omitting them provides slightly conservative (higher) short-circuit currents. However, on low-voltage systems, and particularly at 208 V, there are cases where their inclusion can significantly reduce the calculated short-circuit current.

Also, the normal practice is to disregard the presence of static loads (such as lighting and electric heating) in the network despite the fact that their associated impedance is actually connected in shunt with other network branches. This approach is considered valid since for the usual case the static-load impedances are large compared to the other parallel branches of the network while being approximately 90 deg out of phase with the impedance of fault-generating elements.

In alternating-current circuits the impedance Z is the vector sum of resistance R and reactance X. It is always acceptable to calculate short-circuit currents using vector impedances in the equivalent circuit. For many short-circuit current magnitude calculations at medium or high voltage, and for a few at low voltage, it is sufficiently accurate, conservative, and simpler to ignore resistances and use reactances only. However, for many lower voltage calculations resistance should not be ignored because the calculated currents would be unnecessarily overconservative.

Resistance data are definitely needed for calculations of X/R ratios when applying high-

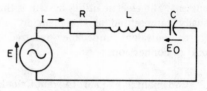

$$E = L \ \frac{dI}{dt} \ + \ RI \ + \ \frac{\int I\,dt}{C} \ + \ E_0$$

$$= L \frac{d^2 Q}{dt^2} \ + \ R \frac{dQ}{dt} \ + \ \frac{Q}{C} \ + \ E_0$$

Fig 68
Series RLC Circuit

and medium-voltage circuit breakers, but they need to be kept as separate entities from the reactance.

5.4 Restraints of Simplified Calculations. The short-circuit calculations described in this chapter are a simple E/Z evaluation of extensive electric power system networks. Before describing the step-by-step procedures in making these calculations, it is appropriate to review some of the restraints imposed by the simplification.

5.4.1 *Impedance Elements.* When an alternating-current electric power circuit contains resistance R, inductance L, and capacitance C, such as the series connection shown in Fig 68, the expression relating current to voltage includes the terms shown in Fig 68. A solution for the current magnitude requires the solution of a differential equation.

If two important restraints are applied to this series circuit, the following simple equation using vector impedances ($X_L = \omega L$ and $X_C = 1/\omega C$) is valid:

$$E = I \left[R + j \left(\omega L - \frac{1}{\omega C} \right) \right]$$

These restraints are that, first, the electric driving force be a sine wave and, second, the impedance coefficients R, L, and C be constants. Unfortunately in short-circuit calculations these restraints may be invalidated. A major reason for this is switching transients.

5.4.2 *Switching Transients.* The vector-impedance analytical tool recognizes only the steady-state sine-wave electrical quantities and does not include the effects of abrupt switching. Fortunately the effects of switching

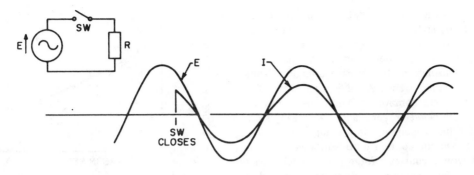

Fig 69
Switching Transient R

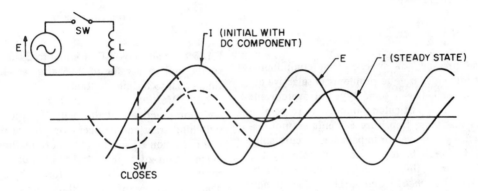

Fig 70
Switching Transient L

transients can be analyzed separately and added. (An independent solution can be obtained from a solution of the formal differential equations, of course.)

In the case of resistance R (Fig 69) the closure of switch SW causes the current to immediately assume the value which would exist in the steady state. No transient adder is needed.

In the case of inductance L (Fig 70) an understanding of the switching transient can best be acquired using the expression

$$E = L \frac{dI}{dt}$$

$$\frac{dI}{dt} = \frac{E}{L}$$

This expression tells us that the application of a driving voltage to an inductance will create a time rate of change in the current magnitude.

The slope of the current–time curve in the inductance will be equal to the quantity E/L.

At the right-hand side of the graph of Fig 70 the steady-state current curve is displayed. It lags the voltage wave by 90 deg and is rising at the maximum rate in the positive direction when the voltage is at the maximum positive value. It holds at a fixed value when the driving voltage is zero. This curve is projected back to the time of circuit switching (dashed curve). Note that at the instant that the switch is to be closed, the steady-state current would have been at a negative value of about 90 percent of crest value. Since the switch has been open prior to this instant, the true circuit current must be zero. After closing the switch, the current wave will display the same slope as the steady-state wave. This is the solid-line current curve beginning at the instant of switch closing.

Note that the difference between this curve and the steady state is a positive direct-current component of the same magnitude as the steady-state wave would have had at the instant of switch closing, in the negative direction. Thus the switching transient takes the form of a direct-current component whose value may be anything between zero and the steady-state crest value, depending on the angle of closing.

If the circuit contained no resistance, the current would continue forever in the displaced form. The presence of resistance causes the direct-current component to be dissipated exponentially. The complete expression for the current would take the form

$$I = \frac{E}{j\omega L} \sin \omega t + I_{dc} \ e^{-Rt/L}$$

The presence of direct-current components may introduce unique problems in selective coordination between some types of overcurrent devices. It is particularly important to bear in mind that these transitory currents are not disclosed by the vector-impedance circuit solution, but must be introduced artifically by the analyst, or by the guide rules which he follows.

5.4.3 *Decrement Factor.* The value at any time of a decaying quantity, expressed in per unit of its initial magnitude, is the decrement factor for that time. Refer to Fig 71 for decrement factors of an exponential decay. The significance of the decrement factor can be understood better if the exponential is expressed in terms of the time constant. If, as indicated in Fig 71, the exponent is expressed as $- t/t'$ with the time variable t in the numerator and the rest combined as a single constant t' (called the time constant) in the denominator, the transitory quantity begins its decay at a rate which would cause it to vanish in one time constant. The exponential character of the decay results in a remnant of 36.8 percent remaining after an elapsed time equal to 1 time constant. Any value of the transitory term selected at, say, time t will be reduced to 0.368 of that value after a subsequent elapsed time equal to 1 time constant. A transitory quantity of magnitude 1.0 at time zero would be reduced to a value of 0.368 after an elapsed time equal to 1 time constant, to a value of

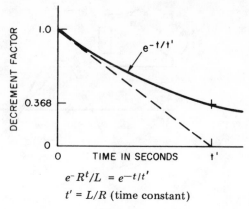

$$e^{-Rt/L} = e^{-t/t'}$$

$$t' = L/R \text{ (time constant)}$$

Fig 71
Decrement Factor

0.135 after an elapsed time equal to 2 time constants, and to a value of 0.05 after an elapsed time equal to 3 time constants.

5.4.4 *Multiple Switching Transients.* It is common practice that the analyst considers the switching transient to occur only once during one excursion of short-circuit current flow. An examination of representative oscillograms of short-circuit currents will often display repeated instances of momentary current interruptions. At times an entire half-cycle of current will be missing. In other cases, especially in low-voltage circuits, there may be present a whole series of chops and jumps in the current pattern. A switching interrupter, especially when switching a capacitor circuit, may be observed to restrike two, or perhaps three times before complete interruption is completed. The restrike generally occurs when the potential difference across the switching contacts is high. It is entirely possible that switching transients, both simple direct- and alternating-current transitory oscillations, may be reinserted in the circuit current a number of times during a single incident of short-circuit current flow. The analyst must remain mindful of possible trouble.

5.4.5 *Practical Impedance Network Synthesis.* One approach to an adequate procedure for computing the phase A current of a three-phase system is indicated in Fig 72. For each physical conducting circuit, the voltage drop is represented as the sum of the self-impedance drop in

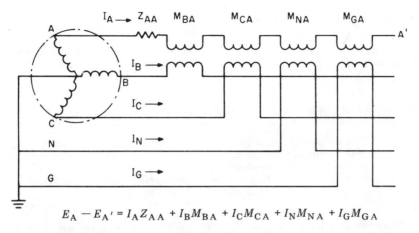

$$E_A - E_{A'} = I_A Z_{AA} + I_B M_{BA} + I_C M_{CA} + I_N M_{NA} + I_G M_{GA}$$

Fig 72
Three-Phase Four-Wire Circuit, Unbalanced Loading

the circuit and the complete array of mutually coupled voltage drops caused by current flow in other coupled circuits. The procedure is complex even in those instances where the current in both the neutral and ground conductors is zero.

The simplified analytical approach to this problem assumes balanced symmetrical loading of a symmetrical polyphase system. With a symmetrical system operating with a symmetrical loading, the effects of all mutual couplings are similarly balanced. What is happening in phase A in the way of self- and mutually coupled voltages is also taking place in phase B with exactly the same pattern, except displaced 120 deg, and it is also taking place in phase C with the same pattern, except displaced another 120 deg. The key to the simplification is the fact that the ratio of the total voltage drop in one phase circuit to the current in that phase circuit is the same in all three phases of the system. Thus it appears that each phase possesses a firm impedance value common with the other phases. This unique impedance quantity is identified as the single-phase line-to-neutral impedance value. Any one line-to-neutral single-phase segment of the system may be sliced out for the analysis, since all are operating with the same load pattern.

The impedance diagram of the simplified concept appears in Fig 73. The need to deal with

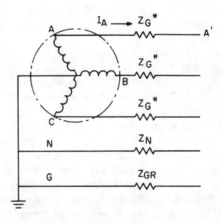

Impedance identity for each symmetrical pattern:

Positive sequence Z_{G1}
Negative sequence Z_{G2}
Zero sequence $Z_{G0} + 3Z_{GR}$*

*Based on zero current in conductor N.

$$E_A - E_{A'} = I_{A1} Z_{G1} + I_{A2} Z_{G2}$$
$$+ I_{A0}(Z_{G0} + 3Z_{GR})$$

Fig 73
Three-Phase Four-Wire Circuit,
Balanced Symmetrical Loading

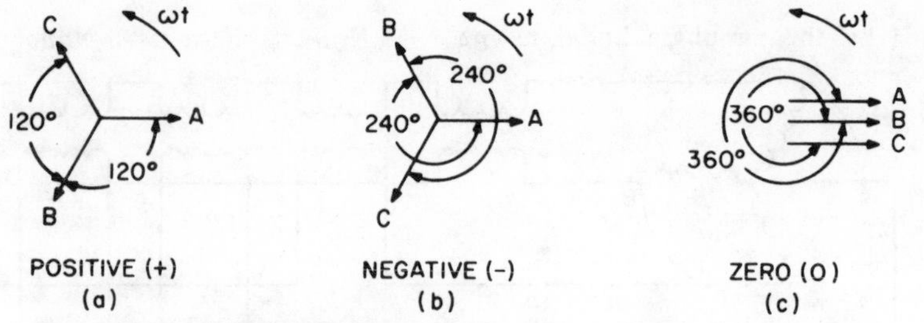

POSITIVE (+)   NEGATIVE (−)   ZERO (0)
(a)   (b)   (c)

The individual phase quantities A, B, and C follow each other
with an angular spacing of

Positive sequence	120°
Negative sequence	240°
Zero sequence	360°

Fig 74
Three-Phase Symmetrical Load Patterns Applicable to a Three-Phase System

mutual coupling has vanished. Since each phase circuit presents identically the same information, it is common to show only a single phase segment of the system in a one-line diagram. The expressions below the sketch in Fig 73 contain some unfamiliar terms. Their meaning will be discussed below.

One restraint associated with this simple analytical method is that all phases of the system share symmetrical loading. While it is true that a three-phase short circuit would satisfy this restraint, some short-circuit problems which must be solved are not balanced. For these unbalanced short-circuit problems we use the concept of symmetrical components. This concept discloses that any conceivable condition of unbalanced loading can be correctly synthesized by the use of appropriate magnitudes and phasing of several systems of symmetrical loading. In a three-phase system, with a normal phase separation of 120 deg, there are just three possible symmetrical loading patterns. These can be quickly identified with the aid of Fig 74. Loadings of the three phase windings A, B, and C must follow each other in sequence, separated by some multiple of 120 deg. In Fig 74(a) they follow each other with a 120 deg separation, in Fig 74(b) with a 240 deg separation, and in Fig 74(c) with a 360 deg separation. Note that separation angles of any other multiples of 120 deg will duplicate one of the three already shown. These loading patterns satisfy the restraints demanded by the analytical method to be used.

Note that Fig 74(a), identified as the positive sequence, represents the normal balanced operating mode. Thus there are only two sequence networks which differ from the normal. Fig 74(b), called the negative sequence, identifies a loading pattern very similar to the positive sequence, except that the electrical quantities come up with the opposite sequence. A current of this pattern flowing in a motor stator winding would create a normal-speed rotating field, but with backward rotation. The pattern of Fig 74(c), called the zero sequence, represents the case in which the equal currents in each phase are in phase. Each phase current reaches its maximum in the same direction at the same instant.

It is understandable that machine interwinding mutual coupling and other mutual coupling effects will be different in the different sequence systems. Hence it is likely that the per-phase impedance of the negative- and zero-sequence systems will differ from that of the positive sequence. Currents of zero sequence, being in phase, do not add up to zero at the end terminal as do both the positive- and negative-sequence currents. They add arithmetically and must be returned to the source via an additional circuit conductor. The zero-sequence voltage drop of this return conductor must be accounted for in the zero-sequence impedance value. With this understanding of the three symmetrical loading patterns, the significance of the notes below the sketch in Fig 73 becomes clear.

The simplifications in analytical procedures

accomplished by the per-phase line-to-neutral balanced system concepts carry with them some important restraints.

(1) The electric power system components must be of symmetrical design pattern.

(2) The electric loading imposed on the system must be balanced and symmetrical.

Wherever these restraints are violated, substantially hybrid network interconnections must be constructed which bridge the zones of unbalanced conditions. In the field of short-circuit current calculations the necessary hybrid interconnections of the sequence networks to accommodate the various unbalanced fault connections can be found in a variety of published references. The necessary hybrid interconnections to accommodate a lack of symmetry in the circuit geometry, such as is needed for an open-delta transformer bank, an open-line conductor, etc, are more difficult to find.

5.4.6 *Other Analytical Tools.* A large number of valid network theorems can be used effectively to simplify certain kinds of problems encountered in short-circuit analysis. These are described and illustrated in many standard texts on alternating-current circuit analysis [1, chap 8]. Of exceptional importance in the short-circuit problem area is Thevenin's theorem and the superposition theorem. Thevenin's theorem allows an extensive complex single-phase network to be reduced to a single driving voltage in series with a single impedance, referred to the particular bus under study. The superposition theorem allows the local effect of a remote voltage change in one source machine to be evaluated by impressing the magnitude of the voltage change, at its point of origin, on the complete impedance network; the current reading in an individual circuit branch is treated as an adder to the prior current magnitude in that branch. These analytical tools, like the others, have specific restraints which must be observed to obtain valid results.

5.4.7 *Respecting the Imposed Restraints.* Throughout this discussion, emphasis has been placed on the importance of respecting the restraints imposed by the analytical procedure in order to obtain valid results. Mention has been made of numerous instances in short-circuit analysis where appropriate corrections must be artificially introduced when analytical

restraints have been violated. One remaining area associated with short-circuit analysis involves variable impedance coefficients. When an arc becomes a series component of the circuit impedance, the R which it represents is not constant. At a current of 1A, it is likely to be, say, 100 Ω, yet at a current of 1000 A, it would very likely be about 0.1 Ω. During each half-cycle of current flow, the arc resistance would traverse this range. It is difficult to determine a proper value to insert in the 60 Hz network. Correctly setting this value of R does not compensate for the violation of the restraint that demands that R be a constant. The variation in R lessens the impedance to high-magnitude current, which results in a wave shape of current which is much more peaked that a sine wave. The current now contains harmonic terms. Since they result from a violation of analytical restraints, they will not appear in the calculated results. Their character and magnitude must be determined by other means and the result artificially introduced into the solution for fault current. A similar type of nonlinearity may be encountered in electromagnetic elements in which iron plays a part in setting the value of L. If the ferric parts are subject to large excursions of magnetic density, the value of L may be found to drop substantially when the flux density is driven into the saturation region. The effect of this restraint violation will, like the case of variable R, result in the appearance of harmonic components in the true circuit current.

5.4.8 *Conclusions.* The purpose of this review of fundamentals is to obtain a better understanding of the basic complexities involved in alternating-current system short-circuit calculations. In dealing with the day-to-day practical problems, the analyst should aim at the following.

(1) Select the optimum location and type of fault to satisfy the purpose of the calculation.

(2) Establish the simplest electric-circuit model of the problem which will both accomplish this purpose and yet minimize the complexity of the solution.

(3) Recognize the presence of system conditions which violate the restraints imposed by the analytical methods in use.

(4) Artificially inject corrections in computed

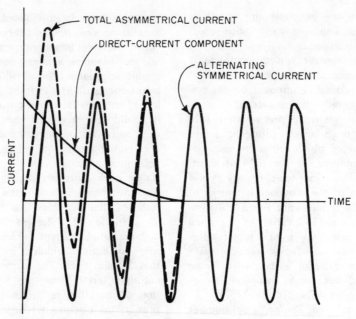

Fig 75
Typical System Fault Current

results to compensate if these conditions are large enough to be significant.

Some of the conclusions of the preceding section apply to the simplified procedures of this chapter. A balanced three-phase fault has been assumed, and a simple equivalent circuit has been described. The current E/Z calculated with the equivalent circuit is an alternating symmetrical rms current, because E is the rms voltage. Within specific constraints to be discussed subsequently, this symmetrical current may be directly compared with equipment ratings, capabilities, or performance characteristics that are expressed as symmetrical rms currents.

The preceding analysis of inductive circuit switching transients indicates that simplified procedures must recognize asymmetry as a system condition and account for it. The correction to compensate for asymmetry considers the asymmetrical short-circuit current wave to be composed of two components. One is the alternating-current symmetrical component E/Z. The other is a direct-current component initially of maximum possible magnitude, equal to the peak of the initial alternating-current

symmetrical component, assuming that the fault occurs at the point on the voltage wave where it creates this condition. At any instant after the fault occurs, the total current is equal to the sum of the alternating- and direct-current components (Fig 75).

Since resistance is present in an actual system, the direct-current component decays to zero as the stored energy it represents is expended in I^2R loss. The decay is assumed to be an exponential, and its time constant is assumed to be proportional to the ratio of reactance to resistance (X/R ratio) of the system from source to fault. As the direct-current component decays, the current gradually changes from asymmetrical to symmetrical.

Asymmetry is accounted for in simplified calculating procedures by applying multiplying factors to the alternating symmetrical current. The resulting estimate of the asymmetrical rms current is used for comparison with equipment ratings, capabilities, or performance characteristics that are expressed as total (asymmetrical) rms currents.

The alternating symmetrical current may also decay with time, as indicated in the discussion of sources of short-circuit current. Changing the impedance representing the machine properly accounts for alternating-current decay of the current to a short circuit at rotating machine terminals. The same impedance changes are assumed to be applicable when representing rotating machines in extensive power systems.

5.5 Detailed Procedure. A significant part of the preparation for a short-circuit current calculation is establishing the impedance of each circuit element, and converting impedances to be consistent with each other for combination in series and parallel. Sources of impedance values for circuit elements are nameplates, handbooks, manufacturers' catalogs, tables included in this chapter, and direct contact with the manufacturer.

Two established consistent forms for expressing impedances are ohms and per unit (per unit differs from percent only by a factor of 100). Individual equipment impedances are often given in percent, which makes comparisons easy, but percent impedances are rarely used without conversion in system calculations. In this chapter the per-unit form of impedance is used because it is more convenient than the ohmic form when the system contains several voltage levels. Impedances expressed as per unit on a defined base can be combined directly, regardless of how many voltage levels exist from source to fault. To obtain this convenience, the base voltage at each voltage level must be related according to the turns ratios of the interconnecting transformers.

In the per-unit system there are four base quantities, base apparent power in volt-amperes, base voltage, base current, and base impedance. The relationship between base, per-unit, and actual quantities is as follows:

per-unit quantity (voltage, current, etc)

$$= \frac{\text{actual quantity}}{\text{base quantity}}$$

Usually a convenient value is selected for base apparent power in volt-amperes, and a base voltage at one level is selected to match the transformer rated voltage at that level. Base voltages at other levels are established by transformer turns ratios. Base current and base impedance at each level are then obtained by standard relationships. The following formulas apply to three-phase systems, where the base voltage is the line-to-line voltage in volts or kilovolts and the base apparent power is the three-phase apparent power in kilovolt-amperes or megavolt-amperes:

base current (amperes)

$$= \frac{\text{base kVA} (1000)}{\sqrt{3} \ (\text{base volts})} = \frac{\text{base kVA}}{\sqrt{3} \ (\text{base kV})}$$

$$= \frac{\text{base MVA} (10^6)}{\sqrt{3} \ (\text{base volts})} = \frac{\text{base MVA} (1000)}{\sqrt{3} \ (\text{base kV})}$$

base impedance (ohms)

$$= \frac{\text{base volts}}{\sqrt{3} \ (\text{base amperes})} = \frac{(\text{base volts})^2}{\text{base kVA} (1000)}$$

$$= \frac{(\text{base kV})^2 (1000)}{\text{base kVA}} = \frac{(\text{base kV})^2}{\text{base MVA}}$$

Impedances of individual power system elements are usually obtained in forms that require conversion to the related bases for a per-unit calculation. Cable impedances are generally expressed in ohms. Converting to per unit using the indicated relationships leads to the following simplified formulas, where the per-unit impedance is Z_{pu}:

$$Z_{\text{pu}} = \frac{\text{actual impedance in ohms (base MVA)}}{(\text{base kV})^2}$$

$$= \frac{\text{actual impedance in ohms (base kVA)}}{(\text{base kV})^2 (1000)}$$

Transformer impedances are in percent of self-cooled transformer ratings in kilovolt-amperes and are converted using

$$Z_{\text{pu}} = \frac{\text{percent impedance (base kVA)}}{\text{kVA rating} (100)}$$

$$= \frac{\text{percent impedance} (10) (\text{base MVA})}{\text{kVA rating}}$$

Motor reactance may be obtained from tables providing per-unit reactances on element ratings in kilovolt-amperes and are converted using

$$X_{\text{pu}} = \frac{\text{per-unit reactance (base kVA)}}{\text{kVA rating}}$$

The procedure for calculating industrial system short-circuit currents consists of the fol-

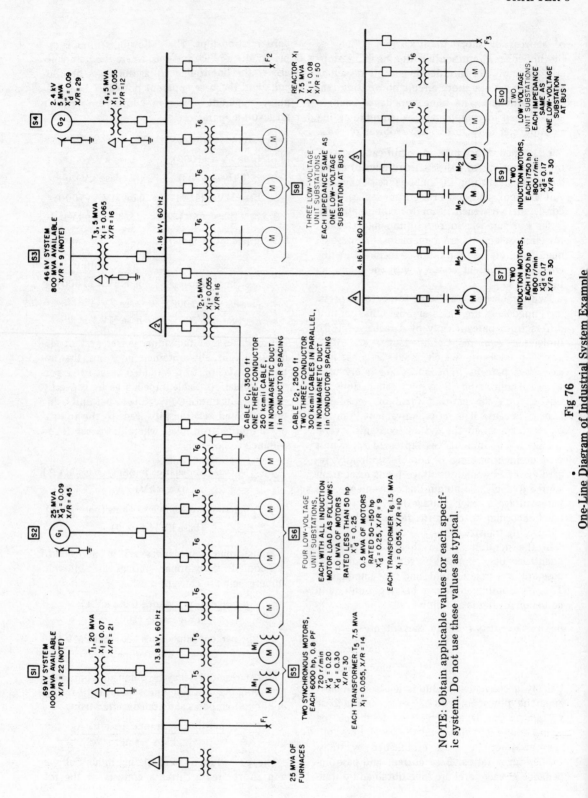

Fig 76
One-Line Diagram of Industrial System Example

NOTE: Obtain applicable values for each specific system. Do not use these values as typical.

Table 18
Typical Reactance Values for Induction and Synchronous
Machines, in Per-Unit of Machine kVA Ratings*

	X_d''	X_d'
Turbine generators†		
2 poles	0.09	0.15
4 poles	0.15	0.23
Salient-pole generators with damper windings†		
12 poles or less	0.16	0.33
14 poles or more	0.21	0.33
Synchronous motors		
6 poles	0.15	0.23
8—14 poles	0.20	0.30
16 poles or more	0.28	0.40
Synchronous condensers†	0.24	0.37
Synchronous converters†		
600 V direct current	0.20	—
250 V direct current	0.33	—
Individual induction motors, usually above 600 V	0.17	—
Groups of motors, each less than 50 hp, usually 600 V and below‡	0.25	—

NOTE: Approximate synchronous motor kVA bases can be found from motor horsepower ratings as follows:
    0.8 power factor motor — kVA base = hp rating
    1.0 power factor motor — kVA base = 0.8 × hp rating

*Use manufacturer's specified values if available.
†X_d' not normally used in short-circuit calculations.
‡The value of X_d'' for groups of motors has been increased slightly to compensate for the very rapid short-circuit current decrement in these small motors. A lower value of X_d'' will normally be appropriate for groups of large motors.

Table 19
Representative Conductor Spacings for
Overhead Lines

Nominal System Voltage (volts)	Equivalent Delta Spacing (inches)
120	12
240	12
480	18
600	18
2400	30
4160	30
6900	36
13 800	42
23 000	48
34 500	54
69 000	96
115 000	204

NOTE to Table 19:

When conductors are not arranged in a delta, the following formula may be used to determine the equivalent delta:

$$d = \sqrt[3]{A \times B \times C}$$

When the conductors are located in one plane and the outside conductors are equally spaced from the middle conductor, the equivalent is 1.26 times the distance between the middle conductor and an outside conductor. For example,

$$\text{equivalent delta spacing} = \sqrt[3]{A \times A \times 2A}$$

$$= 1.26\, A$$

Table 20
Constants of Copper Conductors for 1 ft Symmetrical Spacing*

Size of Conductor (cmil)	(AWG No.)	Resistance R at 50°C, 60 Hz (Ω/conductor/1000 ft)	Reactance X_A at 1 ft Spacing, 60 Hz (Ω/conductor/1000 ft)
1 000 000		0.0130	0.0758
900 000		0.0142	0.0769
800 000		0.0159	0.0782
750 000		0.0168	0.0790
700 000		0.0179	0.0800
600 000		0.0206	0.0818
500 000		0.0246	0.0839
450 000		0.0273	0.0854
400 000		0.0307	0.0867
350 000		0.0348	0.0883
300 000		0.0407	0.0902
250 000		0.0487	0.0922
211 600	4/0	0.0574	0.0953
167 800	3/0	0.0724	0.0981
133 100	2/0	0.0911	0.101
105 500	1/0	0.115	0.103
83 690	1	0.145	0.106
66 370	2	0.181	0.108
52 630	3	0.227	0.111
41 740	4	0.288	0.113
33 100	5	0.362	0.116
26 250	6	0.453	0.121
20 800	7	0.570	0.123
16 510	8	0.720	0.126

NOTE: For a three-phase circuit the total impedance, line to neutral, is

$$Z = R + j(X_A + X_B)$$

*Use spacing factors of Tables 22 and 23 for other spacings.

lowing steps:

(1) Prepare system diagrams
(2) Collect and convert impedance data
(3) Combine impedances
(4) Calculate short-circuit current

Each step will be discussed subsequently in further detail.

5.5.1 *Step 1 — Prepare System Diagrams.* The first step is to prepare a one-line diagram. This diagram should show all sources of short-circuit current and all significant circuit elements. Fig 76 is a one-line diagram of a hypothetical industrial system.

Impedance information may be entered on the one-line diagram after initial data collection and after conversion. Sometimes it is desirable to prepare a separate diagram showing only the impedances after conversion. If the original circuit is complex and several steps of simplification are required, each may be recorded on additional impedance diagrams as the calculation progresses.

The impedance diagram might show reactances only, or it might show both reactances and resistances if a vector calculation is to be made. For calculation of a system X/R ratio, as described later for high-voltage circuit breaker duties, a resistance diagram showing only the resistances of all circuit elements must be prepared.

5.5.2 *Step 2 — Collect and Convert Impedance Data.* Impedance data should be collected for important elements and converted to per unit on bases selected for the study (see

Table 21
Constants of Aluminum Cable, Steel Reinforced, for 1 ft Symmetrical Spacing*

Size of Conductor (cmil)	(AWG No.)	Resistance R at 50°C, 60 Hz (Ω/conductor/1000 ft)	Reactance X_A at 1 ft Spacing, 60 Hz (Ω/conductor/1000 ft)
1 590 000		0.0129	0.0679
1 431 000		0.0144	0.0692
1 272 000		0.0161	0.0704
1 192 500		0.0171	0.0712
1 113 000		0.0183	0.0719
954 000		0.0213	0.0738
795 000		0.0243	0.0744
715 500		0.0273	0.0756
636 000		0.0307	0.0768
556 500		0.0352	0.0786
477 000		0.0371	0.0802
397 500		0.0445	0.0824
336 400		0.0526	0.0843
266 800		0.0662	0.1145
	4/0	0.0835	0.1099
	3/0	0.1052	0.1175
	2/0	0.1330	0.1212
	1/0	0.1674	0.1242
	1	0.2120	0.1259
	2	0.2670	0.1215
	3	0.3370	0.1251
	4	0.4240	0.1240
	5	0.5340	0.1259
	6	0.6740	0.1273

NOTE: For a three-phase circuit the total impedance, line to neutral, is

$$Z = R + j(X_A + X_B)$$

*Use spacing factors of Tables 22 and 23 for other spacings.

Tables 18–24 and Figs 77–79).

Table 18 — Typical Reactance Values for Induction and Synchronous Machines

Table 19 — Representative Conductor Spacings for Overhead Lines

Table 20 — Constants of Copper Conductors, 1 ft Spacing

Table 21 — Constants of Aluminum Cable, Steel Reinforced, 1 ft Spacing

Table 22 — 60 Hz Reactance Spacing Factor X_B, Separation in Inches

Table 23 — 60 Hz Reactance Spacing Factor X_B, Separation in Quarter Inches

Table 24 — 60 Hz Reactance of Typical Three-Phase Cable Circuits

Fig 77 — X/R Ratio of Transformers

Fig 78 — X/R Range for Small Generators and Synchronous Motors

Fig 79 — X/R Range for Three-Phase Induction Motors

The following tables appear in other chapters:

Table 74 — Standard Impedance Values for Three-Phase Transformers (Chapter 8)

Table 87 — Voltage-Drop Values of Three-Phase Busways with Copper Busbars (Chapter 11)

Table 88 — Voltage-Drop Values of Three-Phase Busways with Aluminum Busbars (Chapter 11)

Table 22
60 Hz Reactance Spacing Factor X_B, in Ohms per Conductor per 1000 ft

(feet)	Separation (inches)											
	0	1	2	3	4	5	6	7	8	9	10	11
0	—	−0.0571	−0.0412	−0.0319	−0.0252	−0.0201	−0.0159	−0.0124	−0.0093	−0.0066	−0.0042	−0.0020
1	—	0.0018	0.0035	0.0051	0.0061	0.0080	0.0093	0.0106	0.0117	0.0129	0.0139	0.0149
2	0.0159	0.0169	0.0178	0.0186	0.0195	0.0203	0.0211	0.0218	0.0225	0.0232	0.0239	0.0246
3	0.0252	0.0259	0.0265	0.0271	0.0277	0.0282	0.0288	0.0293	0.0299	0.0304	0.0309	0.0314
4	0.0319	0.0323	0.0328	0.0333	0.0337	0.0341	0.0346	0.0350	0.0354	0.0358	0.0362	0.0366
5	0.0370	0.0374	0.0377	0.0381	0.0385	0.0388	0.0392	0.0395	0.0399	0.0402	0.0405	0.0409
6	0.0412	0.0415	0.0418	0.0421	0.0424	0.0427	0.0430	0.0433	0.0436	0.0439	0.0442	0.0445
7	0.0447	0.0450	0.0453	0.0455	0.0458	0.0460	0.0463	0.0466	0.0468	0.0471	0.0473	0.0476
8	0.0478											

Table 23
60 Hz Reactance Spacing Factor X_B, in Ohms per Conductor per 1000 feet

(inches)	Separation (quarter inches)			
	0	1/4	2/4	3/4
0	—	—	−0.0729	−0.0636
1	−0.0571	−0.0519	−0.0477	−0.0443
2	−0.0412	−0.0384	−0.0359	−0.0339
3	−0.0319	−0.0301	−0.0282	−0.0267
4	−0.0252	−0.0238	−0.0225	−0.0212
5	−0.0201	−0.01907	−0.01795	−0.01684
6	−0.0159	−0.01494	−0.01399	−0.01323
7	−0.0124	−0.01152	−0.01078	−0.01002
8	−0.0093	−0.00852	−0.00794	−0.00719
9	−0.0066	−0.00605	−0.00529	−0.00474
10	−0.0042	—	—	—
11	−0.0020	—	—	—
12	—	—	—	—

Table 24
60 Hz Reactance of Typical Three-Phase Cable Circuits, in Ohms per 1000 ft

		System Voltage			
Cable Size	600 V	2400V	4160V	6900 V	13 800 V
4 to 1					
3 single-conductor cables in magnetic conduit	0.0520	0.0620	0.0618	—	—
1 three-conductor cable in magnetic conduit	0.0381	0.0384	0.0384	0.0522	0.0526
1 three-conductor cable in nonmagnetic duct	0.0310	0.0335	0.0335	0.0453	0.0457
1/0 to 4/0					
3 single-conductor cables in magnetic conduit	0.0490	0.0550	0.0550	—	—
1 three-conductor cable in magnetic conduit	0.0360	0.0346	0.0346	0.0448	0.0452
1 three-conductor cable in nonmagnetic duct	0.0290	0.0300	0.0300	0.0386	0.0390
250—750 kcmil					
3 single-conductor cables in magnetic conduit	0.0450	0.0500	0.0500	—	—
1 three-conductor cable in magnetic conduit	0.0325	0.0310	0.0310	0.0378	0.0381
1 three-conductor cable in nonmagnetic duct	0.0270	0.0275	0.0275	0.0332	0.0337

NOTE: These values may also be used for magnetic and nonmagnetic armored cables.

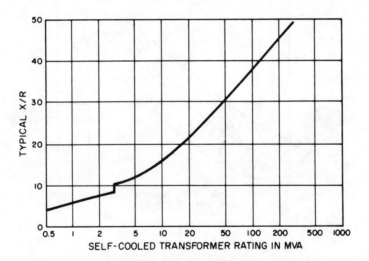

Fig 77
X/R Ratio of Transformers (Based on IEEE Std 320-1972)

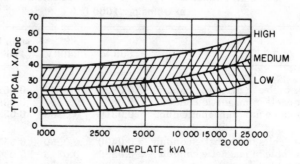

Fig 78
X/R Range for Small Generators and Synchronous Motors
(Solid Rotor and Salient Pole) (From IEEE Std 320-1972)

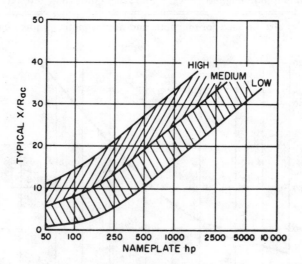

Fig 79
X/R Range for Three-Phase Induction Motors
(From IEEE Std 320-1972)

5.5.3 *Step 3 — Combine Impedances.* The third step is to combine reactances or vector impedances, and resistances where applicable, to the point of fault into a single equivalent reactance or resistance. The three branch elements which form a wye or delta configuration can be converted by the following formulas for further reduction (Fig 80).

(1) Wye to delta [Fig 80(a)]:

$$A = \frac{bc}{a} + b + c$$

$$B = \frac{ac}{b} + a + c$$

$$C = \frac{ab}{c} + a + b$$

(2) Delta to wye [Fig 80(b)]:

$$a = \frac{B \cdot C}{A + B + C}$$

$$b = \frac{A \cdot C}{A + B + C}$$

$$c = \frac{A \cdot B}{A + B + C}$$

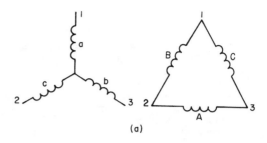

(a)

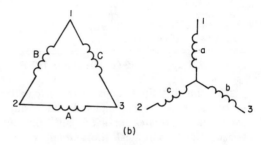

(b)

Fig 80
Wye and Delta Configurations
(a) Wye—Delta. (b) Delta-Wye

5.5.4 *Step 4 — Calculate Short-Circuit Current.* The fourth and final step is to calculate the short-circuit current. The impedance used in the circuit to calculate the short-circuit current depends upon the purpose of the study. This chapter will examine four basic networks of selected impedances used for the results most commonly desired:

(1) First-cycle duties for fuses and low-voltage circuit breakers

(2) First-cycle duties for high-voltage circuit breakers

(3) Contact-parting (interrupting) duties of high-voltage circuit breakers

(4) Short-circuit currents for time-delayed relaying devices

The four networks have the same basic elements except for the impedances of rotating machines. The differing impedances are based on standard application guides where equipment applications are the purpose of the calculation.

(1) *Network 1 — First-Cycle Duties for Fuses and Low-Voltage Circuit Breakers.* Subtransient reactance should be used to represent all rotat-

ing machines in the equivalent network. This will be in accordance with the following standards:

ANSI C37.41-1969 (R 1974), Design Tests for Distribution Cutouts and Fuse Links, Secondary Fuses, Distribution Enclosed Single-Pole Air Switches, Power Fuses, Fuse Disconnecting Switches, and Accessories

IEEE Std 20-1973, Low-Voltage AC Power Circuit Breakers Used in Enclosures (ANSI C37.13-1973)

NEMA AB 1-1975, Molded-Case Circuit Breakers

NEMA SG 3-1975, Low-Voltage Power Circuit Breakers

The standards allow modified reactances for groups of low-voltage induction and synchronous motors fed from a low-voltage substation. If the total motor horsepower rating is equal to the transformer self-cooled rating in kilovoltamperes, a per-unit reactance of 0.25 on the transformer self-cooled rating may be used as a single impedance to represent the group of motors.

Table 25
Rotating-Machine Reactance Multipliers

Type of Rotating Machine	Momentary	Interrupting
All turbine generators; all hydrogenerators with amortisseur windings, all condensers	$1.0\ X_d''$	$1.0\ X_d''$
Hydrogenerators without amortisseur windings	$0.75\ X_d'$	$0.75\ X_d'$
All synchronous motors	$1.0\ X_d''$	$1.5\ X_d''$
Induction motors		
Above 1000 hp at 1800 r/min	$1.0\ X_d''$	$1.5\ X_d''$
Above 250 hp at 3600 r/min	$1.0\ X_d''$	$1.5\ X_d''$
All others, 50 hp and above	$1.2\ X_d''$	$3.0\ X_d''$
All smaller than 50 hp	Neglect	Neglect

From ANSI C37.5-1969 (R 1974) and IEEE Std 320-1972.

After the first three steps of the calculating procedure have been completed and a single equivalent impedance for each fault point obtained, the symmetrical short-circuit current duty is calculated by dividing E_{pu} by Z_{pu} and multiplying by base current:

$$I_{sc} = \frac{E_{pu}}{Z_{pu}} \times I_{base}$$

where I_{sc} is a three-phase zero-impedance (bolted) fault value.

The short-circuit current is now directly applicable for low-voltage equipment whose short-circuit ratings or capabilities are expressed in symmetrical rms currents.

When the equipment ratings or capabilities are expressed as total rms currents, the symmetrical short-circuit current determined as above is multiplied by a multiplying factor found in the applicable standards to arrive at the asymmetrical first-cycle total rms current.

High-voltage short-circuit duties calculated with network 1 are used when high-voltage fuses are applied and when high-voltage system available short circuits are found to be used as factors in subsequent low-voltage calculations.

(2) *Network 2 — First-Cycle Duties for High-Voltage (Above 1kV) Circuit Breakers.* These duties are for comparison with circuit breaker momentary ratings (pre-1964 rating basis) or closing and latching capability (post-1964 rating basis) according to the following standards regardless of circuit breaker age:

ANSI C37.5-1969 (R 1974), Methods for Determining Values of a Sinusoidal Current Wave, a Normal-Frequency Recovery Voltage, and a Guide for Calaculation of Fault Currents for Application of AC High-Voltage Circuit Breakers Rated on a Total Current Basis

IEEE Std 320-1972, Application Guide for AC High-Voltage Circuit Breakers Rated on a Symmetrical Current Basis (ANSI C37.010-1972)

Multiplying factors to apply to reactances of rotating machines are shown in Table 25.

Reduce the equivalent network system to a single equivalent reactance X or Z. Determine the prefault operating voltage E. Divide E by X or Z and multiply by 1.6 to find the first-cycle duty short-circuit total per-unit current, then multiply by base current:

$$I_{sc\ mom} = \frac{E_{pu}}{X_{pu}}\ 1.6\ I_{base}$$

(3) *Network 3 — Contact-Parting (Interrupting) Duties of High-Voltage (Above 1kV) Circuit Breakers.* Consider first the duties for comparison with interrupting ratings of circuit breakers rated on the old total rms current basis (pre-1964 rating basis). ANSI C37.5-1969 (R 1974) procedures apply to calculating duties for these circuit breakers, regardless of age.

Multiplying factors to apply to rotating machine reactances are shown in Table 25. For these calculations the resistance (R) network is also necessary. In the resistance network each

204

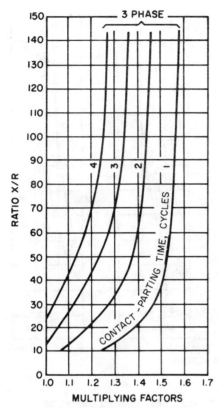

Fig 81
Multiplying Factors (Total Current Rating
Basis) for Three-Phase Faults Fed
Predominantly from Generators Through
No More Than One Transformation (Local)
[From ANSI C37.5-1969 (R 1974)]

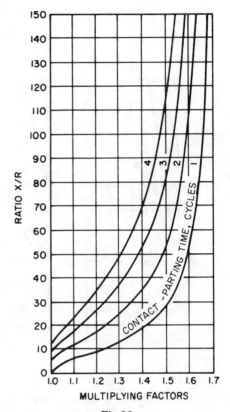

Fig 82
Multiplying Factors (Total Current Rating
Basis) for Three-Phase and Line-to-Ground
Faults Fed Predominantly Through Two or
More Transformations (Remote)
[From ANSI C37.5-1969 (R 1974)]

rotating machine resistance value must be multiplied by the same factor as its corresponding reactance multiplier from Table 25.

Reduce the reactance network to a single equivalent reactance X_{pu} and reduce the resistance network to a single equivalent resistance R_{pu}. Determine the X/R ratio by dividing X_{pu} by R_{pu}; determine E_{pu}, the prefault operating voltage; and determine E/X by dividing E_{pu} by X_{pu} found above.

Select the multiplying factor for E/X correction from curves of ANSI C37.5-1969 (R 1974) (Figs 81 and 82). To use the curves the circuit breaker contact parting time must be known as well as the proximity of generators to the fault

point (local or remote). Local generation applies only when in-plant generators that are predominant contributors to short-circuit currents are not more than one transformation from the fault point.

Minimum contact parting times are usually used and are defined as follows:

8 cycle circuit breaker has 4 cycle contact parting

5 cycle circuit breaker has 3 cycle contact parting

3 cycle circuit breaker has 2 cycle contact parting

2 cycle circuit breaker has 1½ cycle contact parting

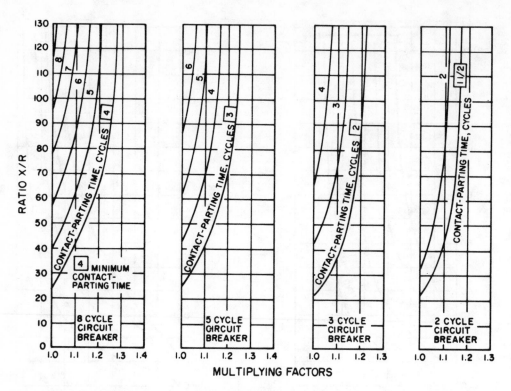

Fig 83
Multiplying Factors for Three-Phase Faults Fed Predominantly from Generators
Through No More Than One Transformation (From IEEE Std 320-1972)

Multiply E_{pu}/X_{pu} by the multiplying factor and the base current:

$$\frac{E_{pu}}{X_{pu}} \times \text{multiplying factor} \times I_{base}$$

This is the calculated total rms current interrupting duty to be compared to the circuit breaker interrupting capability. For older circuit breakers which have three-phase interrupting ratings in megavolt-amperes, the short-circuit current capability is found by dividing the rating in megavolt-amperes by $\sqrt{3}$ and by the operating voltage when the voltage is between the rated maximum and minimum limits. The minimum-limit voltage calculation applies for lower voltages.

Next consider the duties for comparison with the short-circuit (interrupting) capabilities of circuit breakers rated on the new symmetrical rms current basis (post-1964 basis). IEEE Std

320-1972 procedures apply to calculating duties for these circuit breakers.

E/X and the X/R ratio for a given fault point are as already calculated.

Select the multiplying factor for E/X correction from curves of IEEE Std 320-1972 (Figs 83 and 84). To use the curves the circuit breaker contact parting time must be known, as well as the proximity of generators to the fault point (local or remote), as before.

Multiply E_{pu}/X_{pu} by the multiplying factor and the base current:

$$\frac{E_{pu}}{X_{pu}} \times \text{multipling factor} \times I_{base}$$

The result is an interrupting duty to be compared with the rated symmetrical interrupting capability of a circuit breaker, but it is a symmetrical interrupting duty only if the multiplying factor for E/X is 1.0. The rated symmetrical interrupting capability of the circuit

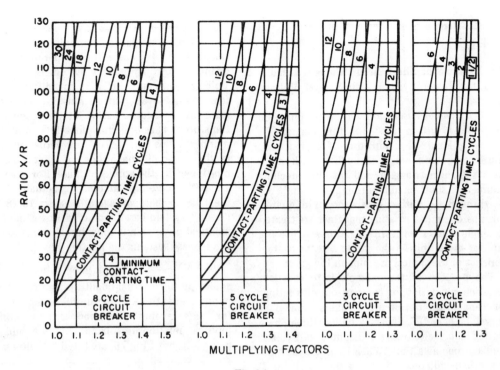

Fig 84
Multiplying Factors for Three-Phase and Line-toGround Faults Fed Predominantly
Through Two or More Transformations (From IEEE Std 320-1972)

breaker is calculated as follows:

symmetrical interrupting capability

$$= \frac{(\text{rated } I_{\text{sc}}) \, (\text{rated maximum } E)}{\text{operating } E}$$

This expression must be restricted to the maximum symmetrical interrupting capability listed for the circuit breaker.

The calculating procedures described for networks 2 and 3 are different in several respects from procedures detailed in earlier editions of this publication, which were based on standards now superseded. The differences are intended to account more accurately for contributions to high-voltage interrupting duty from large induction motors, for the exponential decay of the direct-current component of short-circuit current, and for the alternating-current decay of contributions from nearby generators.

(4) *Network 4 — Short-Circuit Currents for Time-Delayed Relaying Devices.* For the application of instantaneous relays, the value of the first-cycle short-circuit current determined by network 1 or 2 should be used. For an application of time-delay relays beyond 6 cycles, the equivalent system network representation will include only generators and the static equipment between them and the fault point. The generators must be represented by their transient impedance. All motor contributions are omitted. Only the generators that contribute fault current through the relay under consideration to the fault point must be considered for the relay application. The direct-current component will have decayed to near zero and is not considered. The short-circuit symmetrical rms current is $E_{\text{pu}}/X_{\text{pu}}$, where X_{pu} is derived from the equivalent reactance network consisting of generators and passive equipment (cables,

transformers, etc) in the relay protective fault path.

5.6 Example of Short-Circuit Current Calculation for Power System with Several Voltage Levels

5.6.1 *General Discussion.* The three-phase 60 Hz power system used for this example is shown in Fig 76. For purposes of the example, buses are numbered 1 through 4 with numbers shown in triangles, and rotating machine sources of short-circuit currents are numbered S_1 through S_{10} with numbers shown in squares. Groups of similar rotating machines are treated as single sources, each with a rating equal to the sum of the ratings in the group and the characteristics of the typical machine in the group.

The purpose of the example is to calculate short-circuit duties for comparison with ratings or capabilities of circuit breakers applied at buses 1, 2, and 3. Separate faults (three-phase bolted short circuits) are assumed at F_1, F_2, and F_3, one at a time. When F_1 is the fault being calculated, bus 1 is called the "fault bus."

All fault buses are at primary distribution voltage. Duty calculations for their circuit breakers are based on applicable standards, IEEE Std 320-1972 and ANSI C37.5-1969 (R 1974), which describe applications of high-voltage circuit breakers over 1000 V. In addition, first-cycle duties are calculated as described for low-voltage circuit breaker applications in IEEE Std 20-1973. The first-cycle duty described is basically the same as that considered in current standards for fuses.

5.6.2 *Utility System Data.* In-plant generators operating in parallel with utility system ties are the main sources both at bus 1 and at bus 2. The representation of remote utility generators for plant short-circuit calculations is often based on the utility "available," which should be the highest applicable (present or future) magnitude (also specifying the X/R ratio) of short-circuit contribution delivered by the utility from all sources outside the plant (not including contributions from in-plant sources). These data are converted to an equivalent impedance. Obtaining original equivalent impedance data from the utility is equally useful.

5.6.3 *Per-Unit Calculations and Base Quantities.* This example uses per-unit quantities for calculations. The base for all per-unit power quantities throughout the system is 10 MVA. Voltage bases are different for different system voltage levels, but all must be and are related by the turns ratios of interconnecting transformers, as specified in kilovolts at each numbered bus in Fig 76. Any actual quantity is the per-unit magnitude of that quantity multiplied by the applicable base. For example, 1.1 per-unit voltage at bus 1 is actually 1.1 × the 13.8 kV base voltage at bus 1 = 15.18 kV. Per-unit system bases and actual quantities have identical physical relationships. For example, in three-phase systems the relationship that

$$\text{total MVA} = \sqrt{3}\,(E_{L-L} \text{ in kV})\,(I_{\text{line}} \text{ in kA})$$

applies both to actual quantities and to bases of per-unit quantities. Other useful base quantities for this example, derived using common physical relationships from 10 MVA and the base voltages of Fig 76, are listed as follows:

	Base Line-to-Line Voltage E_{L-L}	
	13.8 kV	4.16 kV
Base line current, kA	0.4184	1.388
Base line-to-neutral impedance, ohms	19.04	1.73

This example calculates balanced per-unit three-phase short-circuit current duties using one of three identical per-unit line-to-neutral positive-sequence circuits, energized by per-unit line-to-neutral voltage. Only line-to-line base voltages are listed, however, because line-to-line voltages in per unit of these bases, and line-to-neutral voltages in per unit of their corresponding line-to-neutral base voltages, are identical for balanced three-phase circuits.

5.6.4 *Impedances Represented by Reactances.* The usual calculation of short-circuit duties at voltages over 1000 V involves circuits in which resistance is small with respect to reactance, so computations are simplified by omitting the resistance from the circuit. The slight error introduced makes the solution conservative. This example employs this simplification by using only the reactances of elements when finding the magnitudes of short-circuit duties. Element

Table 26
Passive-Element Reactances, in Per Unit, 10 MVA Base

Transformer T_1, $X = 0.07 (10/20) = 0.035$ per unit
Transformer T_2, $X = 0.055 (10/5) = 0.110$ per unit
Transformer T_3, $X = 0.065 (10/5) = 0.130$ per unit
Transformer T_4, $X = 0.055 (10/5) = 0.110$ per unit
Transformer T_5, $X = 0.055 (10/7.5) = 0.0734$ per unit
Transformer T_6, $X = 0.055 (10/1.5) = 0.367$ per unit
Reactor X_1, $X = 0.08 (10/7.5) = 0.107$ per unit

Cable C_1, from Tables 20 and 22 for 250 kcmil at 1 in spacing,

$$X = 0.0922 - 0.0571 = 0.0351 \ \Omega/1000 \text{ ft}$$

(There are no reactance corrections as this is three-conductor cable in nonmagnetic duct.)

For 3500 ft of cable, the conversion to per unit on a 10 MVA 13.8 kV base is

$$X = (3500/1000) (0.0351/19.04) = 0.0064 \text{ per unit}$$

Cable C_2, 300 kcmil at 1 in spacing,
$$X = 0.0902 - 0.057 = 0.0331 \ \Omega/1000 \text{ ft}$$
For 2500 ft of two cables in parallel at 4.16 kV,
$$X = (2500/1000) (1/2) (0.0331/1.73) = 0.0239 \text{ per unit}$$

resistances are necessary data, however, and play a special role to be defined subsequently.

5.6.5 *Equivalent Circuit Variations Based on Time and Standards.* High-voltage circuit breaker short-circuit current duty calculations may make use of several equivalent circuits for the power system, depending on the time at which duties are calculated and on the procedure described in the standard used as a basis.

The circuit used for calculating first-cycle (momentary) short-circuit current duties uses subtransient reactances for all rotating machine sources of short-circuit current, if based on current fuse and low-voltage circuit breaker standards. If based on IEEE Std 320-1972 and ANSI C37.5-1969 (R 1974), the preceding circuit is changed by representing medium-horsepower induction motors of 50 hp and larger (with some high-rating exceptions) with subtransient reactances multiplied by 1.2, and by omitting induction motors less than 50 hp.

The circuit used for calculating short-circuit (interrupting) duties, at circuit breaker minimum contact-parting times of 1½ to 4 cycles after the short circuit starts, based on IEEE Std 320-1972 and ANSI C37.5-1969 (R 1974), represents synchronous motors with subtransient reactances multiplied by 1.5, large induction motors (over 250 hp at 3600 r/min

or 1000 hp at 1800 r/min) at subtransient reactances multiplied by 1.5, and medium induction motors (all others, 50 hp and larger) at subtransient reactances multiplied by 3.

All equivalent circuits use the same reactances for passive elements.

Resistances are necessary to find fault-point X/R ratios that are used in short-circuit (interrupting) duty calculations based on IEEE Std 320-1972 and ANSI C37.5-1969 (R 1974). The fault-point X/R is the fault-point X divided by the fault-point R. A fault-point X is found by reducing the reactance circuit described in preceding paragraphs to a single equivalent X at the fault point. A fault-point R is found by reducing a related resistance-only circuit. This is derived from the reactance circuit by substituting for the reactance of each element a resistance equal to the reactance divided by the element X/R ratio. For motors whose subtransient reactance is increased by a multiplying factor, the same factor must be applied to the resistance in order to preserve the X/R ratio for the motor.

The X/R data for power system elements of this example, shown in Fig 76, are "medium-typical" data obtained in most cases from tables and graphs that are included in the applicable standards.

Table 27
Subtransient Reactances of Rotating Machines,
in Per Unit, 10 MVA Base

69 kV system,	X =	1.0 (10/1000) = 0.01 per unit
Generator 1,	X_d'' =	0.09 (10/25) = 0.036 per unit
46 kV system,	X =	1.0 (10/800) = 0.0125 per unit
Generator 2,	X_d'' =	0.09 (10/5) = 0.18 per unit

Large synchronous motor M_1, using the assumption that the horsepower rating of an 0.8 power factor machine is its kVA rating,
$X_d'' = 0.20 (10/6) = 0.333$ per unit, each motor
Large induction motor M_2, using the assumption that hp = kVA,
$X_d'' = 0.17 (10/1.75) = 0.971$ per unit
Low-voltage motor group, 1.0 MVA, less than 50 hp,
$X_d'' = 0.25 (10/1) = 2.5$ per unit
Low-voltage motor group, 0.5 MVA, from 50 to 150 hp,
$X_d'' = 0.25 (10/0.5) = 5.0$ per unit

The "approximately 30 cycle" network often is a minimum-source representation intended to investigate whether minimum short-circuit currents are sufficient to operate current-actuated relays. Minimum-source circuits might apply during night time or when production lines are down for any reason. Some of the source circuit breakers may be open and all motor circuits may be off. In-plant generators are represented with transient reactance, assumed at 1.5 times subtransient reactance in the absence of better information.

5.6.6 *Impedance Data and Conversions to Per Unit.* Reactances of passive elements, obtained from Fig 76, are listed in Table 26, along with the conversion of each reactance to per unit on the 10 MVA base.

Most of the data given in Fig 76 are per unit, based on the equipment nameplate rating. Any original percent impedance data were divided by 100 to obtain a per-unit impedance for Fig 76. Conversions are changes of MVA base, multiplied by the ratio of the new MVA base (10 MVA for the example) to the old MVA base (rated MVA). When the equipment rated voltage is not the same as the base voltage, it is also necessary to make voltage base conversions using the square of the ratio of rated voltage to example base voltage as the multiplier. This is not illustrated in this example.

Physical descriptions of cables are used to establish their reactances in ohms based on data in Tables 20 and 22. Dividing an impedance in ohms by the base impedance in ohms converts it to per unit.

5.6.7 *Subtransient Reactances of Rotating Machines, and Reactances for the Circuit to Calculate First-Cycle (Momentary) Short-Circuit Current Duties.* Subtransient reactances of rotating machine sources of short-circuit current are listed in Table 27 togehter with conversions to per unit on the study base.

Reactances representing the rotating machines of utility systems are found by assuming that the "available" megavolt-amperes are 1.0 per unit of a base equal to itself, and 1.0 per-unit megavolt-amperes corresponds to 1.0 per-unit reactance at 1.0 per-unit voltage.

The first-cycle calculation based on the current fuse and low-voltage circuit breaker standards uses a power system circuit with subtransient reactances for all rotating machine sources of short-circuit current. The circuit reactances are available from Tables 26 and 27. The circuit development and impedance simplifications are described subsequently.

5.6.8 *Reactances for the Circuit to Calculate First-Cycle (Momentary) Short-Circuit Current Duties.* The first-cycle calculation based on IEEE Std 320-1972 and ANSI C37.5-1969 (R 1974) uses a power system circuit with subtransient reactances modified as described in Section 5.6.5. The modifications are detailed in Table 28.

Table 28
Reactances for AC High-Voltage Circuit Breaker
First-Cycle (Momentary) Duties

Reactances of source circuits for utility systems, generators,
and large motors (M_1 and M_2) are unchanged

Induction motors rated less than 50 hp are omitted

Low-voltage motor group, 0.5 MVA, from 50 to 150 hp,
$X = 1.2 \, X_d'' = 1.2 \times 5.0 = 6.0$ per unit

Based on ANSI C37.5-1969 (R 1974) and IEEE Std 320-1972.

Table 29
Reactances, X/R Ratios, and Resistances for
AC High-Voltage Circuit Breaker Contact-Parting Time
(Interrupting) Short-Circuit Duties

Transformer T_1, $X/R = 21$, $R = 0.035/21 = 0.001667$ per unit
Transformer T_2, $X/R = 16$, $R = 0.110/16 = 0.00688$ per unit
Transformer T_3, $X/R = 16$, $R = 0.130/16 = 0.00812$ per unit
Transformer T_4, $X/R = 12$, $R = 0.11/12 = 0.00916$ per unit
Transformer T_5, $X/R = 14$, $R = 0.0734/14 = 0.00524$ per unit
Transformer T_6, $X/R = 10$, $R = 0.367/10 = 0.0367$ per unit
Reactor X_1, $X/R = 50$, $R = 0.107/50 = 0.00214$ per unit

Cable C_1, ac resistance at $50°C$ from Table 20 is $0.0487 \, \Omega/1000$ ft,
correction for $75°C = 1.087$

For 3500 ft of cable converted to per unit on a 10 MVA 13.8 kV
base,
$R = (3500/1000)(1.087)(0.0487/19.04) = 0.00972$ per unit

Cable C_2, ac resistance from Table 20 is $0.0407 \, \Omega/1000$ ft

For 2500 ft of two cables in parallel on a 10 MVA 4.16 kV base at
$75°C$,
$R = (2500/1000)(1.087/2)(0.0407/1.73) = 0.0320$ per unit

69 kV system, $X/R = 22$, $R = 0.01/22 = 0.000455$ per unit
Generator 1, $X/R = 45$, $R = 0.036/45 = 0.0008$ per unit

46 kV system, $X/R = 9$, $R = 0.0125/9 = 0.001389$ per unit
Generator 2, $X/R = 29$, $R = 0.18/29 = 0.0062$ per unit

Large synchronous motor M_1, using $X = 1.5 \, X_d'' = 1.5 \, (0.333) = 0.5$ per unit, $X/R = 30$, $R = 0.5/30 = 0.01667$ per unit

Large induction motor M_2, using $X = 1.5 \, X_d'' = 1.5 \, (0.971) = 1.457$
per unit, $X/R = 30$, $R = 1.457/30 = 0.04857$ per unit

Low-voltage motor group below 50 hp is omitted

Low-voltage motor group 50 to 150 hp, using $X = 3.0 \, X_d'' = 3 \times 5 = 15$ per unit, $X/R = 9$, $R = 15/9 = 1.667$ per unit

Based on ANSI C37.5-1969 (R 1974) and IEEE Std 320-1972.

Table 30
Reactances for "Approximately 30 Cycle"
Short-Circuit Currents

Utility system S_1 reactance is unchanged

Generator 2, S_4 is represented with transient reactance, assumed at $1.5\ X_d'' = 1.5 \times 0.18 = 0.27$ per unit

All other sources, S_2, S_3, S_5-S_{10}, are disconnected

5.6.9 *Reactances and Resistances for the Circuit to Calculate Short-Circuit (Interrupting) Current Duties.* Reactances, and resistances derived from them as described previously, are detailed in Table 29.

5.6.10 *Reactances for the Circuit to Calculate "Approximately 30 Cycle" Minimum Short-Circuit Currents.* Minimum generation for this problem occurs with generator 1 down, with the 46 kV utility system connection open, and with all motors disconnected. Reactance details are given in Table 30.

5.6.11 *Circuit and Calculation of First-Cycle (Momentary) Short-Circuit Current Duties for Fuses and Low-Voltage Circuit Breakers.* The circuit used for calculating the alternating currents that establish the first-cycle (momentary) short-circuit duties based on current fuse and low-voltage circuit breaker standards is shown in Fig 85(a). Source circuits S_5 through S_{10} have been simplified using the series and parallel combinations indicated in Table 31, based on the per-unit element impedances obtained directly from Tables 26 and 27. The identities of buses and sources are retained in Fig 85(a), even after the individual element impedances from Fig 76 lose identification when reactances are combined.

The connection of an alternating-current source, the voltage magnitude of which is the prefault voltage at the fault bus, between the dotted "common" connection and the fault at the fault bus causes the flow of per-unit alternating short-circuit current that is being calculated.

The reactances of Fig 85(a) are further simplified as shown in Fig 85(b), without losing track of the three fault locations. The reactance simplifications are summarized in Table 32. The table contains columns of reactances and reciprocals. Arrows are used to indicate the

calculation of a reciprocal. Sums of reciprocals are used to combine reactances in parallel. A dashed line in the reactance column indicates that reactances above the line have been combined in parallel.

The final simplification of reactances to obtain one fault-point X for each fault location is detailed in Table 33. The results for the specified fault buses are the last entries in the reactance columns.

Alternating short-circuit currents are calculated from the circuit reactance reductions assuming the prefault voltage is 1.0 per unit, and alternating rms current is, of course, E/X per unit. Multiplying by base current converts to real units. The resulting symmetrical (alternating only) first-cycle (momentary) short-circuit rms currents are

at F_1, $I_{sym} = (1.0/0.016)\ (0.4184) = 26.15$ kA
at F_2, $I_{sym} = (1.0/0.0423)\ (1.388) = 32.81$ kA
at F_3, $I_{sym} = (1.0/0.1048)\ (1.388) = 13.25$ kA

These currents are useful as primary "available" short-circuit data when following applicable standards to calculate low-voltage short-circuit duties for unit substations supplied from these buses.

Asymmetrical short-circuit duties are necessary for comparison with total rms current ratings of alternating-current high-voltage (and medium-voltage) fuses, such as those in the fused motor-control equipment connected to buses 3 and 4. These are found using multiplying factors from the applicable fuse standards, ANSI C37.41-1969 (R 1974). The applicable standard for the circuit of Fig 85 suggests a general-case multiplying factor of 1.55, but a "special-case" multiplier of 1.2 may be substituted if the voltage is less than 15 kV and if the X/R ratio is less than 4. The circuit of this

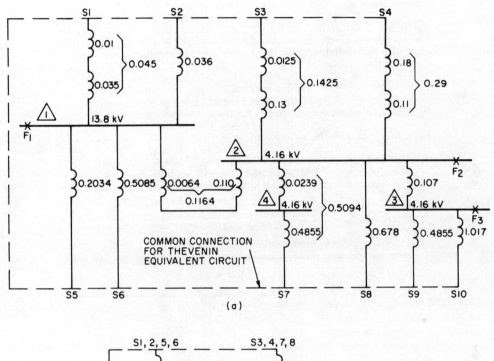

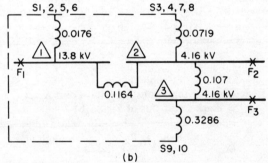

(b)

Fig 85
Circuits of Power System Reactances for Calculation of First-Cycle (Momentary)
Short-Circuit Current Duties for Fuses and Low-Voltage Circuit Breakers
(a) Reactance Diagram. (b) Simplified Reactance Diagram

Table 31
Reactances for Fig 85(a)

S_5 to bus 1, two circuits in parallel, each with M_1 motor X_d'' and T_5 transformer X,

$$X_d'' = (1/2)(0.3333 + 0.0734) = 0.2034 \text{ per unit}$$

S_6 to bus 1 (after combining all the motors of one substation for an equivalent low-voltage motor $X_d'' = 2.5\ (5)/(2.5 + 5) = 1.667$), four circuits in parallel, each with an equivalent low-voltage motor X_d'' in series with a T_6 transformer,

$$X_d'' = (1/4)(1.667 + 0.367) = (1/4)(2.034) = 0.5085 \text{ per unit}$$

S_7 to bus 4, two M_2 induction motors,

$$X_d'' = (1/2)(0.971) = 0.4855 \text{ per unit}$$

S_8 to bus 2, three circuits, each as for the S_6 to bus 1 calculation,

$$X_d'' = (1/3)(2.034) = 0.678 \text{ per unit}$$

S_9 to bus 3, two M_2 induction motors,

$$X_d'' = (1/2)(0.971) = 0.4855 \text{ per unit}$$

S_{10} to bus 3, two circuits, each as for the S_6 to bus 1 calculation,

$$X_d'' = (1/2)(2.034) = 1.017 \text{ per unit}$$

Table 32
Reactance Combinations for Fig 85(a)

S_1, S_2, S_5, S_6			S_3, S_4, S_7, S_8			S_9, S_{10}		
X		$1/X$	X		$1/X$	X		$1/X$
0.045	→	22.22	0.1425	→	7.02	0.4855	→	2.060
0.036	→	27.78	0.29	→	3.45	1.017	→	0.983
0.2034	→	4.91	0.5094	→	1.96	0.3286	←	3.043
0.5085	→	1.97	0.678	→	1.47			
0.0176	←	56.88	0.0719	←	13.90			

Table 33
Reactance Combinations for Fault-Point X at
Each Fault Bus of Fig 85(b)

Fault at F_1			Fault at F_2			Fault at F_3		
X		$1/X$	X		$1/X$	X		$1/X$
0.3286			0.0176			0.1340	→	7.46
0.107			0.1164			0.0179	→	13.90
0.4356	→	2.30	0.1340	→	7.46	0.0468	←	21.36
0.0719	→	13.90	0.107			0.107		
0.0617	←	16.20	0.3286			0.1538	→	6.502
0.1164			0.4356	→	2.30	0.3286	→	3.043
0.1781	→	5.62	0.0719	→	13.90	0.1048	←	9.545
0.0176	→	56.82	0.0423	←	23.66			
0.016	←	62.44						

example will not have X/R ratios as low as 4. The asymmetrical (total) first-cycle (momentary) short-circuit rms currents for fuse applications are

at F_1, I_{tot} = 1.55(26.15) = 40.53 kA
at F_2, I_{tot} = 1.55(32.81) = 50.86 kA
at F_3, I_{tot} = 1.55(13.25) = 20.54 kA

5.6.12 *Circuit and Calculation of First-Cycle (Momentary) Short-Circuit Current Duties for High-Voltage Circuit Breakers.* The circuit used for calculating the alternating currents that establish the first-cycle (momentary) short-circuit duties based on IEEE Std 320-1972 and ANSI C37.5-1969 (R 1974) is shown by Fig 86(a). The rotating machine reactance changes from subtransient are shown in Table 28. These changes affect the Table 31 reactance combination simplifications of source circuits S_5 through S_{10}, as detailed in Table 34.

Fig 86(b) shows the last step of reactance simplification before the several fault location identities are lost. Table 35 details the reactance simplifications.

The final simplification of reactances to obtain one fault-point X for each fault is detailed in Table 36. The symmetrical first-cycle (momentary) short-circuit rms currents obtained from the fault-point values of Table 36 are

at F_1, I_{sym} = (1.0/0.0164) (0.4184) = 25.5 kA
at F_2, I_{sym} = (1.0/0.045) (1.388) = 30.9 kA
at F_3, I_{sym} = (1.0/0.1139) (1.388) = 12.2 kA

Asymmetrical short-circuit duties for comparison with alternating-current high-voltage (over 1000 V) circuit breaker momentary ratings (former rating basis) or closing and latching capabilities (present rating basis) are found using a 1.6 multiplying factor according to the reference standards. These total (asymmetrical) first-cycle (momentary) short-circuit rms currents are

at F_1, I_{tot} = 1.6(25.5) = 40.8 kA
at F_2, I_{tot} = 1.6(30.9) = 49.4 kA
at F_3, I_{tot} = 1.6(12.2) = 19.5 kA

5.6.13 *Circuit and Calculation of Contact-Parting Time (Interrupting) Short-Circuit Current Duties for High-Voltage Circuit Breakers.* In addition to a circuit of power system reactances for calculating alternating currents (I_{pu} = E/X), a resistance-only circuit is needed to es-

tablish fault-point X/R ratios. Duties are calculated by applying multiplying factors to E/X. The multiplying factors depend on the fault-point X/R and also on other factors defined subsequently.

The circuits used for calculating E/X and fault-point R are shown in Figs 87(a) and 88(a), respectively. The rotating machine reactances for the circuit of Fig 87(a), if changed from subtransient, are shown in Table 29. Table 37 details how these changes affect the Table 34 simplifications of source circuits S_5 through S_{10}. Table 37 also includes resistance simplifications of source circuits for Fig 88(a).

Figs 87(b) and 88(b) show the last steps of reactance and resistance simplifications, respectively, before the several fault-location identities are lost. Tables 38 and 39 detail the reactance and resistance simplifications starting from Figs 87(a) and 88(a), respectively. The final simplifications of reactances and resistances to obtain one fault-point X and one fault-point R for each fault location are detailed in Tables 40 and 41, respectively.

Values of per-unit E/X for each fault bus are readily obtained from Table 40 when E = 1.0 (as for this example); they are the final entries in the $1/X$ columns, opposite the fault-point X entries. Values converted to actual currents are

at F_1, E/X = 58.96(0.4184) = 24.67 kA
at F_2, E/X = 20.69(1.388) = 28.72 kA
at F_3, E/X = 7.808(1.388) = 10.84 kA

Values of X/R for each fault bus are obtained from the fault-point X and R entries of Tables 40 and 41 as follows:

at F_1, X/R = 0.071/0.000537 = 31.6
at F_2, X/R = 0.0483/0.00349 = 13.9
at F_3, X/R = 0.1281/0.00489 = 26.2

The reference standards contain graphs of multiplying factors which determine calculated short-circuit current duties when applied to E/X values. The proper graph is selected knowing the following information:

(1) Three-phase or single-phase short-circuit current (three-phase for this example)

(2) Rating basis of the circuit breaker being applied (previous total-current short-circuit ratings or present symmetrical-current short-circuit ratings)

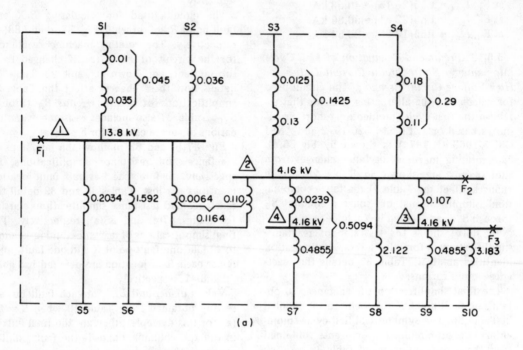

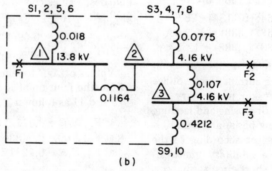

Fig 86
Circuits of Power System Reactances for Calculation of First-Cycle (Momentary)
Short-Circuit Current Duties for High-Voltage Circuit Breakers
(a) Reactance Diagram. (b) Simplified Reactance Diagram

Table 34
Reactances for Fig 86(a), Detailed as Changes from Table 31

S_5, S_7, S_9 no change
S_6 to bus 1, four circuits in parallel, each with 1.2 X_d'' of 50 to 150 hp
   induction motors and transformer T_6,
      $X_d'' = (1/4)(6.0 + 0.367) = (1/4)(6.367) = 1.592$ per unit
S_8 to bus 2, three circuits, each as for the S_6 to bus 1 calculation,
      $X_d'' = (1.3)(6.367) = 2.122$ per unit
S_{10} to bus 3, two circuits, each as for the S_6 to bus 1 calculation,
      $X_d'' = (1.2)(6.367) = 3.183$ per unit

Table 35
Reactance Combinations for Fig 86(a)

| S_1, S_2, S_5, S_6 | | S_3, S_4, S_7, S_8 | | S_9, S_{10} | |
X	$1/X$	X	$1/X$	X	$1/X$
0.045 →	22.22	0.1425 →	7.018	0.4855 →	2.060
0.036 →	27.78	0.29 →	3.448	3.183 →	0.314
0.2034 →	4.92	0.5094 →	1.963	0.4214 ←	2.374
1.592 →	0.63	2.122 →	0.471		
0.018 ←	55.55	0.0775 ←	12.90		

Table 36
Reactance Combinations for Fault-Point X at
Each Fault Bus of Fig 86(b)

| Fault at F_1 | | Fault at F_2 | | Fault at F_3 | |
X	$1/X$	X	$1/X$	X	$1/X$
0.4212		0.018		0.1344 →	7.441
0.107		0.1164		0.0755 →	12.90
0.5282 →	1.893	0.1344 →	7.441	0.0492 ←	20.34
0.0775 →	12.90	0.5282 →	1.893	0.107	
0.0676 ←	14.79	0.0775 →	12.90	0.1562 →	6.402
0.1164		0.45 ←	22.23	0.4212 →	2.374
0.184 →	5.43			0.1139 ←	8.776
0.018 →	55.55				
0.0164 ←	60.98				

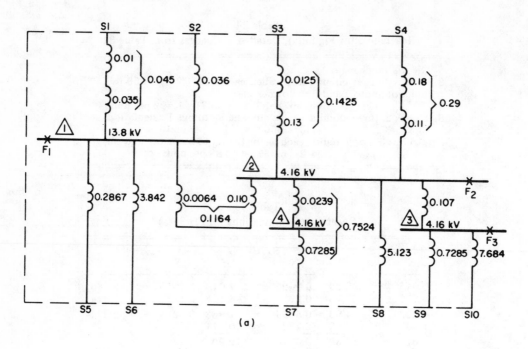

(a)

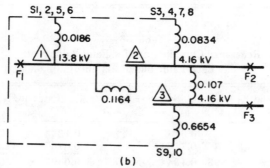

(b)

Fig 87
Circuits of Power System Reactances for Calculation of E/X and Fault-Point X
for Contact-Parting Time (Interrupting) Short-Circuit Current Duties
for High-Voltage Circuit Breakers
(a) Reactance Diagram. (b) Simplified Reactance Diagram

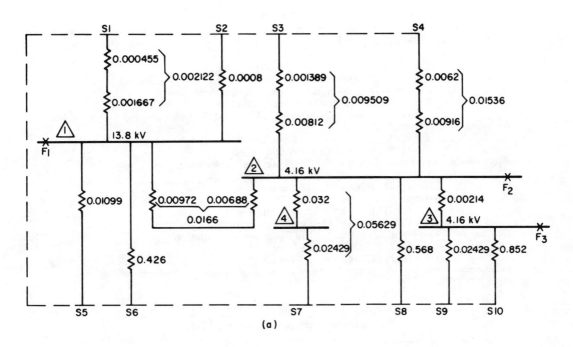

(a)

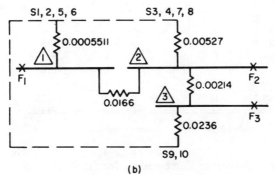

(b)

Fig 88
Circuits of Power System Resistances for Calculation of Fault-Point R
for Contact-Parting Time (Interrupting) Short-Circuit Current Duties
for High-Voltage Circuit Breakers
(a) Resistance Diagram. (b) Simplified Resistance Diagram

Table 37
Reactances for Fig 87(a) and Resistances for Fig 88(a)

S_5 to bus 1, two circuits in parallel, each with 1.5 X_d'' of synchronous motor M_1
and transformer T_4,

$X = (1/2)(0.5 + 0.0734) = 0.2867$ per unit
$R = (1/2)(0.01667 + 0.00531) = (1/2)(0.02198) = 0.01099$ per unit

S_6 to bus 1, four circuits in parallel, motor group and transformer T_6,

$X = (1/4)(15 + 0.367) = (1/4)(15.367) = 3.842$ per unit
$R = (1/4)(1.667 + 0.0367) = (1/4)(1.704) = 0.426$ per unit

S_7 to bus 4, two motors M_2,

$X = (1/2)(1.457) = 0.7285$ per unit
$R = (1/2)(0.04857) = 0.02429$ per unit

S_8 to bus 2, three circuits, each as for the S_6 to bus 1 calculation,

$X = (1/3)(15.367) = 5.123$ per unit
$R = (1/3)(1.704) = 0.568$ per unit

S_9 to bus 3, two motors M_2,

$X = (1/2)(1.457) = 0.7285$ per unit
$R = (1/2)(0.04857) = 0.02429$ per unit

S_{10} to bus 3, two circuits, each as for the S_6 to bus 1 calculation,

$X = (1/2)(15.367) = 7.684$ per unit
$R = (1/2)(1.704) = 0.852$ per unit

Table 38
Reactance Combinations for Fig 87(a)

| S_1, S_2, S_5, S_6 | | S_3, S_4, S_7, S_8 | | S_9, S_{10} | |
X	$1/X$	X	$1/X$	X	$1/X$
0.045	→ 22.22	0.1425	→ 7.018	0.7285	→. 1.373
0.036	→ 27.78	0.29	→ 3.448	7.684	→ 0.130
0.2867	→ 3.49	0.7524	→ 1.329	0.6653	← 1.503
3.842	→ 0.26	5.123	→ 0.195		
0.0186	← 53.75	0.0834	← 11.99		

Table 39
Resistance Combinations for Fig 88(a)

| S_1, S_2, S_5, S_6 | | S_3, S_4, S_7, S_8 | | S_9, S_{10} | |
R	$1/R$	R	$1/R$	R	$1/R$
0.002122	→ 471.3	0.009509	→ 105.2	0.02429	→ 41.17
0.0008	→ 1250.0	0.01536	→ 65.10	0.852	→ 1.174
0.01099	→ 90.99	0.05629	→ 17.77	0.02362	← 42.34
0.426	→ 2.35	0.568	→ 1.76		
0.0005511	← 1815	0.00527	← 189.8		

Table 40
Reactance Combinations for Fault-Point X at Each Fault Bus of Fig 87(b)

Fault at F_1		Fault at F_2		Fault at F_3	
X	$1/X$	X	$1/X$	X	$1/X$
0.6654		0.0186		0.135 →	7.407
0.107		0.1164		0.0834 →	11.99
0.7724 →	1.295	0.135 →	7.407	0.0516 ←	19.40
0.0834 →	11.99	0.7724 →	1.295	0.107	
0.0753 ←	13.29	0.0834 →	11.99	0.1586 →	6.305
0.1164		0.0483 ←	20.69	0.6654 →	1.503
0.1917 →	5.217			0.1281 ←	7.808
0.0186 →	53.76				
0.017 ←	58.98				

Table 41
Resistance Combinations for Fault-Point R at Each Fault Bus of Fig 88(b)

Fault at F_1		Fault at F_2		Fault at F_3	
R	$1/R$	R	$1/R$	R	$1/R$
0.02362		0.0005511		0.01715 →	58.31
0.00214		0.0166		0.00527 →	189.8
0.02576 →	38.82	0.01715 →	58.31	0.00403 ←	248.1
0.00527 →	189.8	0.00527 →	189.8	0.00214	
0.00437 ←	228.6	0.02576 →	38.82	0.00617 →	162.0
0.0166		0.00349 ←	286.9	0.02362 →	42.34
0.02097 →	47.68			0.00489 ←	204.4
0.0005511 →	1815				
0.000537 ←	1863				

(3) Rating interrupting time of the circuit breaker being applied

(4) Fault-point X/R ratio

(5) Proximity of generators

The proximity of generators determines the choice between graphs for faults fed predominantly from generators through not more than one transformation ("local" in this example) and for faults fed predominantly through two or more transformations ("remote" in this example). Utility contributions are considered to be from remote generators in most industrial system duty calculations.

For many systems having only remote sources and no in-plant generators, it is clear that the remote multiplying factor is the only choice. For the fewer systems that have in-plant generator primary power sources, both multiplying factors may be necessary, as explained subsequently.

In this example short-circuit duties are calculated for total-current (TOT) short-circuit rated (previous basis) circuit breakers with 8 cycle and 5 cycle rated interrupting times (TOT 8 and TOT 5) and symmetrical-current (SYM) short-circuit rated (present basis) circuit breakers with 5 cycle rated interrupting times (SYM 5).

Multiplying factors obtained from both the local and remote graphs of IEEE Std 320-1972 and ANSI C37.5-1969 (R 1974) are shown in Table 42 for the other conditions previously established in this example. These graphs are shown as Figs 81—84. The local multiplying factors are smaller because they include the effects of generator alternating (symmetrical) current decay. Remote multiplying factors are

Table 42
Three-Phase Short-Circuit Current Multiplying Factors for E/X for Example Conditions

Fault Location	Fault-Point X/R Ratio	Circuit Breaker Type	Multiplying Factor	
			Local	Remote
F_1	31.6	TOT 8	1.05	1.18
		TOT 5	1.14	1.27
		SYM 5	1.04	1.15
F_2	13.9	TOT 8	1.0*	1.0*
		TOT 5	1.01	1.06
		SYM 5	1.0*	1.0*
F_3	26.2	TOT 8	1.02	1.14
		TOT 5	1.11	1.21
		SYM 5	1.01	1.10

*Standards indicate that a 1.0 multiplying factor applies without further checking when X/R = 15 or less for SYM circuit breakers of all rated interrupting times and for TOT 8 circuit breakers.

based on no decay of the generator alternating (symmetrical) current up to circuit breaker contact-parting time.

It is not immediately apparent for any of the fault buses of this example which of the two multiplying factors applies. The standards leave the definitions of the words "predominantly" and "transformation" to the interpretation of the user.

The remote multiplying factor is always the conservative selection because it is the larger. However, the difference can be appreciable. For example, Table 42 indicates calculated short-circuit duties for TOT 8 applications at bus 1 that differ by about 13 percent. Since bus 1 has both remote and local generator contributions, each of about the same magnitude, this example uses the remote multiplying factors to calculate short-circuit duties.

The calculated short-circuit interrupting duty rms currents for three-phase faults at bus 1, using remote multiplying factors, are

for TOT 8 circuit breakers,

$$1.18(24.67) = 29.11 \text{ kA-T}$$

for TOT 5 circuit breakers,

$$1.27(24.67) = 31.33 \text{ kA-T}$$

for SYM 8 circuit breakers,

$$1.15(24.67) = 28.37 \text{ kA-S}$$

The kA-T designation denotes an rms current duty in kiloamperes to be compared with the total-current short-circuit (interrupting) capability of a total rated circuit breaker. This is a total (asymmetrical) rms current duty.

The kA-S designation denotes an rms current duty in kiloamperes to be compared with the symmetrical-current short-circuit (interrupting) capability of a symmetrical rated circuit breaker. This is a symmetrical rms current duty only if the multiplying factor for E/X is 1.0; otherwise it is neither symmetrical nor asymmetrical, but part way in between.

The F_2 fault calculation for a TOT 5 circuit breaker is not detailed in this example. TOT 8 and SYM 5 duties at bus 2 are already available, since 1.0 multiplying factors apply, as follows:

for TOT 8 circuit breakers,

$$1.0(28.72) = 28.72 \text{ kA-T}$$

for SYM 5 circuit breakers,

$$1.0 (28.72) = 28.72 \text{ kA-S}$$

The calculated short-circuit (interrupting) duty rms currents for three-phase faults at bus 3, using remote multiplying factors, are

for TOT 8 circuit breakers,

$$1.14 (10.84) = 12.36 \text{ kA-T}$$

for TOT 5 circuit breakers,

$$1.21(10.84) = 13.12 \text{ kA-T}$$

for SYM 5 circuit breakers,

$$1.10(10.84) = 11.92 \text{ kA-S}$$

Table 43
AC High-Voltage Circuit Breaker Short-Circuit Ratings or Capabilities, in Kiloamperes

Circuit Breaker Nominal Size Identification	Example Maximum System Operating Voltage (kV)	8 Cycle Total-Rated Circuit Breakers		5 Cycle Symmetrical-Rated Circuit Breakers	
		Momentary Rating (Total First-Cycle RMS Current)	Interrupting Rating (Total RMS Current at 4 Cycle Contact-Parting Time)	Closing and Latching Capability (Total First-Cycle RMS Current)	"Short-Circuit" Capability (Symmetrical RMS Current at 3 Cycle Contact-Parting Time)
4.16—75	4.16	20	10.5	19	10.1
4.16—250	4.16	60	35	58	33.2
4.16—350	4.16	80	48.6	78	46.9
13.8—500	13.8	40	21	37	19.6
13.8—750	13.8	60	31.5	58	30.4
13.8—1000	13.8	80	42	77	40.2

Table 44
Calculated Bus 1 Short-Circuit Duties
Compared with Ratings or Capabilities of
AC High-Voltage Circuit Breakers

Type of Circuit Breaker	TOT 8 (8 Cycle Total Rated)	SYM 5 (5 Cycle Symmetrical Rated)
First-cycle duty, total rms current	40.8 kA	40.8 kA
Short-circuit (interrupting) duty, rms current	29.1 kA-T	28.4 kA-S
Circuit breaker nominal size	13.8—750	13.8—750
Momentary current rating, or closing and latching capability	60 kA	58 kA
Interrupting rating, or short-circuit current capability	31.5 kA	30.4 kA

Based on ANSI C37.5-1969 (R 1974) and IEEE Std 320-1972.

Table 45
Calculated Bus 2 Short-Circuit Duties
Compared with Ratings or Capabilities of
AC High-Voltage Circuit Breakers

Type of Circuit Breaker	TOT 8 (8 Cycle Total Rated)	SYM 5 (5 Cycle Symmetrical Rated)
First-cycle duty, total rms current	49.4 kA	49.4 kA
Short-circuit (interrupting) duty, rms current	28.7 kA-T	28.7 kA-S
Circuit breaker nominal size	4.16—250	4.16—250
Momentary current rating, or closing and latching capability	60 kA	58 kA
Interrupting rating, or short-circuit current capability	35 kA	33.2 kA

Based on ANSI C37.5-1969 (R 1974) and IEEE Std 320-1972.

5.6.14 *Circuit Breaker Short-Circuit Capabilities Compared with Calculated Short-Circuit Current Duties.* Short-circuit ratings, or capabilities derived from them, for circuit breakers that might be applied in the example system are listed in Table 43. The capabilities which are derived from symmetrical short-circuit ratings using a ratio of rated maximum voltage to operating voltage are computed using the example operating voltages listed in the table.

Circuit breakers for bus 1 application, both TOT 8 and SYM 5 types, having short-circuit ratings or capabilities no less than the calculated duties at bus 1, are listed in Table 44 with the calculated duties for comparison. Circuit breakers for bus 2 and 3 applications are listed in Tables 45 and 46, respectively, with short-circuit ratings or capabilities and calculated duties.

5.6.15 *Circuit and Calculation of "Approximately 30 Cycle" Minimum Short-Circuit Currents.* The circuit used is shown in Fig 89. The

rotating machine reactances are shown in Table 30. Table 47 details the reactance simplifications starting from Fig 89(b).

A prefault voltage of 1.0 per unit is assumed, I is calculated at E/X per unit, and the conversion is made to amperes. There is no direct-current component remaining to cause asymmetry. The resulting symmetrical "approximately 30 cycle" short-circuit currents are

at $F_1, I = (1.0/0.0413) (0.4184) = 10.14$ kA
at $F_2, I = (1.0/0.1133) (1.388) = 12.25$ kA
at $F_3, I = (1.0/0.2203) (1.388) = 6.30$ kA

5.7 Example of Short-Circuit Current Calculation for a System under 1000 V. As in portions of a power system over 1000 V, calculation of short-circuit currents at various locations in a system under 1000 V is essential for proper application of circuit breakers, fuses, buses, and cables. All must withstand momentarily the thermal and magnetic stresses imposed by the maximum possible short-circuit current. In ad-

Table 46
Calculated Bus 3 Short-Circuit Duties
Compared with Ratings or Capabilities of
AC High-Voltage Circuit Breakers

Type of Circuit Breaker	TOT 8 (8 Cycle Total Rated)	SYM 5 (5 Cycle Symmetrical Rated)
First-cycle duty, total rms current	19.5 kA	19.5 kA
Short-circuit (interrupting) duty, rms current	12.4 kA-T	11.9 kA-S
Circuit breaker nominal size	4.16—250	4.16—250
Momentary current rating, or closing and latching capability	60 kA	58 kA
Interrupting rating, or short-circuit current capability	35 kA	33.2 kA

Based on ANSI C37.5-1969 (R 1974) and IEEE Std 320-1972.

dition, circuit breakers and fuses must safely interrupt these maximum fault currents.

For the three-phase system, the three-phase fault will usually produce the maximum fault current. On a balanced three-phase system, the line-to-line fault current will never exceed 87 percent of the three-phase value. The line-to-ground fault current could exceed the three-phase short-current by a small percentage; however, this is apt to occur only when there is little or no motor load and the primary system fault contribution is small.

The calculation of symmetrical short-circuit current duties is normally sufficient for the application of circuit breakers and fuses under 1000 V, since they have published symmetrical current interrupting ratings. The ratings are based on the first-cycle symmetrical rms current, determined at 1/2 cycle after fault inception and incorporate an asymmetrical capability as necessary for a circuit X/R ratio of 6.6 or less (short-circuit power factor of 15 per-

cent or greater). A typical system served by a transformer rating of 1000 or 1500 kVA will usually have a short-circuit X/R ratio within these limits. For larger or multitransformer systems it is advisable to check the X/R ratio, and if greater than 6.6, then the circuit breaker or fuse application should be based on asymmetrical current limitations not further covered here (see IEEE Std 20-1973).

The low-voltage fault calculation procedure differs very little from that used for finding first-cycle short-circuit duties in higher voltage systems. All connected motor ratings are included as fault contributing sources, and this contribution is based on the subtransient reactance of the machines. The contribution from the primary system should be equivalent to that calculated for its "momentary duty." Due to the quantity and small ratings of motors usually encountered in low-voltage systems, it is customary to use an assumed typical value for their equivalent reactance in the low-voltage

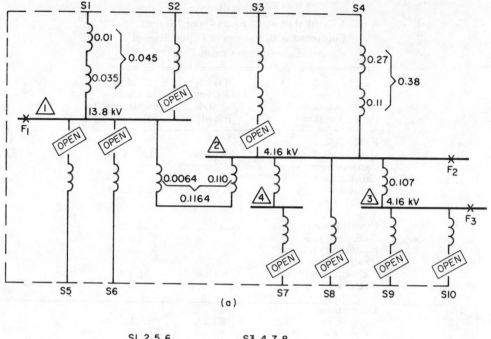

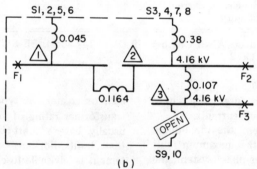

**Fig 89
Circuits of Power System Reactances for Calculation of
"Approximately 30 Cycle" Minimum Short-Circuit Currents
(a) Reactance Diagram. (b) Simplified Reactance Diagram**

**Table 47
Reactance Combinations for Fault-Point X at Each Fault Bus of Fig 89(b)**

Fault at F_1		Fault at F_2		Fault at F_3
X	$1/X$	X	$1/X$	X
0.38		0.045		0.1133
0.1164		0.1164		0.107
0.4964	→ 2.0145	0.1614	→ 6.1958	0.2203
0.045	→ 22.2222	0.38	→ 2.6316	
0.04126	← 24.2367	0.1133	← 8.8274	

226

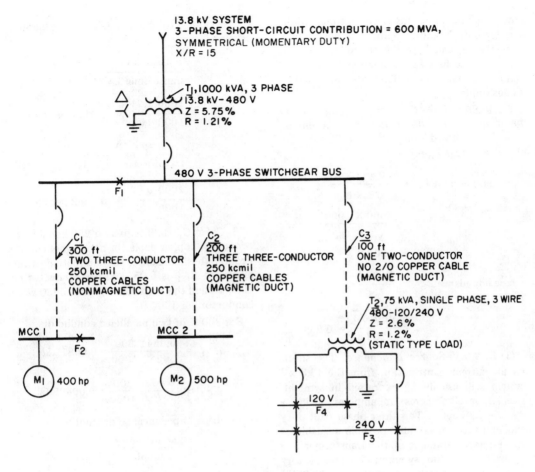

**Fig 90
Low-Voltage System**

NOTE: The motor horsepower indicated at MCC 1 and 2 represents a lumped total of small induction three-phase machines ranging in size from 10 to 150 hp.

short-circuit network. This typical value is 25 percent on the individual motor rating or the total rating of a group of motors, both in kilovolt-amperes.

The example fault calculation presented here is for a 480 V three-phase system, illustrated by the single-line diagram of Fig 90. The system data shown are typical of that required to perform the calculations.

Bolted three-phase faults F_1 and F_2 are assumed at each of the bus locations, and zero-

impedance (bolted) line-to-line faults F_3 and F_4 are assumed at the 120/240 V single-phase locations. Both resistance and reactance components of the circuit element impedances are used in order to illustrate the more precise procedure and to obtain X/R ratios.

Resistances are usually significant in low-voltage short-circuit current calculations. Their effect may be evaluated either by a complex impedance reduction or by separate X and R reductions. The complex reduction leads to the most accurate short-circuit current magnitude. The separate X and R reductions are simpler, conservative, and have the added benefit that they give the best approximation for the X/R ratio at the fault point. They are illustrated by this example.

5.7.1 *Step 1 — Convert All Element Impedances to Per-Unit Values on a Common Base.* The assumed base power is 1000 kVA and the base voltage E_b = 480 V:

$$\text{base current } I_b = \frac{\text{kVA }(1000)}{\sqrt{3} \times E_b}$$

$$= \frac{1000 \times 1000}{\sqrt{3} \times 480}$$

$$= 1202.8 \text{ A}$$

$$\text{base impedance } Z_b = \frac{E_b/\sqrt{3}}{I_b}$$

$$= \frac{480/\sqrt{3}}{1202.8} = 0.2304 \text{ } \Omega$$

(1) *13.8 kV Source Impedance.* The short-circuit current contribution from the 13.8 kV system will usually be expressed in symmetrical megavolt-amperes or amperes, giving a specific X/R ratio. This three-phase fault duty should be the maximum possible available at the primary terminals of the transformer and equivalent to the symmetrical "momentary duty" level. For this example, the 13.8 kV fault duty is 600 MVA or 25 102 A symmetrical at an X/R ratio of 15. The R_s and X_s values comprising the equivalent impedance Z_s can be obtained as follows:

$$Z_s = \frac{\text{base kVA}}{\text{short-circuit kVA}}$$

$$= \frac{1000}{600\,000} = 0.00166 \text{ per unit}$$

Since $Z_s = \sqrt{(R_s)^2 + (X_s)^2}$ and $X_s/R_s = 15$, the value of $R_s = Z_s/\sqrt{1 + (15)^2} = 0.00011$ per unit, and the value of $X_S = 15R_S = 0.00165$ per unit.

(2) *1000 kVA Transformer Impedance.* The transformer manufacturer provides the information that the impedance is 5.75 percent on the self-cooled base rating of 1000 kVA, and the resistance is 1.21 percent (R_{T1}). Reactance $X =$

$\sqrt{Z^2 - R^2}$ = 5.62 percent (X_{T1}). The per-unit values are

$$R_{T1} = \frac{\text{base kVA}}{\text{transformer kVA}} \frac{\%R_{T1}}{100}$$

$$= \frac{1000}{1000} \frac{1.21}{100} = 0.0121 \text{ per unit}$$

$$X_{T1} = \frac{\text{base kVA}}{\text{transformer kVA}} \frac{\%X_{T1}}{100}$$

$$= \frac{1000}{1000} \frac{5.62}{100} = 0.0562 \text{ per unit}$$

(3) *Cable C_1* (300 ft of two 250 kcmil three-conductor copper cables in nonmagnetic duct). From published tables, the alternating-current resistance R_{C1} is 0.0541 Ω per conductor per 1000 ft, and the reactance X_{C1} is 0.0330 Ω per conductor per 1000 ft.

For 300 ft of two paralleled conductors,

$$R_{C1} = \frac{0.0541}{2} \frac{300}{1000} = 0.00812 \text{ } \Omega$$

$$X_{C1} = \frac{0.0330}{2} \frac{300}{1000} = 0.00495 \text{ } \Omega$$

Converting impedances to per unit,

$$R_{C1} = \frac{\text{actual ohms}}{\text{base ohms}}$$

$$= \frac{0.00812}{0.2304} = 0.0352 \text{ per unit}$$

$$X_{C1} = \frac{\text{actual ohms}}{\text{base ohms}}$$

$$= \frac{0.00495}{0.2304} = 0.0215 \text{ per unit}$$

(4) *Cable C_2* (200 ft of three 250 kcmil three-conductor copper cables in magnetic duct). From published tables, the alternating-current resistance R_{C2} is 0.0552 Ω per conductor per 1000 ft, and the reactance X_{C2} is 0.0379 Ω per conductor per 1000 ft.

For 200 ft of three paralleled conductors,

$$R_{C2} = \frac{0.0552}{3} \frac{200}{1000} = 0.00368 \text{ } \Omega$$

$$X_{C2} = \frac{0.0379}{3} \frac{200}{1000} = 0.00253 \text{ } \Omega$$

Converting impedances to per unit,

$$R_{C2} = \frac{0.00368}{0.2304} = 0.01597 \text{ per unit}$$

$$X_{C2} = \frac{0.00253}{0.2304} = 0.01098 \text{ per unit}$$

(5) *Cable* C_3 (100 ft of one no. 2/0 two-conductor copper cable in magnetic duct). From published tables, the alternating-current resistance R_{C3} is 0.102 Ω/1000 ft, and the reactance X_{C3} is 0.0407 Ω/1000 ft.

For 100 ft,

$$R_{C3} = 0.102 \frac{100}{1000} = 0.0102 \ \Omega$$

$$X_{C3} = 0.0407 \frac{100}{1000} = 0.00407 \ \Omega$$

Converting impedances to per unit,

$$R_{C3} = \frac{0.0102}{0.2304} = 0.0443 \text{ per unit}$$

$$X_{C3} = \frac{0.00407}{0.2304} = 0.01766 \text{ per unit}$$

(6) *Motor Contribution.* The running motor loads at motor-control center 1 and 2 buses total 400 hp and 500 hp, respectively. Typical assumptions made for 480 V small motor groups are that 1 hp = 1 kVA, and the average subtransient reactance is 25 percent. The resistance is 4.167 percent based on a typical X/R ratio of 6.

Converting impedances to per unit on the 1000 kVA base,

$$R_{M1} = \frac{\text{base kVA}}{\text{motor kVA}} \frac{R_{M1}}{100}$$

$$= \frac{1000}{400} \frac{4.167}{100} = 0.1042 \text{ per unit}$$

$$X_{M1} = \frac{\text{base kVA}}{\text{motor kVA}} \frac{X_{M1}}{100}$$

$$= \frac{1000}{400} \frac{25}{100} = 0.625 \text{ per unit}$$

$$R_{M2} = \frac{1000}{500} \frac{4.167}{100} = 0.0833 \text{ per unit}$$

$$X_{M2} = \frac{1000}{500} \frac{25}{100} = 0.500 \text{ per unit}$$

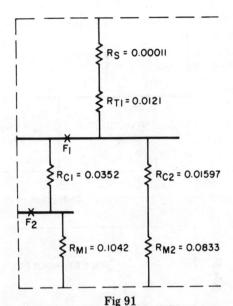

Fig 91
Resistance Network for Faults at F_1 and F_2

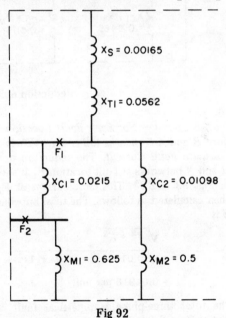

Fig 92
Reactance Network for Faults at F_1 and F_2

5.7.2 *Step 2 — Draw Separate Resistance and Reactance Diagrams Applicable for Fault Locations F_1 and F_2* (Figs 91 and 92). Since the single-phase 120/240 V system has no fault-contributing sources, it will not be represented in these diagrams.

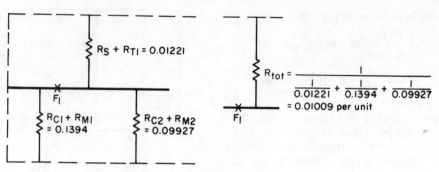

Fig 93
Reduction of R Network for Fault at F_1

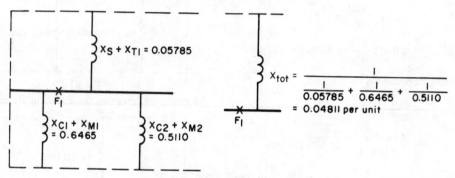

Fig 94
Reduction of X Network for Fault at F_1

5.7.3 *Step 3 — For Each Fault Location Reduce R and X Networks to Per-Unit Values and Calculate Fault Current.* The reduction of the R and X networks at fault location F_1 is shown in Figs 93 and 94. The fault current at F_1 is then calculated as follows. The total impedance Z is

$$Z = \sqrt{R^2 + X^2}$$

$$= \sqrt{(0.01009)^2 + (0.04811)^2}$$

$$= 0.04916 \text{ per unit}$$

The total three-phase symmetrical fault current at F_1 is $E/Z \times$ base current, that is,

$$\frac{\text{base amperes}}{\text{per unit } Z} = \frac{1202.8}{0.04916} = 24\ 470 \text{ A}$$

and the X/R ratio of the system impedance for the fault at F_1 is

$$X/R = \frac{0.04811}{0.01009} = 4.77$$

The reduction of the R and X networks at fault location F_2 is shown in Figs 95 and 96. The fault current at F_2 is then calculated as follows. The total impedance Z is

$$Z = \sqrt{R^2 + X^2}$$

$$= \sqrt{(0.0319)^2 + (0.0657)^2} = 0.073 \text{ per unit}$$

The total three-phase symmetrical fault current at F_2 is

$$\frac{\text{base amperes}}{\text{per unit } Z} = \frac{1202.8}{0.073} = 16\ 480 \text{ A}$$

and the X/R ratio of the system impedance for the fault at F_2 is

$$X/R = \frac{0.0657}{0.0319} = 2.06$$

5.7.4 *Step 4 — Draw Separate Resistance and Reactance Diagrams Applicable for Faults at the 120/240 V Single-Phase Secondary of the 75 kVA Transformer, and Calculate Fault Cur-*

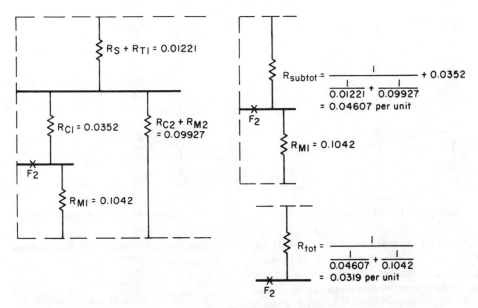

**Fig 95
Reduction of R Network for Fault at F_2**

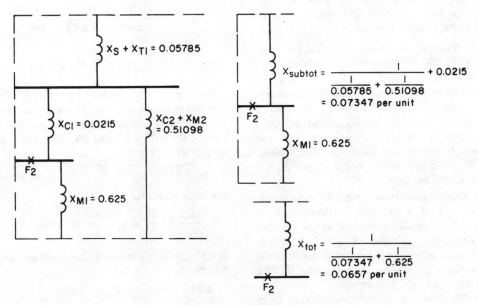

**Fig 96
Reduction of X Network for Fault at F_2**

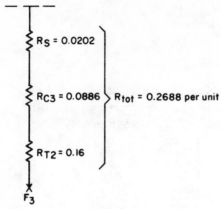

Fig 97
Resistance Network for Fault at F_3

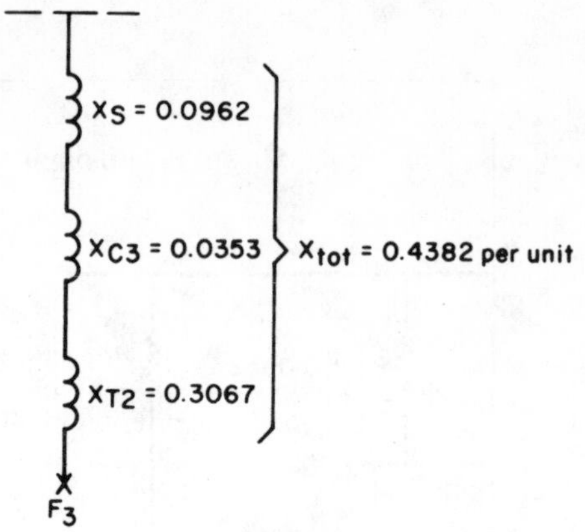

Fig 98
Reactance Network for Fault at F_3

rents. Per-unit calculations of short-circuit currents at the low-voltage side of a single-phase transformer connected line to line to a three-phase system may continue to use the same base, in this example 1000 kVA, but as a single-phase base. Impedances in the primary system connected to the transformer have double the values used for three-phase calculations to account for both outgoing and return paths of single-phase primary currents.

NOTE: This procedure assumes that the positive- and negative-sequence impedances are equal.

The total system three-phase fault impedance, as calculated above for fault at F_1, consists of $R_s = 0.0101$ per unit and X_s 0.0481 per unit. Since these are line-to-neutral values, they are doubled to obtain the line-to-line equivalents. Thus R_s becomes 0.0202 per unit and X_s becomes 0.0962 per unit.

The single-phase cable circuit C_3 was determined to have a per-unit line-to-neutral resistance R_{C3} equal to 0.0443 and a per-unit line-to-neutral reactance X_{C3} of 0.01766. These values must also be doubled for the line-to-line fault calculation, and become 0.0886 and 0.0353 per unit, respectively.

The 75 kVA transformer impedance, from published tables, is 2.6 percent on the base rating of 75 kVA and including the full secondary winding. The impedance components are 1.2 percent resistance R_{T2} and 2.3 percent reactance X_{T2}.

The per-unit values on the common 1000 kVA base are

$$R_{T2} = \frac{\text{base kVA}}{\text{transformer kVA}} \frac{\%R_{T2}}{100}$$

$$= \frac{1000}{75} \frac{1.2}{100} = 0.16 \text{ per unit}$$

$$X_{T2} = \frac{\text{base kVA}}{\text{transformer kVA}} \frac{\%X_{T2}}{100}$$

$$= \frac{1000}{75} \frac{2.3}{100} = 0.3067 \text{ per unit}$$

For a line-to-line fault at F_3 across the 240 V secondary winding of the 75 kVA transformer, the applicable resistance and reactance diagrams are shown in Figs 97 and 98. The total impedance Z is

$$Z = \sqrt{(0.2688)^2 + (0.4382)^2} = 0.5141 \text{ per unit}$$

the total short-circuit kVA is

$$\frac{\text{base kVA}}{\text{per unit } Z} = \frac{1000}{0.5141} = 1945 \text{ kVA}$$

and the total symmetrical rms short-circuit current is

$$\frac{\text{kVA} (1000)}{E_{L-L}} = \frac{1945 \times 1000}{240} = 8104 \text{ A}$$

For a line-to-line fault across the 120 V secondary of the 75 kVA transformer, the transformer resistance and reactance values are modified to compensate for the half-winding

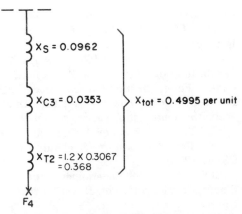

Fig 99
Resistance Network for Fault at F_4

Fig 100
Reactance Network for Fault at F_4

effect. On the same 75 kVA base rating, impedances of one 120 V winding are obtained from those of the 240 V winding using a resistance multiplier of approximately 1.5 and a reactance multiplier of approximately 1.2. These multipliers are typical for a single-phase distribution class transformer. However, for greater accuracy, the transformer manufacturer should be consulted.

For a fault at F_4 the resistance and reactance diagrams are shown in Figs 99 and 100. The total impedance Z is

$$Z = \sqrt{(0.3488)^2 + (0.4995)^2} = 0.6092 \text{ per unit}$$

the total short-circuit kVA is

$$\frac{\text{base kVA}}{\text{per unit } Z} = \frac{1000}{0.6092} = 1642 \text{ kVA}$$

and the total symmetrical rms short-circuit current is

$$\frac{\text{kVA}(1000)}{E_{\text{L}-\text{L}}} = \frac{1642 \times 1000}{120} = 13\,683 \text{ A}$$

5.8 Calculation of Fault Currents for DC Systems. The calculation of direct-current fault values is essential in the design and application of distribution and protective apparatus used in direct-current systems. A knowledge of mechanical stresses imposed by these fault currents is also important in the installation of cables, buses, and their supports.

As in the application of alternating-current protective devices, the magnitude of the available direct-current short-circuit current is the prime consideration. Since high-speed or semi-high-speed direct-current protective devices can interrupt the flow of fault current before the maximum value is reached, the rate of rise of the fault current must be considered, along with the interruption time, in order to determine the maximum current that will actually be obtained. Lower speed protective devices will generally permit the maximum value to be reached before interruption.

The sources of direct fault currents are
(1) Generators
(2) Synchronous converters
(3) Motors
(4) Electronic rectifiers
(5) Semiconductor rectifiers
(6) Batteries
(7) Electrolytic cells

Simplified procedures for the calculation of direct-current short-circuit currents are not well established, and therefore this chapter can only provide reference to publications containing helpful information (see IEEE Std 66-1969, Determination of Short-Circuit Characteristics of Direct-Current Machinery, and [2]−[6].)

5.9 Standards References. The following standards publications were used as references in preparing this chapter.

ANSI C37.5-1969 (R 1974), Methods for Determing Values of a Sinusoidal Current Wave, a Normal-Frequency Recovery Voltage, and a

Guide for Calculation of Fault Currents for Application of AC High-Voltage Circuit Breakers Rated on a Total Current Basis

ANSI C37.41-1969 (R 1974), Design Tests for Distribution Cutouts and Fuse Links, Secondary Fuses, Distribution Enclosed Single-Pole Air Switches, Power Fuses, Fuse Disconnecting Switches, and Accessories

IEEE Std 20-1973, Low-Voltage AC Power Circuit Breakers Used in Enclosures (ANSI C37.13-1973)

IEEE Std 66-1969, Determination of Short-Circuit Characteristics of Direct-Current Machinery

IEEE Std 320-1972, Application Guide for AC High-Voltage Circuit Breakers Rated on a Symmetrical Current Basis (ANSI C37.010-1972)

NEMA AB 1-1975, Molded-Case Circuit Breakers

NEMA SG 3-1975, Low-Voltage Power Circuit Breakers

5.10 References and Bibliography

5.10.1 *References*

[1] STEVENSON, W. D., JR. *Elements of Power System Analysis.* New York: McGraw-Hill, 1962.

[2] AIEE COMMITTEE REPORT. Protection of Electronic Power Converters. *AIEE Transactions*, vol 69, 1950, pp 813-829.

[3] CRITES, W. R., and DARLING, A. G. Short-Circuit Calculating Procedure for DC Systems with Motors and Generators. *AIEE Transactions (Power Apparatus and Systems)*, pt III, vol 73, Aug 1954, pp 816-825.

[4] DORTORT, I.K. Extended Regulation Curves for Six-Phase Double-Way and Double-Wye Rectifiers. *AIEE Transactions (Communications and Electronics)*, pt I, vol 72, May 1953, pp 192-202.

[5] DORTORT, I.K. Equivalent Machine Constants for Rectifiers. *AIEE Transactions (Communications and Electronics)*, pt I, vol 72, Sept 1953, pp 435-438.

[6] HERSKIND, C. C., SCHMIDT, A., JR., and RETTIG, C. E. Rectifier Fault Currents — II. *AIEE Transactions*, vol 68, 1949, pp 243-252.

5.10.2 *Bibliography*

[7] BEEMAN, D. L., Ed. *Industrial Power Systems Handbook.* New York: McGraw-Hill, 1955, chap 2.

[8] *Electrical Transmission and Distribution Reference Book.* East Pittsburgh, PA: Westinghouse Electric Corporation, 1964.

[9] GREENWOOD, A. Basic Transient Analysis for Industrial Power Systems. *Conference Record, 1972 IEEE Industrial and Commerical Power Systems and Electric Space Heating Joint Technical Conference*, IEEE 72CHO600-7-IA, pp 13-20.

[10] REED, M. B. *Alternating Current Circuit Theory*, 2nd ed. New York: Harper and Brothers, 1956.

[11] ST. PIERRE, C. R. *Time-Sharing Computer Programs (DATUM$) for Power System Data Reduction.* Schenectady, NY: General Electric Company, 1973.

[12] WAGNER, C. F., and EVANS, R. D. *Symmetrical Components.* New York: McGraw-Hill, 1933.

6. Grounding

6.1 Introduction. All phases of the subject of grounding applicable to the scope of the IEEE Industrial and Commercial Power Systems Committee have been studied and documented in IEEE Std 142-1972, Grounding of Industrial and Commercial Power Systems (ANSI C114.1-1973), and that standard will constitute the basic source of technical guidance for this chapter.

This chapter, therefore, will simply identify and discuss those facets of the grounding technology which relate to industrial plants. The subject matter will be subdivided into the following four categories:

(1) System grounding
(2) Equipment grounding
(3) Static and lightning protection grounding
(4) Connection to earth

6.2 System Grounding. Electric power distribution system grounding is concerned with the nature and location of an intentional electric interconnection between the electric system conductors and ground (earth). The common classifications of grounding to be found in industrial plant power distribution systems are

(1) Ungrounded
(2) Resistance grounded
(3) Reactance grounded
(4) Solidly grounded

The nature of electric system grounding may have a striking effect on the magnitude of line-to-ground voltages which must be endured under both steady-state and transient conditions. Electric systems which allow severe overvoltage can expect reduced useful life of insulation, which will appear as frequent insulation failures (circuit faults). In rotating electric machines where insulation space is limited, this conflict between voltage stress and the useful life is particularly acute.

In addition to the control of system overvoltages, intentional electric system neutral grounding makes possible sensitive and speedy fault protection based on detection of ground-current flow. Grounded systems in most cases are arranged so that circuit protective devices will remove a faulty circuit from the system regardless of the type of fault. Any contact from phase to ground in the grounded system thus results in immediate isolation of the faulty circuit and its loads. However, the experience of some engineers has been that greater service continuity can be obtained with grounded-neutral than with ungrounded-neutral systems. Furthermore, a very high order of rotating machine fault protection may be acquired by a simple inexpensive ground overcurrent relay. It will likely be found that the protective qualities of rotating machine differential protection will be enhanced by grounding the power supply system.

The following practice is recommended for establishing the system grounding connection.

(1) Systems used to supply phase-to-neutral loads that must be solidly grounded:

   120/240 V, single-phase, three-wire

   208Y/120 V, three-phase, four-wire

   480Y/277 V, three-phase, four-wire

(2) Systems that should be resistance grounded:

   480 V, three-phase, three-wire

   600 V, three-phase, three-wire

   2400 V, three-phase, three-wire

   4160 V, three-phase, three-wire

   6900 V, three-phase, three-wire

   13 800 V, three-phase, three-wire

There is divided opinion among engineers as to the seriousness of the overvoltage problem on ungrounded systems (600 V and less), and the likelihood of its affecting the electric-service continuity. Some engineers feel that service continuity is improved, and insulation failures are reduced, by using the grounded system. Others feel that, under proper operating conditions, the ungrounded system offers an added degree of service continuity not jeopardized by any serious likelihood of dangerous transient overvoltages. A detailed discussion of the factors influencing a choice of the grounded or ungrounded system is given in Chapter 1 of IEEE Std 142-1972.

The (so-called) ungrounded system is actually high-reactance capacitively grounded as a result of the capacitance coupling to ground of every energized conductor. The operating advantage sometimes claimed for the ungrounded system stems from the fact that a single line-to-ground fault, if sustained, will not result in an automatic tripout of the circuit. This results merely in the flow of a small charging current to ground. It is quite generally conceded that this practice introduces potential hazards to other apparatus [1].

For the duration of the ground fault on one conductor, the other two phase conductors throughout the entire metallic system are subjected to 73 percent overvoltage. It is, therefore, extremely important to locate the faulty circuit promptly and repair or remove it before the abnormal voltage stresses produce breakdown on other machines or circuits. Because of the capacitance coupling to ground, the ungrounded system is subject to dangerous overvoltages (five times normal or more) as a

result of an intermittent contact ground fault (arcing ground), or a high inductive reactance connected from one line to ground. So long as no disturbing influences occur on the system, the line-to-ground potentials (even on an ungrounded system) remain steady at about 58 percent of the line-to-line voltage value. Accumulated operating experience indicates that, in general-purpose industrial power distribution systems, the overvoltage incidents associated with ungrounded operation diminish the useful life of insulation in such a way that electric circuit and machine failures occur more frequently than they do on grounded systems. The advantage of an ungrounded system, in not immediately dropping load upon the occurrence of a ground fault, may be largerly destroyed by the practice of ignoring a ground fault, and allowing it to remain on the system until a second one occurs causing a power interruption. An adequate detection system together with an organized program for removing ground faults is considered essential for operation of the ungrounded system. These observations are limited to alternating-current systems. Direct-current system operation is not subject to many of the overvoltage hazards present in alternating-current systems.

Resistance-grounded systems employ an intentional resistance connection between the electric system neutral and ground. This resistance appears in parallel with the system-to-ground capacitive reactance and makes this circuit behave more like a resistor than a capacitor. Even a high-resistance connection ($R \leqslant X_{\text{co}}/3$, where R is the intentional resistance between the electric system neutral and ground, and $X_{\text{co}}/3$ is the total system-to-ground capacitive reactance) will suffice to curb the overvoltage-producing tendencies of a pure capacitively grounded system. A low-resistance connection will exercise a rigid control of line-to-ground potentials and also make available a substantial (controlled) line-to-ground fault current for securing selective ground-fault relaying.

High-resistance grounding, for effective control of the severe transitory overvoltages, should introduce between the electric system and ground a resistance whose ohmic value is of the same order as (or lower than) the total system-to-ground capacitive reactance ($X_{\text{co}}/3$).

This will limit to moderate value the overvoltages created by an inductive reactance connection from one phase to ground or from an intermittent-contact line-to-ground short circuit. It will not avoid the sustained 73 percent overvoltage on two phases during the presence of a ground fault on the third phase. Nor will it have much effect on a low-impedance overvoltage source such as an interconnection with conductors of a higher voltage system, a ground fault on the outer end of an extended winding transformer or stepup autotransformer, or a ground fault at the transformer—capacitor junction connection of a series capacitor welder.

Low-resistance grounding requires a grounding connection of very much lower resistance. The resistance value is tailored to provide a ground-fault current acceptable for relaying purposes. Typical current values used range from 400 A on modern systems using sensitive window current transformer ground-sensor relaying up to perhaps 2000 A in the larger systems using ground-responsive relays connected in current transformer residual circuits. In mobile electric shovel application, much lower levels of ground-fault current (50—25 A) are dictated by the acute shock-hazard considerations.

Reactance-grounded systems are not ordinarily employed in industrial power systems. The permissible reduction in available ground-fault current without risk of transitory overvoltages is limited. The criterion for curbing the overvoltages is that the available ground-fault current be at least 25 percent of the three-phase fault current ($X_0/X_1 \leqslant 10$, where X_0 is the zero-sequence inductive reactance and X_1 the positive-sequence inductive reactance of the system). The resulting fault current can be high and present an objectionable degree of arcing damage at the fault, leading to a preference for resistance grounding. Much greater reduction in fault-current value is permissible with resistance grounding without risk of overvoltage.

Solidly grounded systems exercise the greatest control of overvoltages but result in the highest magnitudes of ground-fault current. These high-magnitude fault currents may introduce new problems and intensify design problems in the equipment grounding system.

Solidly grounded systems are used extensively at operating voltages of 600 V and less. A large-magnitude available ground-fault current is desirable to secure optimum performance of phase-overcurrent trips or interrupters. The low line-to-neutral driving voltage of the supply system (346 V in the 600 V system and 277 V in the 480 V system) lessens the likelihood of dangerous voltage gradients in the ground return circuits even when higher than normal ground-return impedances are present.

6.2.1 *System-Grounding Design Deviations.* The intent of the preceding advisory recommendations is to promote broad application of the fewest variety of system-grounding patterns that will satisfy the operational requirements of industrial plant electric power systems in general. Even minor deviations in design practice within a particular variety are to be avoided as far as possible. Nonetheless, it is admitted that the list of recommended patterns is not all inclusive, and hence is not to be regarded as mandatory.

Local problems within individual electric power systems will evolve here and there, which will justify a deviation from the patterns listed in previous paragraphs. Circumstances can arise which may well justify "solid grounding" with circuit patterns other than named in the table of recommendations. For example, in case the power supply is obtained from the utility company via feeders from a 4.16Y/2.4 kV solidly grounded substation bus, the user will be justified in adopting that pattern. Further, should the user desire to serve 120 V single-phase one-side-grounded load circuits, there could be firm justification for solidly grounding the midpoint of one phase of a 240 V delta system to obtain a 240 V three-phase four-wire delta pattern. Also, the use of a 120 V three-phase delta system for general-purpose power could well justify solid corner-of-the-delta grounding.

In designing the electric power supply system to serve electrically operated excavating machinery, the existence of a greatly accentuated degree of electric-shock-voltage exposure may justify the use of a system-grounding pattern employing a 25 A resistive grounding connection (to establish a 25 A level of available ground-fault current). The achievement of security from dangerous electric-shock injury for operators and bystanders may be judged to over-shadow the reduction in rotating-machine fault-detection sensitivity which is sacrificed. This illustrates the economic tradeoff function.

There may be sound justification for the insertion of a reactor in the neutral connection of a generator which is to be connected to a solidly grounded three-phase four-wire distribution system in order to avoid excessive generator-winding current in response to a line-to-ground fault on the system. This connection should not be referred to as a "reactance-grounded" pattern, even though it might so appear when viewed from the generator unit.

The foregoing examples illustrate clearly the need for design flexibility to tailor the system grounding pattern to cope with the unique and the unusual. However, the decision to deviate from the advisory recommendations should be based on a specific engineering evaluation of a need for that deviation (in contrast with a mere desire to do things in a different way).

6.3 Equipment Grounding. Equipment grounding pertains to that system of electric conductors by which all metallic structures through which energized conductors run will be interconnected.

The main purposes of equipment grounding are as follows.

(1) To maintain low potential difference between nearby metallic members in any area to ensure freedom from electric-shock hazard to personnel in the area [2].

(2) To provide an effective electric conductor system over which short-circuit currents involving ground can flow without sparking or other evidence of thermal distress to avoid fire hazard to combustible material or gas ignition in combustible atmospheres [3].

All electric conductor housings, equipment enclosures, and motor frames shall be interconnected by an equipment grounding system which will satisfy the foregoing requirements. The rules for achieving these objectives are given in NFPA No 70, National Electrical Code (1975), (ANSI C1-1975), Article 250, and in ANSI C2-1973, National Electrical Safety Code.

In case of an insulation failure along a power conductor of an electric power circuit, allowing an electric connection between the energized conductor and some portion of a metal enclosure, there exists the tendency to impart to the metal enclosure the same electric potential as exists on the power conductor. Unless all such conductive enclosures have been intentionally grounded in an approved manner, the occurrence of an insulation breakdown on the power conductor may cause to appear on such an enclosure a voltage of sufficient magnitude to constitute a dangerous electric-shock hazard to anyone who touches it. A round-about grounding connection to the enclosure may be insufficient to avoid dangerous electric-shock hazard, yet it may permit a substantial amount of ground-fault current to flow. The heat released consequently can be responsible for a definite fire, or explosion hazard in addition.

Only by intentionally grounding the metallic enclosures in a manner which assures the presence of adequate current-carrying capability and an adequately low value of ground-fault circuit impedance can both electric-shock hazard and fire hazard be avoided. In many application areas the applicable electrical codes prescribe such grounding practices as a mandatory requirement.

Industry electric accident statistics compiled by the State of California indicate that a great many personnel injuries are occurring as a result of electric shock when making contact with metallic parts that are normally not energized and should have been expected to remain unenergized. Effective equipment grounding practices would eliminate these personnel injuries.

Fire insurance companies, when presenting summaries of industrial experience, have indicated that, in their judgment, approximately one out of every seven fires in industrial establishments owes its origin to the electric system. While these reports undoubtedly contain some unjustified assignments under the category of defective wiring, nonetheless there is reason to suspect that difficulties in electric system operation are responsible for a greater number of fires than would be first imagined. Perhaps the development and adoption of more effective practices in equipment grounding systems can effect a marked reduction in fire hazard.

Recent technical investigations [3] [4] point out the necessity of making good electric junctions between sections of conduit or metal raceways and ensuring adequate cross-sectional area and conductivity of these bonds. In power circuit wiring associated with electric conductors of no. 1/0 cross section or larger, there is unmistakable evidence that the short-circuit current flowing by way of ground will seek a path close to the outgoing conductor that will usually consist of the conduit or the metal enclosing raceway. This is true even though the conduit or raceway may be repeatedly bonded to heavy steel building members or nearby piping systems. In order to provide an effective current-carrying path, either the cross-sectional area of the conduit or raceway must be increased, or a paralleling grounding conductor must be run inside the conduit or raceway. Paralleling conductors or circuits running external to the conduit or raceway are quite ineffective in shunting short-circuit currents away from the metallic conductor enclosure.

Preferred methods of grounding the following types of equipment are given in detail in IEEE Std 142-1972, Chapter 2:

(1) Structures
(2) Outdoor stations
(3) Large generators and motor rooms
(4) Conductor enclosures
(5) Miscellaneous motors
(6) Portable equipment

6.4 Static and Lightning Protection Grounding

6.4.1 *Static Grounding.* Industrial plants handling solvents, dusty materials, or other flammable products often have a potentially hazardous operating condition because of static electricity accumulating on equipment, on materials being handled, or even on operating personnel.

The discharge of a static charge to ground or to other equipment in the presence of flammable or explosive materials is often the cause of fires and explosions which result in the loss of many lives each year and an accompanying large property loss.

The simple expedient of grounding equipment is not always the solution of the problem. Each installation must be studied in order that an adequate method of control may be selected.

The protection of human life is a prime objective in the control of electrostatic charges.

In addition to the direct danger to life from explosion or fire created by an electrostatic discharge, there is the possibility of personal injury from being startled by an electric shock, which may, in turn, induce an accident, such as a fall from a ladder or platform.

Another objective in controlling static electricity is the avoidance of

(1) Investment losses, buildings, contained equipment, or stored materials
(2) Lost production

The avoidance of losses in this manner represents good insurance.

An additional need for effective electrostatic control may be that of improved product quality. For example, in grinding operations, an electrostatic charge may prevent the achievement of the degree of fineness desired in the finished product, or in certain textile operations, it may cause fibers to stand on end instead of lying flat, resulting in an inferior product.

Material handled by chutes or ducts have been known to accumulate static charges causing the material to cling to the inside surfaces of the chutes or ducts and, thus, clog them.

IEEE Std 142-1972, Chapter 3, contains a detailed treatment of the topics

(1) Fundamental causes of static
(2) Methods of testing for static
(3) Hazards in various types of industries
(4) Methods of static control

6.4.2 *Lightning Protection Grounding.* Lightning protection grounding is concerned with the conduction to earth of current discharges in the atmosphere originating in cloud formations. The function of the lightning grounding system is to convey these lightning discharge currents safely to earth without incurring damaging potential differences across electrical insulation in the industrial power system, without overheating lightning grounding conductors, and without the disrupting breakdown of air between the lightning ground conductors and other metallic members of the structure [2], [5].

Lightning represents a vicious source of overvoltage. It is capable of imparting a potential of one half million volts or more to a stricken object. The current in the direct discharge may be more than 100 000 A. The rate at which this current builds up may be as much as 10 000 A/µs.

The presence of such high-magnitude fast-rising surge current emphasizes the need for high discharge capability in surge arresters and low impedance in the connecting leads.

For example, should a direct stroke which has made contact with a rod or mast on an industrial building encounter an inductance of as little as 1 µH with a current buildup rate of 10 000 A/µs, there would result a 10 000 V potential drop across this inductance. It is this inductive voltage drop which is responsible for application rules which demand that the lightning down-leads either be bonded to the building structure or separated therefrom by several feet. Surge protective grounding circuits should be as short and direct as possible with no unnecessary turns or corners.

Dry-type transformers generally possess basic insulation level values below those of standard liquid-filled units of the same voltage class. The application of such transformers to industrial systems subject to surge voltages, requiring surge arresters for protection, warrants a careful check on protection adequacy. In some instances assured protection cannot be achieved unless the supply system is of solid-multi-grounded design to allow the use of grounded-neutral arresters. In less critical areas protection may be achieved with low-cost line-to-line rated surge arresters with a limit on the separation distance between the arrester and the transformer. The type and rating of the surge arrester needed to protect the transformer's basic insulation level, as influenced by the system-grounding design and surge arrester location restrictions, justifies a careful review.

The effective resistance of a grounding electrode to a surge current may be quite different from that determined by direct current or commercial-frequency alternating current. For example, an insulated conductor running parallel to a high-voltage line near ground level represents a usable grounding electrode for discharge of lightning currents, although a direct-current resistance measurement between this conductor and earth would show infinite resistance. An extreme in the other direction

might be represented by the running of a grounding cable, perhaps 500 ft, to metallic plates buried in a lake or pond. Such a circuit arrangement might show very low direct-current ground resistance, yet present a very high effective resistance to lightning discharge currents controlled by the characteristics of the long interconnecting grounding cable.

6.5 Connection to Earth

6.5.1 *General Discussion.* Connections to earth having acceptably low values of resistance are needed to discharge lightning stroke currents, dissipate the released bound charge resulting from nearby strokes, and drain off static voltage accumulations (IEEE Std 142-1972).

The presence of overhead high-voltage transmission circuits may introduce a requirement for a connection to earth to pass safely the ground-fault current which would result from a broken line conductor falling on some part of the building structure.

To a large extent the internal electric distribution system installed within commercial buildings and industrial plants is entirely enclosed in grounded metal. Except for cable tray systems, conductors are enclosed in conduit, metallic armor, or metal raceway. The other electric elements of equipments and machines can be expected to be encased in metal cabinets or metallic machine frames. All of these metallic enclosures and cable trays will be intentionally interconnected and, in turn, will be bonded to other metallic components within the area, such as building structural members, piping systems, messenger cables, etc. Thus the local electric system will be self-contained within its own shell of conducting metal and can be designed to operate adequately and safely without any connection to earth itself. This can be likened to the electric distribution system as installed on an airplane. The airplane structure constitutes an adequate ground system. No connection to earth is needed to achieve an adequate, safe electric system.

6.5.2 *Recommended Acceptable Values.* The most elaborate grounding system that can be designed may prove to be inadequate unless the connection of the system to the earth is adequate and has a low resistance [2]. The earth connection is one of the most important parts of the grounding system. It is also the most difficult part to design and to obtain.

The perfect connection to earth would have zero resistance, but this is impossible to obtain. Ground resistances of less than 1 Ω can be obtained, although this low a resistance may not be necessary. Since the resistance required varies inversely with the fault current to ground, the larger the fault current, the lower must be the resistance.

For larger substations and generating stations, the earth resistance should not exceed 1 Ω. For smaller substations and for industrial plants, in general, a resistance of less than 5 Ω should be obtained if practicable. The National Electrical Code, Section 250-84, approves the use of a single made electrode for the system grounding electrode, if its resistance does not exceed 25 Ω.

6.5.3 *Resistivity of Soils.* The resistivity of the earth is a prime factor in establishing the resistance of a grounding electrode. The resistivity of soil varies with the depth from the surface, with the moisture and chemical content, and with the soil temperature. For representative values of resistivity for general types of soils and the effects of moisture and temperature, see IEEE Std 142-1972, Chapter 4.

6.5.4 *Soil Treatment.* Soil resistivity may be reduced anywhere from 15 to 90 percent, depending upon the kind and texture of the soil, by chemical treatment. There are a number of chemicals suitable for this purpose, including sodium chloride, magnesium sulfate, copper sulfate, and calcium chloride. Common salt and magnesium sulfate are most commonly used.

Chemicals are generally applied by placing them in the circular trench around the electrode in such a manner as to prevent direct contact with the electrode. While the effects of treatment will not become apparent for a considerable period, they may be accelerated by saturating the area with water. Also, such treatment is not permanent and must be renewed periodically, depending on the nature of chemical treatment and the characteristics of the soil.

6.5.5 *Existing Electrodes.* All grounding electrodes fall into one of two catagories, (1) those

which are an inherent part of the establishment, and (2) those which have been "made" for electrical grounding purposes.

The National Electrical Code, Section 250-81, designates that underground metal water piping, available on the premises, be the preferred grounding electrode. This preference prevails regardless of length, except that when the effective length of buried pipe is less than 10 ft, it shall be supplemented with an electrode of the type named in Section 250-82 or 83.

For safety grounding and for small distribution systems where the ground currents are of relatively low magnitude, such buried metal-pipe electrodes are usually preferred because they are economical in first cost. However, before reliance can be placed on any electrodes of this group, it is essential that their resistance to earth be measured to ensure that some unforeseen discontinuity has not seriously affected their suitability. The use of plastic pipe in new, and of wooden ones in older, water systems may seriously impair its value as a grounding electrode. Even iron or steel piping may include gaskets which act as insulators. Sometimes small metal (brass) wedges are used to ensure electrical continuity. These wedges must be replaced when repairs are made. Interior piping systems which are likely to become energized must be bonded to the electric system grounding conductor. If the piping system contains a member designed to permit easy removal, a bonding jumper must be installed bridging the removable member.

6.5.6 *Concrete-Encased Grounding Electrodes.* Concrete below ground level is a good electrical conductor, as good as moderately low-resistivity earth. Consequently metal electrodes encased in such concrete will function as excellent grounding electrodes [5], [6], (see also National Electrical Code). In areas of poor soil conductivity, the beneficial effects of the concrete encasement are most pronounced.

To create a "made" electrode by encasement of a metal electrode in concrete would likely not be economically attractive; but most industrial establishments employ much concrete-encased metal below grade for other purposes. The reinforcing steel in concrete foundations and footings is a good example. The concrete encasement of steel, in addition to contributing to low grounding resistance, serves to immunize the steel against corrosive disintegration, such as would take place if the steel were in direct contact with the earth [5]. Even though copper and steel are in contact with each other within the bed of moist concrete, destructive disintegration of the steel member does not take place.

Steel reinforcing bars (re-bars) in foundation piers consist usually of groups of four or more vertical members held by horizontal spacer square "rings" at regular intervals. The vertical members are wired to heavy horizontal members in the spread footing at the base of the pier. Meausrements show that such a pier has an electrode resistance of about half the resistance of a simple ground rod driven to the same depth in earth. Electrical connection to the re-bar system is conveniently made by a bar welded to one vertical re-bar and a J-bolt for the column base plate. The J-bolt then becomes the electrode connection. A weld to a re-bar at a point where the bar is in appreciable tension is to be avoided.

Usually such footings appear every 15 to 20 ft in all directions in industrial buildings. A good rule of thumb for determining the effective overall resistance of the grounding mat is to divide the resistance of one typical footing by half the number of footings around the outside wall of the buildings. (Inner footings aid little in lowering the overall resistance.)

Copper cable embedded in concrete is similarly benefited, a fact that may be of particular value under circumstances of high earth resistivity.

The National Electrical Code, Sections 250-82 and 83, recognizes concrete-encased copper and steel as an effective grounding electrode under either the catagory of an "available" or a "made" electrode.

6.5.7 *Made Electrodes.* Made electrodes may be subdivided into driven electrodes, buried strips or cables, grids, buried plates, and counterpoises. The type selected will depend upon the type of soil encountered and the available depth. Driven electrodes are generally more satisfactory and economical where bedrock is 10 ft or more below the surface, while grids, buried strips, or cables are preferred for lesser

depths. Grids are frequently used for substations or generating stations to provide equipotential areas throughout the entire station where hazards to life and property would justify the higher cost. They also require the least amount of buried material per mho of ground conductance. Buried plates have not been used extensively in recent years because of the higher cost as compared to rods or strips. Also, when used in small numbers, they are the least reliable type of made electrode. The counterpoise is a form of the buried cable electrode, and its use is generally confined to locations having high-resistance soils, such as sand or rock, where other methods are not satisfactory.

6.5.8 *Galvanic Corrosion.* There has developed an increased awareness of possible aggravated galvanic corrosion of buried steel members if crossbonded to buried dissimilar metal, such as copper [7]−[9].

The result has been a trend to seek a design of electrical grounding electrode which is, galvanically, neutral with respect to the steel structure. In some cases the grounding electrode design employs steel-exposed metal electrodes with insulated copper cable interconnections [7].

The corrosion of buried steel takes place even in the absence of a crossconnection to buried dissimilar metal. The exposed surface of the buried steel inherently contains bits of dissimilar conducting material, foreign metal fragments, or slag inclusions, which create local galvanic cells and local circulating currents. At spots where current leaves the metal surface, metal ions leave the parent metal and account for destructive corrosion. The cross bonding to dissimilar metal may aggravate the rate of corrosion, but is not the only cause for the action.

Electrical engineering technology must recognize the problem and seek grounding electrode designs which will produce no observable increase in the rate of corrosive disintegration of nonelectrical buried metal members. An overriding priority dictates that the electrical grounding electrode itself not suffer destruction by galvanic corrosion. Relative economics will be an inevitable factor in the design choice.

The recent increase in the use of plastic pipes for water supply to buildings removes one of the most common sources of complaint. The absence of buried metal piping, however, demands that some other suitable grounding electrode be located, or created.

A timely release of new knowledge bearing on this problem is the electrical behavior pattern of concrete-encased metal below grade (see Section 6.5.6). The relationship to galvanic corrosion is (1) there is generally an extensive array of concrete-encased-steel reinforcing members within the foundations and footings, which collectively account for a huge total surface area, resulting in extremely low current density values at the steel surface, (2) the concrete-encased re-bars themselves constitute an excellent permanent low-resistance earthing connection with little or no economic penalty, and (3) current flow across the steel—concrete boundary does not disintegrate the steel as it would if the steel were in contact with earth.

A discussion on methods of calculating and measuring resistance to earth, current loading capacity of soils, as well as recommended methods and techniques of constructing connections to earth may be found in Chapter 4 of IEEE Std 142-1972.

6.6 Standards References. The following standards publications were used as references in preparing this chapter.

ANSI C2 (1977 ed), National Electrical Safety Code

IEEE Std 80-1976, Safety in AC Substation Grounding

IEEE Std 142-1972, Grounding of Industrial and Commercial Power Systems (ANSI C114.1-1973)

NFPA No 70, National Electrical Code (1975), (ANSI C1-1975)

NFPA No 78, Lightning Protection Code (1968), (ANSI C5.1-1969)

6.7 References and Bibliography

6.7.1 *References*

[1] BEEMAN, D. L., Ed. *Industrial Power Systems Handbook.* New York: McGraw-Hill, 1955, chap 5—7.

[2] AIEE COMMITTEE REPORT. Voltage Gradients Through the Ground Under Fault Conditions. *AIEE Transactions (Power Apparatus and Systems),* pt III, vol 77, Oct 1958, pp 669-692.

[3] KAUFMANN, R. H. Some Fundamentals of Equipment-Grounding Circuit Design. *AIEE Transactions (Applications and Industry),* pt II, vol 73, Nov 1954, pp 227-232.

[4] BISSON, A. J., and ROCHAU, E. A. Iron Conduit Impedance Effects in Ground Circuit Systems. *AIEE Transactions (Applications and Industry),* pt II, vol 73, Jul 1954, pp 104-107.

[5] FAGAN, E.J., and LEE, R. H. The Use of Concrete-Encased Reinforcing Rods as Grounding Electrodes. *IEEE Transactions on Industry and General Applications,* vol IGA-6, Jul/Aug 1970, pp 337-348.

[6] WIENER, P. A Comparison of Concrete-Encased Grounding Electrodes to Driven Ground Rods. *IEEE Transactions on Industry and General Applications,* vol IGA-6, May/Jun 1970, pp 282-287.

[7] COLMAN, W. E., and FROSTICK, H. G. Electrical Grounding and Cathodic Protection at the Fairless Works. *AIEE Transactions (Applications and Industry),* pt II, vol 74, Mar 1955, pp 19-24.

[8] HERTZBERG, L. B. The Water Utilities Look at Electrical Grounding. *IEEE Transactions on Industry and General Applications,* vol IGA-6, May/Jun 1970, pp 278-281.

[9] ZASTROW, O. W. Underground Corrosion and Electrical Grounding. *IEEE Transactions on Industry and General Applications,* vol IGA-3, May/Jun 1967, pp 237-243.

6.7.2 *Bibliography*

[10] BRERETON, D.S., and HICKOCK, H. N. System Neutral Grounding for Chemical Plant Power Systems. *AIEE Transactions (Applications and Industry),* pt II, vol 74, Nov 1955, pp 315-320.

[11] BULLARD, W. R. Grounding Principles and Practice. Part IV — System Grounding. *Electrical Engineering,* vol 64, Apr 1945, pp 145-151.

[12] DALZIEL, C. F. Effects of Electric Shock on Man. *IRE Transactions on Medical Electronics,* vol PGME-5, Jul 1956, pp 44-62.

[13] GIENGER, J. A., DAVIDSON, O. C., and BRENDEL, R. W. Determination of Ground Fault Current on Common AC Grounded-Neutral Systems in Standard Steel or Aluminum Conduit. *AIEE Transactions (Applications and Industry),* pt II, vol 79, May 1960, pp 84-90.

[14] HARVIE, R. A. Hazards During Ground Faults on 480-V Grounded Systems. *IEEE Transactions on Industry Applications*, vol IA-10, Mar/Apr 1974, pp 190-196.

[15] JENSEN, C. Grounding Principles and Practice. Part II — Establishing Grounds. *Electrical Engineering*, vol 64, Feb 1945, pp 68-74.

[16] JOHNSON, A. A. Grounding Principles and Practice. Part III — Generator-Neutral Grounding Devices. *Electrical Engineering*, vol 64, Mar 1945, pp 92-99.

[17] KAUFMANN, R. H. Let's Be More Specific About Equipment Grounding. *Proceedings of the American Power Conference*, vol 24, 1962, pp 913-922.

[18] THACKER, H. B. Grounded Versus Ungrounded Low-Voltage Alternating Current Systems. *Iron and Steel Engineer*, Apr 1954.

[19] WEST, R. B. Equipment Grounding for Reliable Ground-Fault Protection in Electrical Systems Below 600 V. *IEEE Transactions on Industry Applications*, vol IA-10, Mar/Apr 1974, pp 175-189.

7. Power Factor

7.1 General. This chapter provides some basic information about power-factor improvement, including common and important applications. The approach is to cover the necessary fundamentals and include selected references for appropriate detailed information on switching, harmonics, resonance, automatic control, measurements, metering arrangements, and capacitor protection. High-frequency systems and series capacitors are excluded because their applications are limited.

7.1.1 *Emphasis on Capacitors.* Adding capacitors is generally the most economical way to improve the plant power factor, especially in existing plants. Therefore the application emphasis will be on capacitors. However, there are some cases where synchronous motors are most economical, and they will also be covered.

In addition to their relatively low cost, capacitors have other properties, such as ease of installation, practically no maintenance, and very low losses. Capacitors are manufactured in relatively small ratings (kvar) for economical manufacturing and engineering reasons. The individual units can, however, be combined into suitable banks to obtain a large range of ratings. Thus capacitors can be added in small or large units to meet current operating requirements and avoid a larger than necessary installation in anticipation of increased future requirements.

7.1.2 *Benefits of Power-Factor Improvement.* Most benefits provided by a power-factor improvement stem from the reduction of reactive power in the system. This may result in (1) lower purchased-power costs if the utility enforces a power-factor clause, (2) release of system electrical capacity, (3) voltage improvement, and (4) lower system losses. Maximum benefits are obtained when capacitors or synchronous motors are located at low-power-factor loads.

Although reducing the power bill is still the primary reason for improving the power factor, and it is becoming more important because of conservation of energy, the function of releasing system capacity is sometimes the decisive factor.

7.2 Typical Plant Power Factor. The unimproved power factor in plants depends on equipment design and operating conditions. Thus in many cases it is not possible to predict accurately what should be expected in a new plant. The typical figures listed in this section are drawn from operating experience.

7.2.1 *General Industry Applications.* The power-factor values in Table 48 are by industry category and are representative of the nor-

Table 48
Typical Unimproved Power-Factor Values
by Industries

Industry	Power Factor
Auto parts	75—80
Brewery	75—80
Cement	80—85
Chemical	65—75
Coal mine	65—80
Clothing	35—60
Electroplating	65—70
Foundry	75—80
Forge	70—80
Hospital	75—80
Machine manufacturing	60—65
Metalworking	65—70
Office building	80—85
Oil-field pumping	40—60
Paint manufacturing	55—65
Plastic	55—70
Stamping	60—70
Steel works	65—80
Textile	65—75
Tool, die, jig	60—65

Table 49
Typical Operating Power-Factor Values
by Plant Operations

Operation	Power Factor
Air compressor	
external motors	75—80
hermetic motors	50—80
Metalworking	
arc welding	35—60
with commercially	
furnished capacitors	70—80
Machining	40—65
Melting	
arc furnace	75—90
induction furnace, 60 Hz	100
Stamping	
standard	60—70
high speed	45—60
Spraying	60—65
Welding	
arc	35—60
resistance	40—60
Weaving	
individual	60
multiple	70
brind	70—75

mal part of industrial or commercial equipment assemblage. The values can be expected to change in some cases where solid-state converters are used.

7.2.2 *Plant Applications.* The values in Table 49 are those that may be expected per shift. The overall plant power factor will be different if there are partly loaded motors or idle shifts.

7.2.3 *Utilization Applications*

(1) *Motors.* The power factor of a partly loaded induction motor is poor, as indicated in Fig 108. This figure also shows the attractive improvement over the entire load range by using a capacitor of proper rating.

The T-frame motor, available since about 1964, generally has a lower power factor than the U-frame design. The comparison is shown in Fig 109.

Hermetic and wound-rotor type motors have a lower operating power factor than other induction motors of the same power and speed ratings.

(2) *Rectifiers*

(a) *Diode type with no phase control.* Small single-phase units have about 50 percent power factor at full load, while larger multiphase units may have 95—98 percent power factor.

(b) *Thyristor drives.* The power factor is roughly proportional to the ratio of direct-current output voltage to rated voltage. At partial loads the power factor is poor. A step-by-step procedure for determining the capacitor rating required for power-factor improvement is given in [1].

(3) *Electric Furnaces.* Arc furnaces have a poor power factor, typically 65—75 percent. Power-factor improvement may be a system problem. Induction furnaces have a power factor of typically 30—70 percent; switched capacitors are used to maintain near unity power factor.

(4) *Lamps.* Incandescent lamps operate at unity power factor. Fluorescent and other discharge lamps have a poor operating power factor of approximately 70 percent. With a corrected ballast the power factor will range from about 90 percent lagging to slightly leading.

(5) *Transformers.* These are not ordinarily considered to be loads, but they do contribute to lowering the system power factor. The trans-

former exciting current is usually 1—2 percent of the transformer rating in kilovolt-amperes and is independent of load. Reactive power is also required by the transformer leakage reactance. Such power varies as the square of the load current. At rated current the leakage reactance requires reactive power equal in magnitude to the transformer rating in kilovolt-amperes times the nameplate impedance in per unit.

7.3 Instruments and Measurements for Power-Factor Studies. When power-factor studies are made, sufficient data should be obtained to select the proper rating and location of capacitors or synchronous motors. If the study is for power billing rate purposes, then the power bills usually furnish sufficient information to determine the capacitor rating required.

Most utility rates include a demand charge, usually obtained from a demand register attachment on the watthour meter or by recording or printing-type instruments.

The power factor may be measured directly or obtained from other indications such as kilowatt, kilovolt-ampere, or kilovar readings. Average values may be obtained from kilowatt-hour, kilovar-hour, or kilovolt-ampere-hour readings.

Measurements by recording or graphic instruments are most desirable and useful because they provide a permanent record.

Indicating instruments are satisfactory for spot checking individual feeder circuits or loads. They can be used also in place of recording instruments if readings are taken at frequent intervals.

The preferred measurements are power in kilowatts, reactive power in kilovar, and voltage in volts, from which the apparent power in kilovolt-amperes and the power factor can be calculated. Voltage readings are especially desirable if automatic capacitor control with a voltage-responsive master element is contemplated. Direct measurement of the power factor is not particularly desirable, since power-factor interpretation may be misleading; for example, even at 95 percent power factor, the reactive power component (in kilovar) of the load is 33 percent as great as the active power component (in kilowatts).

When a power-factor meter is used, it is generally a polyphase instrument. Even then, power-factor readings are accurate only on balanced loads and become increasingly inaccurate as the load unbalance increases.

For metering arrangement and connection diagrams refer to [2, chap 3].

If switchboard meters are provided, their readings may be taken as applying to that portion of the load that each serves.

Readings from such instruments often disclose great variations in the power factor between different parts of a plant. Such knowlege can be valuable in locating equipment for power-factor improvement to best advantage.

7.4 Basic Power-Factor Economics. High plant power factors can yield direct savings. Some, such as reduced power bills and release of system capacity, are quite visible; others, such as decreased I^2R losses, are less visible but nonetheless real as are many indirect savings as a result of more efficient performance.

The cost of improving the power factor in existing plants, and of maintaining proper levels as load is added, depends on the power-factor value selected and upon the equipment chosen to supply the compensating reactive power. In general, medium-voltage capacitors cost less per kilovar than low-voltage capacitors. However, the cost of installation may change the comparison. Industrial capacitor additions should give a large payback through power-bill savings in two to four years.

The combination of reduced power billing and released system capacity by improving the power factor is very attractive economically. Fig 101 is a nomogram that gives the power factor at which substation equipment costs S and capacitor costs C on an installation basis are equal. Thus only relative cost values S/C are necessary in exploring the economic application. Later absolute values can be used for an accurate comparison (see [3]).

Some utility rates specify a minimum operating power factor which may be as high as 90 or 95 percent.

Optimizing the system power-factor level consists of combining an engineer's sense of system operating efficiency with a businessman's sense of investment profitability.

7.5 Power-Factor Fundamentals. Most utilization devices require two components of current.

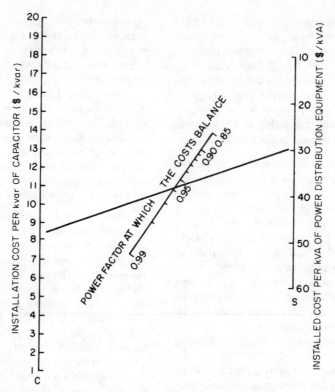

Fig 101
Nomogram for Determining Cost Balance for
Power-Factor Improvement

(1) The power-producing current or working current is that current which is converted by the equipment into useful work, usually in the form of heat, light, or mechanical power. The unit of measurement of power is the watt.

(2) Magnetizing current, also known as wattless, reactive, or nonworking current, is the current required to produce the flux necessary to the operation of electromagnetic devices. Without magnetizing current, energy could not flow through the core of a transformer or across the air gap of an induction motor. The unit of measurement of reactive power is the var.

The normal phasor relationship of these two components of current to each other, to the total current, and to the system voltage is illustrated in Fig 102. It shows that the active current and the reactive current add vectorically to form the total current which can

be determined from the expression

total current

$$= \sqrt{(\text{active current})^2 + (\text{reactive current})^2}$$

(Eq 1a)

$$I = \sqrt{(I \cos \phi)^2 + (I \sin \phi)^2} \qquad \text{(Eq 1b)}$$

At a given voltage V, the active, reactive, and apparent power are proportional to current and are related as follows:

apparent power

$$= \sqrt{(\text{active power})^2 + (\text{reactive power})^2}$$

(Eq 2a)

$$VI = \sqrt{(VI \cos \phi)^2 + (VI \sin \phi)^2} \qquad \text{(Eq 2b)}$$

As evidenced when Fig 102 is compared with Fig 103, the phasor diagram for power is similar in form to that for current.

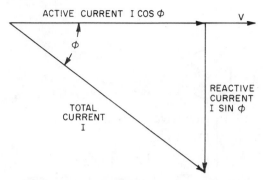

Fig 102
Angular Relationship of Current and Voltage in Alternating-Current Circuits

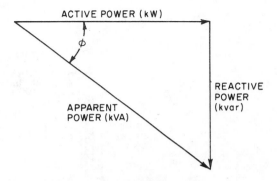

Fig 103
Relationship of Active, Reactive, and Total Power

7.5.1 *Definition of Power Factor.* The power factor in this text is defined as the ratio of active power to apparent power in a circuit. It varies from one to zero, but is generally given in percent:

$$\text{power factor} = \frac{\text{active power}}{\text{apparent power}} = \frac{kW}{kVA} \quad \text{(Eq 3a)}$$

power factor = cosine of angle between active power and apparent power

$$= \cos \phi \quad \text{(Eq 3b)}$$

active power = apparent power × power factor

$$= (kVA)(pf) \quad \text{(Eq 4)}$$

7.5.2 *Leading and Lagging Power Factor.* The power factor may be leading or lagging, depending on the direction of both the active and reactive power flows. If these flows are in the same direction, the power factor at that point of reference is lagging. If either power component flow is in an opposite direction, the power factor at that point of reference is leading. Since capacitors are a source of reactive power only, their power factor is always leading.

An induction motor has a lagging power factor as it requires both active and reactive power to flow into the motor (same direction). An overexcited synchronous motor can supply reactive power to the system. The active power component flows into the motor and the reactive power flows into the power system (opposite directions), so the power factor is leading. In an actual power system the system pow-

er factor may be lagging, even though some apparatus such as overexcited synchronous motors may have a leading power factor.

7.5.3 *How to Improve the Power Factor.* When the reactive power component in a circuit is reduced, the total current is reduced. If the active power component does not change, as is usually true, the power factor will improve as the reactive power component is reduced. When the reactive power component becomes zero, the power factor will be unity or 100 percent.

This is shown pictorially in Fig 104. The load requires an active current of 80 A, but because the motor requires a reactive current of 60 A, the supply circuit must carry the vector sum of these two components, which is 100 A. After a capacitor is installed to supply the motor reactive-current requirements, the supply circuit needs to deliver only 80 A to do exactly the same work. The supply circuit is now carrying only active power, so no system capacity is wasted in carrying reactive power.

Thus for all practical purposes, the only way to improve the power factor is to reduce the reactive power component. This is usually done with capacitors. Synchronous motors may also be used for power-factor improvement. The reactive power output that they are capable of supplying to the line is a function of excitation and motor load. The curves of Fig 105 show the reactive power that a typical synchronous motor is capable of delivering under various load conditions with normal excitation.

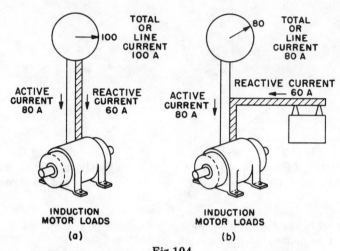

Fig 104
Schematic Arrangement Showing How Capacitors Reduce
Total Line Current by Supplying Reactive Power Requirements Locally

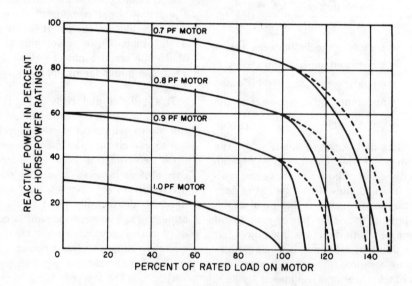

Fig 105
Leading Reactive Power in Percent of Motor Horsepower Ratings
for Synchronous Motors at Part Load and at Various Power-Factor Ratings
Solid Lines Are Based on Reduction in Excitation at Overload to Maintain Rated
Full-Load Current; Dashed Lines Represent Rated Excitation to
Maintain Rated Pullout Torque

At high overloads a synchronous motor may take reactive power from the line.

7.5.4 *Calculation Methods for Power-Factor Improvement.* From the right triangle relationship of Fig 102 several simple and useful mathematical expressions may be written:

$$\cos \phi = \frac{\text{active power}}{\text{apparent power}} = \frac{\text{kW}}{\text{kVA}} \quad \text{(Eq 3)}$$

$$\tan \phi = \frac{\text{reactive power}}{\text{active power}} = \frac{\text{kvar}}{\text{kW}} \quad \text{(Eq 5)}$$

$$\sin \phi = \frac{\text{reactive power}}{\text{apparent power}} = \frac{\text{kvar}}{\text{kVA}} \quad \text{(Eq 6)}$$

Because the active power component usually remains constant and the apparent power and reactive power components change with the power factor, the expression involving the active power component is the most convenient to use. This expression may be rewritten as

$$\text{reactive power} = \text{active power} \times \tan \phi \quad \text{(Eq 7a)}$$

$$\text{kvar} = (\text{kW})(\tan \phi) \quad \text{(Eq 7b)}$$

where the value of $\tan \phi$ corresponds to the power factor angle.

For example, assume that it is necessary to determine the capacitor rating to improve the load power factor:

$$\begin{array}{l}\text{reactive power at} \\ \text{original power factor}\end{array} = \text{active power} \times \tan \phi_1$$

$$= (\text{kW})(\tan \phi_1) \quad \text{(Eq 8)}$$

$$\begin{array}{l}\text{reactive power at} \\ \text{improved power factor}\end{array} = \text{active power} \times \tan \phi_2$$

$$= (\text{kW})(\tan \phi_2) \quad \text{(Eq 9)}$$

where ϕ_1 is the angle of the original power factor and ϕ_2 is the angle of the improved power factor. Therefore the capacitor rating required to improve the power factor is

$$\text{reactive power} = \text{active power}$$

$$\times (\tan \phi_1 - \tan \phi_2) \quad \text{(Eq 10a)}$$

$$\text{kvar} = (\text{kW})(\tan \phi_1 - \tan \phi_2) \quad \text{(Eq 10b)}$$

For simplification $(\tan \phi_1 - \tan \phi_2)$ is often written as $\Delta\tan$. Therefore,

$$\text{reactive power} = \text{active power} \times \Delta\tan \quad \text{(Eq 11a)}$$

$$\text{kvar} = (\text{kW})(\Delta \tan) \quad \text{(Eq 11b)}$$

All tables, charts, and curves which have a kW multiplier for determining the reactive power requirements are based on Eq 11 (see Table 50).

Example. Using Table 50, find the capacitor rating required to improve the power factor of a 500 kW load from 0.76 to 0.93:

$$\text{kvar} = \text{kW} \times \text{multiplier}$$

$$= 500 \times 0.46$$

$$= 230$$

7.5.5 *Location of Reactive Power Supply.* The benefits obtained by installing capacitors or synchronous machines for power-factor improvement result from the reduction of reactive power in the system. They should in general be installed as close to the load as practical. Fig 106 shows four possible capacitor locations. The most desirable location for power-factor improvement is C_1, then, in the following order, C_2 (typically a plug-in busway), C_3, and C_4. The same principle applies to the location of synchronous motors as far as power-factor improvement is concerned.

Economics must be considered when determining the capacitor location. The cost of a switching device, where required, should be included in the cost comparison.

It is common practice to connect capacitors ahead of individual motors as shown in location C_1 in Fig 106. This provides power-factor improvement at the load and permits switching the capacitor and motor as a unit. More details are given in Section 7.9.

Power-factor improvement for small loads, or for those units that for some other reason may not lend themselves to having capacitors directly associated with the load, may be accomplished by connecting capacitors at a distribution point such as a panel-board or plug-in busway at location C_2, or at a substation at location C_3. Automatic switching, controlled either by electrical sensors or by interlocks with other equipment, has been successfully used at both locations C_2 and C_3.

A combination of capacitors at locations C_1 and C_2, and with bus-switched units as at location C_3, may prove the economical overall choice.

Table 50
kW Multipliers to Determine Reactive-Power Requirements for Power-Factor Improvement

Original Power Factor	Corrected Power Factor																				
	0.80	0.81	0.82	0.83	0.84	0.85	0.86	0.87	0.88	0.89	0.90	0.91	0.92	0.93	0.94	0.95	0.96	0.97	0.98	0.99	1.0
0.50	0.982	1.008	1.034	1.060	1.086	1.112	1.139	1.165	1.192	1.220	1.248	1.276	1.306	1.337	1.369	1.403	1.440	1.481	1.529	1.589	1.732
0.51	0.937	0.962	0.989	1.015	1.041	1.067	1.094	1.120	1.147	1.175	1.203	1.231	1.261	1.292	1.324	1.358	1.395	1.436	1.484	1.544	1.687
0.52	0.893	0.919	0.945	0.971	0.997	1.023	1.050	1.076	1.103	1.131	1.159	1.187	1.217	1.248	1.280	1.314	1.351	1.392	1.440	1.500	1.643
0.53	0.850	0.876	0.902	0.928	0.954	0.980	1.007	1.033	1.060	1.088	1.116	1.144	1.174	1.205	1.237	1.271	1.308	1.349	1.397	1.457	1.600
0.54	0.809	0.835	0.861	0.887	0.913	0.939	0.966	0.992	1.019	1.047	1.075	1.103	1.133	1.164	1.196	1.230	1.267	1.308	1.356	1.416	1.559
0.55	0.769	0.795	0.821	0.847	0.873	0.899	0.926	0.952	0.979	1.007	1.035	1.063	1.093	1.124	1.156	1.190	1.227	1.268	1.316	1.376	1.519
0.56	0.730	0.756	0.782	0.808	0.834	0.860	0.887	0.913	0.940	0.968	0.996	1.024	1.054	1.085	1.117	1.151	1.188	1.229	1.277	1.337	1.480
0.57	0.692	0.718	0.744	0.770	0.796	0.822	0.849	0.875	0.902	0.930	0.958	0.986	1.016	1.047	1.079	1.113	1.150	1.191	1.239	1.299	1.442
0.58	0.655	0.681	0.707	0.733	0.759	0.785	0.812	0.838	0.865	0.893	0.921	0.949	0.979	1.010	1.042	1.076	1.113	1.154	1.202	1.262	1.405
0.59	0.619	0.645	0.671	0.697	0.723	0.749	0.776	0.802	0.829	0.857	0.885	0.913	0.943	0.974	1.006	1.040	1.077	1.118	1.166	1.226	1.369
0.60	0.583	0.609	0.635	0.661	0.687	0.713	0.740	0.766	0.793	0.821	0.849	0.877	0.907	0.938	0.970	1.004	1.041	1.082	1.130	1.190	1.333
0.61	0.549	0.575	0.601	0.627	0.653	0.679	0.706	0.732	0.759	0.787	0.815	0.843	0.873	0.904	0.936	0.970	1.007	1.048	1.096	1.156	1.299
0.62	0.516	0.542	0.568	0.594	0.620	0.646	0.673	0.699	0.726	0.754	0.782	0.810	0.840	0.871	0.903	0.937	0.974	1.015	1.063	1.123	1.266
0.63	0.483	0.509	0.535	0.561	0.587	0.613	0.640	0.666	0.693	0.721	0.749	0.777	0.807	0.838	0.870	0.904	0.941	0.982	1.030	1.090	1.233
0.64	0.451	0.474	0.503	0.529	0.555	0.581	0.608	0.634	0.661	0.689	0.717	0.745	0.775	0.806	0.838	0.872	0.909	0.950	0.998	1.068	1.201
0.65	0.419	0.445	0.471	0.497	0.523	0.549	0.576	0.602	0.629	0.657	0.685	0.713	0.743	0.774	0.806	0.840	0.877	0.918	0.966	1.026	1.169
0.66	0.388	0.414	0.440	0.466	0.492	0.518	0.545	0.571	0.598	0.626	0.654	0.682	0.712	0.743	0.775	0.809	0.846	0.887	0.935	0.995	1.138
0.67	0.358	0.384	0.410	0.436	0.462	0.488	0.515	0.541	0.568	0.596	0.624	0.652	0.682	0.713	0.745	0.779	0.816	0.857	0.905	0.965	1.108
0.68	0.328	0.354	0.380	0.406	0.432	0.458	0.485	0.511	0.538	0.566	0.594	0.622	0.652	0.683	0.715	0.749	0.786	0.827	0.875	0.935	1.078
0.69	0.299	0.325	0.351	0.377	0.403	0.429	0.456	0.482	0.509	0.537	0.565	0.593	0.623	0.654	0.686	0.720	0.757	0.798	0.846	0.906	1.049
0.70	0.270	0.296	0.322	0.348	0.374	0.400	0.427	0.453	0.480	0.508	0.536	0.564	0.594	0.625	0.657	0.691	0.728	0.769	0.817	0.877	1.020
0.71	0.242	0.268	0.294	0.320	0.346	0.372	0.399	0.425	0.452	0.480	0.508	0.536	0.566	0.597	0.629	0.663	0.700	0.741	0.789	0.849	0.992
0.72	0.214	0.240	0.266	0.292	0.318	0.344	0.371	0.397	0.424	0.452	0.480	0.508	0.538	0.569	0.601	0.635	0.672	0.713	0.761	0.821	0.964
0.73	0.186	0.212	0.238	0.264	0.290	0.316	0.343	0.369	0.396	0.424	0.452	0.480	0.510	0.541	0.573	0.607	0.644	0.685	0.733	0.793	0.936
0.74	0.159	0.185	0.211	0.237	0.263	0.289	0.316	0.342	0.369	0.397	0.425	0.453	0.483	0.514	0.546	0.580	0.617	0.658	0.706	0.766	0.909
0.75	0.132	0.158	0.184	0.210	0.236	0.262	0.289	0.315	0.342	0.370	0.398	0.426	0.456	0.487	0.519	0.553	0.590	0.631	0.679	0.739	0.882
0.76	0.105	0.131	0.157	0.183	0.209	0.235	0.262	0.288	0.315	0.343	0.371	0.399	0.429	0.460	0.492	0.526	0.563	0.604	0.652	0.712	0.855
0.77	0.079	0.105	0.131	0.157	0.183	0.209	0.236	0.262	0.289	0.317	0.345	0.373	0.403	0.434	0.466	0.500	0.537	0.578	0.626	0.685	0.829
0.78	0.052	0.078	0.104	0.130	0.156	0.182	0.209	0.235	0.262	0.290	0.318	0.346	0.376	0.407	0.439	0.473	0.510	0.551	0.599	0.659	0.802
0.79	0.026	0.052	0.078	0.104	0.130	0.156	0.183	0.209	0.236	0.264	0.292	0.320	0.350	0.381	0.413	0.447	0.484	0.525	0.573	0.633	0.776
0.80	0.000	0.026	0.052	0.078	0.104	0.130	0.157	0.183	0.210	0.238	0.266	0.294	0.324	0.355	0.387	0.421	0.458	0.499	0.547	0.609	0.750
0.81		0.000	0.026	0.052	0.078	0.104	0.131	0.157	0.184	0.212	0.240	0.268	0.298	0.329	0.361	0.395	0.432	0.473	0.521	0.581	0.724
0.82			0.000	0.026	0.052	0.078	0.105	0.131	0.158	0.186	0.214	0.242	0.272	0.303	0.335	0.369	0.406	0.447	0.495	0.555	0.698
0.83				0.000	0.026	0.052	0.079	0.105	0.132	0.160	0.188	0.216	0.246	0.277	0.309	0.343	0.380	0.421	0.469	0.529	0.672
0.84					0.000	0.026	0.053	0.079	0.106	0.134	0.162	0.190	0.220	0.251	0.283	0.317	0.354	0.395	0.443	0.503	0.646
0.85						0.000	0.027	0.053	0.080	0.108	0.136	0.164	0.194	0.225	0.257	0.291	0.328	0.369	0.417	0.477	0.620
0.86							0.000	0.026	0.053	0.081	0.109	0.137	0.167	0.198	0.230	0.264	0.301	0.342	0.390	0.450	0.593
0.87								0.000	0.027	0.055	0.083	0.111	0.141	0.172	0.204	0.238	0.275	0.316	0.364	0.424	0.567
0.88									0.000	0.028	0.056	0.084	0.114	0.145	0.177	0.211	0.248	0.289	0.337	0.397	0.540
0.89										0.000	0.028	0.056	0.086	0.117	0.149	0.183	0.220	0.261	0.309	0.369	0.512
0.90											0.000	0.028	0.058	0.089	0.121	0.155	0.192	0.233	0.281	0.341	0.484
0.91												0.000	0.030	0.061	0.093	0.127	0.164	0.205	0.253	0.313	0.456
0.92													0.000	0.031	0.063	0.097	0.134	0.175	0.223	0.283	0.426
0.93														0.000	0.032	0.066	0.103	0.144	0.192	0.252	0.395
0.94															0.000	0.034	0.071	0.112	0.160	0.220	0.363
0.95																0.000	0.037	0.079	0.126	0.186	0.329
0.96																	0.000	0.041	0.089	0.149	0.292
0.97																		0.000	0.048	0.108	0.251
0.98																			0.000	0.060	0.203
0.99																				0.000	0.143
																					0.000

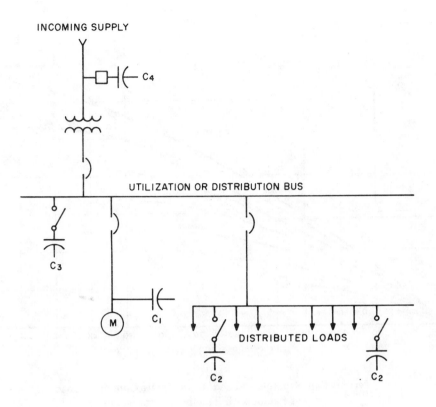

Fig 106
Possible Shunt Capacitor Locations

Large plants with extensive primary distribution systems often install capacitors at the primary voltage bus at location C_4 when utility billing encourages the user to improve the power factor.

7.6 Release of System Capacity. The expression "release of capacity" means that as the power factor is improved, the current in the existing system will be reduced, permitting additional load to be served by the same system. Equipment such as transformers, cables, and generators may be thermally overloaded. Frequently the active power rating of the prime mover corresponds to the apparent power rating of the generator. Thus improvement of the power factor can release both active power and apparent power capacity.

Various expressions for determining the amount of capacity released by power-factor improvement, along with actual examples,

curves, charts, and the economics, are covered in [2]–[4]. Fig 107 shows curves to determine the capacity released.

Example. If a plant has a load of 1000 kVA at 70 percent power factor and 480 kvar of capacitors are added, the system electric capacity released is approximately 28.5 percent, that is, the system can carry 28.5 percent more load (at 70 percent power factor) without exceeding the apparent power rating before the power factor was improved. The final power factor of the original load plus the additional load is approximately 90 percent.

7.7 Voltage Improvement. Although capacitors raise a circuit voltage, it is rarely economical to apply them in industrial plants for that reason alone. The voltage improvement may, therefore, be regarded as an additional benefit.

The following approximate expression shows the importance of reducing the reactive power

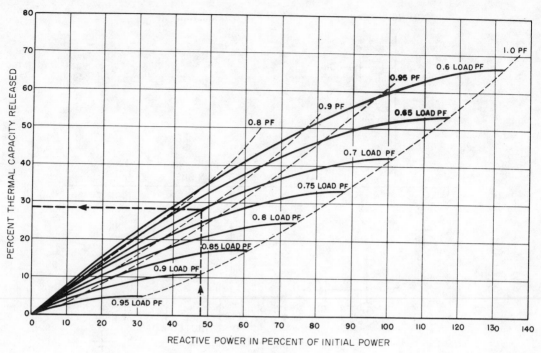

Fig 107
Percent Capacity Released and Approximate Combined Load
Power Factor with Reactive Compensation
————— Original Load Power Factor (cos ϕ_1); — — — Final Power Factor (cos ϕ_2)

component of current in order to reduce the voltage drop:

$$\Delta V \approx RI \cos \phi \pm XI \sin \phi \qquad \text{(Eq 12)}$$

where ΔV is the voltage change, which may be a drop or rise in voltage, ΔV, R, X, and I may be in absolute values of ΔV in volts, R and X in ohms, and I in amperes; or they may be in per-unit values with ΔV in per-unit volts. (Refer to Chapter 5 for an explanation of per-unit quantities.) ϕ is the power-factor angle. Plus is used when the circuit power factor is lagging and minus when it is leading.

ΔV is always positive (voltage drop) for a circuit having a lagging power factor and usually negative (voltage rise) for the typical industrial circuit having a leading power factor. This may be rewritten as

$$\Delta V \approx R \times \text{active power current}$$
$$\pm X \times \text{reactive power current} \quad \text{(Eq 13)}$$

Perhaps the most useful form of Eq 12 is

$$\Delta V \approx I(R \cos \phi \pm X \sin \phi) \qquad \text{(Eq 14)}$$

where $R \cos \phi$ reflects the active power contribution to voltage drop per ampere of total current and $X \sin \phi$ similarly reflects the reactive power contribution to voltage drop. Typically $X \sin \phi$ is many times greater than $R \cos \phi$, say 5—10 times greater. Thus, typically, reactive power flow produces a voltage drop magnitude that is several times greater than that produced by actual power flow. Since the power factor acts directly to reduce reactive power flow, it is most effective in reducing voltage drop.

An examination of Eq 13 shows that it is only necessary to know the system reactance and the capacitor rating to predict the voltage change due to the change in reactive power. Eq 13 may therefore be rewritten in a simple form to determine the voltage change due to capaci-

254

tors at a transformer secondary bus:

$$\%\Delta V = \frac{\text{capacitor kvar} \times \% \text{ transformer impedance}}{\text{transformer kVA}}$$

$$(\text{Eq 15})$$

The voltage increases when a capacitor is switched on and decreases when it is switched off. A capacitor permanently connected to the bus will provide a permanent boost in voltage.

Example. The percent change in voltage at the bus when the transformer is rated 1000 kVA with 6 percent impedance and with a capacitor bank rated 300 kvar, using Eq 13, is calculated as

$$\%\Delta V = \frac{300}{1000} \times 6$$

$$= 1.8\% \text{ voltage rise}$$

If excessive voltage becomes a problem, it is suggested that the transformer taps should be changed.

The voltage regulation of a system from no load to full load is practically unaffected by the amount of capacitors, unless the capacitors are switched; however, the addition of capacitors can raise the voltage level. The voltage rise due to capacitors in most industrial plants with modern power distribution systems and a single transformation is rarely more than a few percent.

7.8 Power System Losses. Although the financial return from conductor loss reduction alone is seldom sufficient to justify the installation of capacitors, it is an attractive additional benefit, especially in old plants with long feeders or in irrigation or other field pumping operations.

System conductor losses are proportional to current squared, and since current is reduced in direct proportion to power-factor improvement, the losses are inversely proportional to the square of the power factor:

$$\% \text{ power loss} \propto 100 \left(\frac{\text{original pf}}{\text{improved pf}} \right)^2 \quad (\text{Eq 16})$$

$$\% \text{ loss reduction} = 100 \left[1 - \left(\frac{\text{original pf}}{\text{improved pf}} \right)^2 \right]$$

$$(\text{Eq 17})$$

7.9 Selection of Capacitors with Induction Motors. Even where economics may not favor the individual motor–capacitor method because of a diversity among motors in operation and the higher unit cost of capacitors in small ratings, this method is gaining in popularity because of its operational advantages. It attains good power factors from the beginning of operations, without the need for power-factor surveys; it puts the right amount of capacitors at the correct location as production equipment is added, taken away, or moved about the plant; it unloads distribution facilities; and it assures that the capacitors are always on the line when (and only when) the motor is energized and therefore when the power-factor improvement is needed.

7.9.1 *Effectiveness of Capacitors.* The power factor of a squirrel-cage motor at full load is usually between 80 and 90 percent, depending upon the motor speed and type of motor. At light loads, however, the power factor drops rapidly as illustrated in Fig 108. Generally induction motors do not operate at full load (often the drive is "overmotored"), and consequently they have low operating power factors. Even though the power factor of an induction motor varies materially from no load to full load, the motor reactive power does not change very much. This characteristic makes the squirrel-cage motor a particularly attractive application for capacitors. With a properly selected capacitor, the operating power factor is excellent over the entire load range of the motor, as shown in Fig 108. It is generally in excess of 95 percent of full load and higher at partial loads.

7.9.2 *Limitations of Capacitor–Motor Switching.* Capacitors have been applied to induction motors and switched with the motor as a unit with good results, except in a few cases. Experience has shown that, when difficulties are encountered, it is usually because too large a capacitor rating has been used. The three factors which limit the value of capacitors to be switched with a motor are (1) excessive inrush current or reclosing, (2) transient torques, and (3) overvoltage due to self-excitation. These limitations apply when the capacitor is connected to the load side of the motor starter as

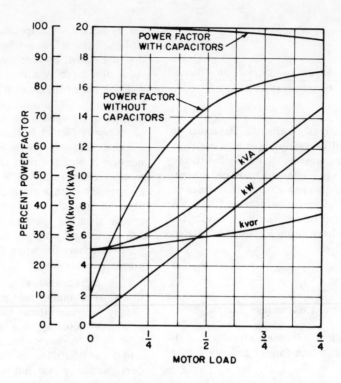

Fig 108
Motor Characteristics for Typical Medium-Size
and Medium-Speed Induction Motor

shown in Fig 110(a) and (b), and the capacitor and motor are switched as a unit.

One factor frequently not considered is that the capacitors can materially change the motor time constant which influences the safe time for reconnecting the motor—capacitor combination to the line.

Example. A 700 hp, 4000 V, 900 r/min motor had an open-circuit time constant of 0.675 s. With the proper capacitor rating of 155 kvar, the new time constant was 4.45 s. The time for the residual voltage to decay to 25 percent, a commonly accepted safe value before reconnection to the power source, was about 6 s. Thus this fact becomes important in high-inertia drives and fast reclosing switching (see [5], [6]).

7.9.3 *Selection of Capacitor Ratings.* NFPA No 70, the National Electrical Code (1975), (ANSI C1-1975), Section 460-7, states that "The total kvar rating of capacitors that are connected on the load side of a motor controller shall not exceed the value required to raise the no-load power factor of the motor to unity." [See Fig 110(a) and (b).]

The NEC has now properly omitted tables of capacitor values for motors because of the difference in motor designs among manufacturers. A good rule to follow is to measure the motor no-load current in selecting the capacitor rating. This current can be generally measured by a hook-on type ammeter.

There may be great differences in the capacitor ratings to use for a given motor rating, depending primarily on the motor speed; the slower the speed, the larger the capacitor rating which can be used. This is well demonstrated in the literature [2], [4]—[6] for a wide range of motor horsepower and speed ratings. There will also be large differences in the recommended capacitor ratings for motors

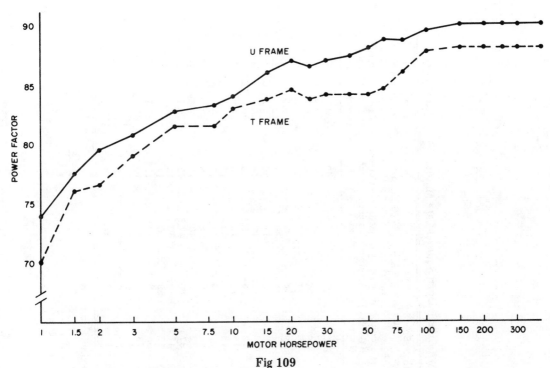

Fig 109
Power Factor Versus Motor Horsepower Rating for U-Frame and T-Frame Designs

NOTE: Based on compilation of data from six major manufacturers for
three-phase NEMA class B 1800 r/min totally enclosed 460 V motors at full load.

of different design generations,[11] such as

Pre-U frame, generally before 1955
U frame, since 1955
T frame, since 1964

Fig 109 shows the differences in power factor for U- and T-frame designs.

When the motor capacitor rating is not known or when measurement of the motor no-load current is impractical, Tables 51—53 will serve as guides.

Hermetic motors are built with a minimum of copper and iron and have quite different

characteristics from standard motors. The power factor is so poor that it is not unusual for no-load current to be about half the full-load value. Therefore capacitor ratings are larger than for standard NEMA class B motors.

7.9.4 *Location of Motor Capacitors.* Capacitors may be connected to each motor and switched with the motor, as shown in Fig 110(a) and (b), or capacitors may be permanently connected to the feeder circuit at selected starters for convenience, as shown in Fig 110(c).

The preferred location from an overall standpoint for applications not involving repetitive switching is that of Fig 110(a) or (b). In either case the capacitor and motor are switched as a unit by the motor starter, so the capacitor is always in service when the motor is in operation. The connection in Fig 110(a) may be used for new installations, as the motor overload relay can be selected at the time of purchase on the basis of the reduced line current due to the capacitors.

---

[11] For pre-U-frame designs see [2, chap 9] and [4, chap 4, table 11, p 86]. For U-frame designs see IEEE Std 141, Electric Power Distribution for Industrial Plants, 1964 edition, Tables 6.4—6.11, pp 125-129. In addition capacitor manufacturers have available publications that provide tables recommending capacitor ratings for motors.

Table 51
Suggested Capacitor Ratings for Pre-U-Frame NEMA Class B Open Squirrel-Cage Motors

Induction Motor Rating (hp)	3600 r/min		1800 r/min		Nominal Motor Speed 1200 r/min		900 r/min		720 r/min		600 r/min	
	Capacitor Rating (kvar)	Line Current Reduction (%)	Capacitor Rating (kvar)	Line Current Reduction (%)	Capacitor Rating (kvar)	Line Current Reduction (%)	Capacitor Rating (kvar)	Line Current Reduction (%)	Capacitor Rating (kvar)	Line Current Reduction (%)	Capacitor Rating (kvar)	Line Current Reduction (%)
3	1.5	14	1.5	15	1.5	20	2	27	2.5	35	3.5	41
5	2	12	2	13	2	17	3	25	4	32	4.5	37
7½	2.5	11	2.5	12	3.5	15	4	22	5.5	30	6	34
10	3	10	3	11	5	14	5	21	6.5	27	7.5	31
15	4	9	4	10	6.5	13	6.5	18	8	23	9.5	27
20	5	9	5	10	7.5	12	7.5	16	9	21	12	25
25	6	9	6	10	9	11	9	15	11	20	14	23
30	7	8	7	9	11	11	10	14	12	18	16	22
40	9	8	9	9	13	10	12	13	15	16	20	20
50	12	8	11	9	15	10	15	12	19	15	24	19
60	14	8	14	8	18	10	18	11	22	15	27	19
75	17	8	16	8	25	9	21	10	26	14	32.5	18
100	22	8	21	8	30	9	27	10	32.5	13	40	17
125	27	8	26	8	35	9	32.5	10	40	13	47.5	16
150	32.5	8	30	8	42.5	9	37.5	10	47.5	12	52.5	15
200	40	8	37.5	8	52.5	8	47.5	9	60	12	65	14
250	50	8	45	8	60	8	57.5	9	70	11	77.5	13
300	57.5	8	52.5	7	67.5	8	65	9	80	11	87.5	12
350	65	8	60	7	75	8	75	9	87.5	10	95	11
400	70	8	65	7	80	8	85	9	95	10	105	11
450	75	8	67.5	6	82.5	8	92.5	9	100	9	110	11
500	77.5	8	72.5	6			97.5	9	107.5	9	115	10

Applies to three-phase, 60 Hz motors when switched with capacitors as a single unit (from [2]).

Table 52
Suggested Capacitor Ratings for T-Frame NEMA Class B Motors

Induction Motor Rating (hp)	3600 r/min		1800 r/min		1200 r/min		900 r/min		720 r/min		600 r/min	
	Capacitor Rating (kvar)	Line Current Reduction (%)	Capacitor Rating (kvar)	Line Current Reduction (%)	Capacitor Rating (kvar)	Line Current Reduction (%)	Capacitor Rating (kvar)	Line Current Reduction (%)	Capacitor Rating (kvar)	Line Current Reduction (%)	Capacitor Rating (kvar)	Line Current Reduction (%)
3	1.5	14	1.5	23	2.5	28	3	38	3	40	4	40
5	2	14	2.5	22	3	26	4	31	4	40	5	40
7½	2.5	14	3	20	4	21	5	28	5	38	6	45
10	4	14	4	18	5	21	6	27	7.5	36	8	38
15	5	12	5	18	6	20	7.5	24	8	32	10	34
20	6	12	6	17	7.5	19	9	23	10	29	12	30
25	7.5	12	7.5	17	8	19	10	23	12	25	18	30
30	8	11	8	16	10	19	14	22	15	24	22.5	30
40	12	12	13	15	16	19	18	21	22.5	24	25	30
50	15	12	18	15	20	19	22.5	21	24	24	30	30
60	18	12	21	14	22.5	17	26	20	30	22	35	28
75	20	12	23	14	25	15	28	17	33	14	40	19
100	22.5	11	30	14	30	12	35	16	40	15	45	17
125	25	10	36	12	35	12	42	14	45	15	50	17
150	30	10	42	12	40	12	52.5	14	52.5	14	60	17
200	35	10	50	11	50	10	65	13	68	13	90	17
250	40	11	60	10	62.5	10	82	13	87.5	13	100	17
300	45	11	68	8	75	12	100	14	100	13	120	17
350	50	12	75	8	90	12	120	13	120	13	135	15
400	75	10	80	8	100	12	130	13	140	13	150	15
450	80	8	90	8	120	10	140	12	160	14	160	15
500	100	8	120	9	150	12	160	12	180	13	180	15

Applies to three-phase, 60 Hz motors when switched with capacitors as a single unit.

Table 53
Suggested Capacitor Ratings, in Kilovar, for NEMA Class C, D, and Wound-Rotor Motors

Induction Motor Rating (hp)	Design C Motor 1800 and 1200 r/min	900 r/min	Design D Motor 1200 r/min	Wound-Rotor Motor
15	5	5	5	5.5
20	5	6	6	7
25	6	6	6	7
30	7.5	9	10	11
40	10	12	12	13
50	12	15	15	17.5
60	17.5	18	18	20
75	19	22.5	22.5	25
100	27	27	30	33
125	35	37.5	37.5	40
150	37.5	45	45	50
200	45	60	60	65
250	54	70	70	75
300	65	90	75	85

Applies to three-phase, 60 Hz motors when switched with capacitors as a single unit.

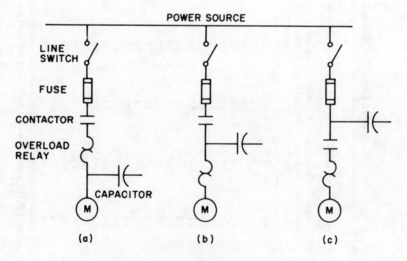

Fig 110
Electrical Location of Capacitors when Used with Inductor Motors
for Power-Factor Improvement

Fig 110(b) may be preferred for existing installations as no change in the overload relay is required because the current through the overload relay is the motor current.

The arrangement shown in Fig 110(c) is used when capacitors are permanently connected to the system. Its main advantage is the elimination of a separate switching device for the capacitors.

When the capacitor is connected as in Fig 110(a), the current through the overload relay is less than the motor current alone. The percent line current reduction may range from 10 to 25 percent.

With less margin in present motor designs and with motors usually applied closer to the actual load requirements, there is less margin for overloading. Therefore the motor overload relay should be selected or changed for the lower motor current with capacitors.

The percent line current reduction may be approximated from the following expression:

$$\%\Delta I = 100 \left(1 - \frac{\cos \phi_1}{\cos \phi_2}\right) \quad \text{(Eq 18)}$$

where

$\%\Delta I$ = Percent line current reduction
$\cos \phi_1$ = Power factor before installation of capacitor
$\cos \phi_2$ = Power factor after installation of capacitor

7.9.5 *Selection of Capacitors for Motors.* The following points should be considered when selecting motors and accompanying capacitors.

(1) Select motors that have long hours of use, so that each capacitor has a high duty factor and is likely to be on the line when needed.

(2) Choose large motors first.

(3) Limit capacitor ratings to the values recommended by the NEC and the motor manufacturer.

(4) Power-factor improvement for a multiplicity of small loads or general load compensation may be accomplished by locating capacitors at a distribution point such as a panelboard, a plug-in busway, at a substation, or on a feeder. These capacitors are usually permanently connected through a proper disconnecting means.

The method of allocating capacitors to various sizes of motors is, (1) determine the total capacitance needed; (2) list the motors in the portion of the plant involved; and (3) assign capacitors to motors in descending order of horsepower rating until the total required amount of capacitance has been accumulated.

Since it is not generally practical or economical to use capacitors with all motors, additional plant capacitor requirements will have to be supplied by capacitors at other locations.

A few such studies for each type of plant operation and desired power factor will yield a figure of the minimum size motor to which capacitors should be economically assigned.

7.9.6 *Motor-Capacitor Applications to Avoid*

(1) Motors should not be subject to reversing or plugging [7].

(2) Motors should not be restarted while still running and generating substantial back voltage.

(3) Capacitors should not be used with crane or elevator motors where the load may drive the motor, or on multispeed motors.

(4) Open-transition reduced-voltage starters should be avoided with wye—delta connections and capacitors. Capacitors should be connected on the line side of contactors involved in any open-circuit transition for voltage change or speed.

Although it is possible to apply capacitors in these cases, technical investigations are required. It is suggested that the capacitors be switched separately or assigned with other loads.

7.9.7 *Induction Versus Synchronous Motors.* Induction motors with capacitors are often more economical than synchronous motors alone. Sometimes the type of drive will dictate the use of one type of motor, but where a free choice is available, an economic comparison should be made. The capacitor—induction-motor combination generally has the advantage of lower maintenance.

Generally synchronous motors used for system power-factor improvement are of the 0.8 power-factor type, because the incremental cost of the reactive power produced is low. Synchronous motors are especially attractive for slow-speed drives.

The following expression may be used as a guide to obtain the capacitor rating required to deliver the same net reactive power as an

0.8 power-factor synchronous motor of the same horsepower rating at full load:

capacitor rating

$$\approx 1.12 \times \text{induction-motor rating} \quad \text{(Eq 19a)}$$

$$\text{kvar} \approx 1.12 \times \text{hp} \quad \text{(Eq 19b)}$$

For induction-motor applications of up to 500 hp it will be found necessary to add approximately 1.1—1.2 kvar of capacitors per horsepower to make the combination comparable to an 0.8 power-factor synchronous-motor application.

From the previous discussions relating to the selection of the proper capacitor rating it is evident that not all the capacitors can be switched with the motor. Therefore in making an economic comparison, a separate switching device must be included in the cost comparisons.

The equipment comparisons for initial costs should include:

synchronous motor
+ starter
+ exciter

versus

induction motor
+ starter
+ capacitors
+ separate capacitor switching devices, when needed

These economic comparisons may be summarized as follows.

(1) *Low-Voltage Systems (1000 V or Less).* The induction-motor method is more economical up to about 200 hp with full-voltage starters and to about 350 hp with reduced-voltage starters.

(2) *Medium-Voltage Systems.* The synchronous motor costs less than the induction-motor equipment over the entire horsepower and speed ranges if a power circuit breaker or a contactor of adequate interrupting capacity is used to switch the excess capacitors.

7.10 Automatic Control Equipment. Automatic switching of capacitors is seldom used in industrial plants, but when used, it is usually required for one or more of these reasons:

(1) To control circuit loading

(2) To reduce voltage during light-load conditions and to improve voltage regulations under all load conditions

(3) To meet the requirements of a rate clause that sets maximum or minimum reactive power demand; or to comply with other utility requirements

The most common control types are

Current control	Single step (ON or OFF)
Voltage control	Generally single step (one or more capacitor blocks)
Reactive power control	Generally multistep (usually a series of capacitor blocks)
Time control	Time clock

These various control systems, the basis of selection, the characteristics of associated master elements, and the construction and use of load diagrams to select the bandwidth and amount of capacitors to be switched per step are described in the literature [2], [4].

More automatic switching may be required in plants having a large portion of load in thyristor motor drives because of the drive characteristics of reactive power versus active power.

7.11 Capacitor Standards and Operating Characteristics

7.11.1 *Capacitor Ratings.* There are ratings for shunt capacitor units from 240 through 21 600 V (See NEMA CP1-1976, Shunt Capacitors). Although this NEMA standard lists only a limited number of ratings for low-voltage service (240—600 V), several manufactuers have additional ratings of 1—15 kvar to cover applications with motors. Larger units are available in ratings up to 600 kvar, 13 200 V, three-phase units.

7.11.2 *Maximum Voltage.* Where capacitors are to be operated above 100 percent of rated voltage, including harmonics, refer the application to the capacitor manufacturer.

7.11.3 *Temperature.* Capacitors are more sensitive to temperature limits than are many other devices. Therefore manufacturers' precautions on installation must be followed. Capacitors should not be placed in hot locations near furnaces or resistors, exposed to sunshine in hot climates, or placed where air cannot circulate unless special provision is made for cooling or for operating the capacitors below nameplate voltage. Neglect of these points will shorten capacitor life. [See IEEE Std 18-1968, Shunt Power Capacitors (ANSI C55.1-1968).]

7.11.4 *Time to Discharge.* The NEC, Section 460-6, requires capacitors to be discharged to a residual voltage of 50 V or less in 1 min for capacitors rated 600 V or less, and Section 460-28 requires discharge to 50 V or less in 5 min for those rated above 600 V. This is usually accomplished with built-in discharge resistors. However, they are not required when capacitors are connected without disconnecting means directly to other discharge paths such as motors or transformers.

7.11.5 *Effect of Harmonics on Capacitors.* Capacitors have a substantial margin for harmonic currents and voltages. IEEE Std 18-1968 and NEMA CP1-1976 require capacitors to carry 135 percent of rating in kvar, including that of the fundamental and harmonics.

If the voltage is approximately normal, it is unlikely that a capacitor would be overloaded by harmonics, although it can happen, as illustrated in Fig 111.

For specific effects of harmonics on capacitors consult the capacitor manufacturer.

7.11.6 *Operating Characteristics.* The following relationships apply when capacitors are operated at other than their design rated operating conditions.

(1) The reactive power varies approximately as the square of the applied voltage.

(2) The reactive power varies approximately as the frequency.

7.11.7 *Overcurrent Protection.* The NEC, Section 460-8(b), requires overcurrent protection for capacitors under 600 V and Section 460-25 requires overcurrent protection for capacitors over 600 V nominal. An exception is provided for capacitors under 600 V which are protected by the motor overcurrent device. Fuses are generally used and may be applied to each individual capacitor unit or to small groups of units in parallel. The blowing of a fuse of properly selected rating may provide an indication of overcurrent operating conditions as well as an actual fault in the capacitor.

A fuse for a capacitor is not for overload protection in the same sense as it is used for other

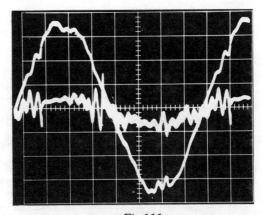

Fig 111
Oscillogram of 480 V Line Voltage and
Capacitor Current near a Thyristor-Controlled
Furnace, 150 ft away from a 1000 kVA
Transformer; Capacitor Overheated

electric apparatus, such as a motor. The current ratings of capacitor fuses range from 165 to 250 percent of the capacitor current rating to allow for inrush current; therefore their overload protection is quite limited.

The type and rating of the fuse must be carefully selected for the proper time–current characteristic and voltage, so only the fuse rating recommended by the capacitor manufacturer should be used. The fuse time–current characteristic is the most important factor, as it is a measure of fuse performance, whereas the current rating may be only a nominal value established for the particular type of fuse.

7.11.8 *Low-Voltage Switching Devices.* There is rarely any problem encountered in the interruption, closing, or repetitive operation of low-voltage air circuit breakers, molded-case circuit breakers, contactors, or switches associated with capacitor equipments for industrial service.

The manufacturer of the capacitor switching device should be consulted concerning the device capacitance current switching capability. Vacuum circuit breakers are derated considerably when used in capacitor switching applications (see IEEE Std 342-1973, Application Guide for Capacitance Current Switching for AC High-Voltage Circuit Breakers Rated on a Symmetrical Current Basis (ANSI C37.0731-1973)].

The NEC, Section 460-8, requires switching

Table 54
Capacitor Rating Multipliers to Obtain Switching-Device* Rating

Type of Switching Device	Multiplier to Obtain Equivalent Capacitor Rating	Equivalent Current per kvar		
		240 V	480 V	600 V
Magnetic-type power circuit breaker	1.35	3.25	1.62	1.30
Molded-case circuit breakers				
magnetic type	1.35	3.25	1.62	1.30
others	≈ 1.5	≈ 3.61	≈ 1.8	≈ 1.44
Contactors, enclosed†	1.5	3.61	1.8	1.44
Safety switch‡	1.35	3.25	1.62	1.30
Safety switch (fusible)‡	1.65	3.98	1.98	1.58

*Switching device must have a continuous-current rating that is equal to or exceeds the current associated with the capacitor kvar rating times the indicated multiplier. Enclosed switch ratings at 40°C (104°F) ambient temperature.

†If contactor manufacturers give specific ratings for capacitors, these should be followed.

‡This requirement is given in NEMA CP1-1976, Section 4.09, page 15.

devices to be selected for at least 135 percent of the continuous-current rating of the capacitor and to have the proper interrupting rating for the system short-circuit capacity.

Table 54 is a convenient reference in selecting the various switching devices for low-voltage systems.

7.11.9 *Medium-Voltage Switching Devices.* The phenomenon of reconduction of the arc in capacitor switching devices sometimes creates a problem on high-voltage transmission lines or where large blocks of capacitors are switched, such as ratings of 10 000–15 000 kvar on 13.8 kV and higher voltage systems. Very few industrial plants have capacitors of such large blocks, and very few plants have many capacitors located on the primary distribution systems. The capacitors are usually located on the lower voltage utilization circuits.

The repetitive duty of switching devices is generally not a problem either, since there are very few industrial applications where capacitor banks must be switched so often that contact or mechanism maintenance is frequently required.

7.11.10 *Selection of Cable Sizes.* In selecting cables, allowance must be made for the same 1.35 multiplier and any additional allowance or derating for temperature.

7.11.11 *Inspection and Testing of Capacitors.* Capacitors should be inspected at intervals de-

termined by experience. Ventilation, ambient temperature, and applied voltage should be checked as well as capacitor appearance. Refer to manufacturers' manuals for detailed information and guidance.

7.12 Transients. The industry has not had much difficulty with transients considering the extensive number of capacitor installations in service. However, with the increased use of thyristor motor drives and inverters along with the need for capacitors to improve their power factor, which is very poor at low speed [1], more problem areas can be expected. This is especially true in plants having many thyristor drives scattered throughout and where the total drive load is an appreciable part of the total plant load.

A rectifier performs its function by commutating (or switching) the load selectively from one phase (of the alternating-current system) to the next at a particular point on the alternating-current wave. Such commutation produces switching transients and, depending on associated phase controlling and overlap, harmonics of various magnitudes. In some cases capacitors may improve the voltage wave shape that is distorted by transients associated with rectifier commutation. Fig 112 shows where capacitors have nearly eliminated a transient voltage dip of approximately 50 percent of crest value of the voltage wave.

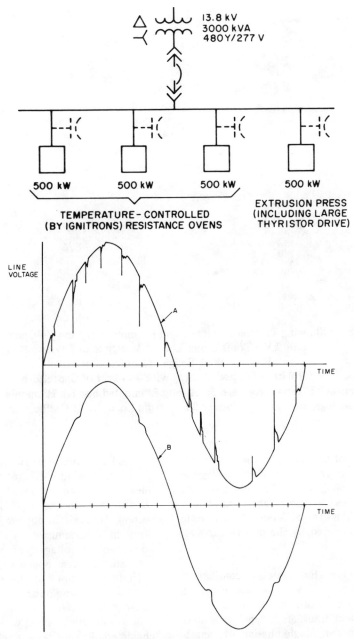

Fig 112
Illustration of the Reduction of Rectifier Commutation Switching Transients by the
Application of Power Capacitors. Before the Addition of Capacitors the Total Plant Peak
Load is 2500 kW at 0.83 Power Factor. Capacitors (Shown by Dashed Lines)
Totaling 130 kvar Increase the Power Factor to 0.95
A — Line Voltage Distorsions Caused by Chopped-Wave Loads (Notches Varied with
Loading, but dV/dt was Hundreds of Volts per Microsecond);
B — Improvement with Capacitors Installed at Loads

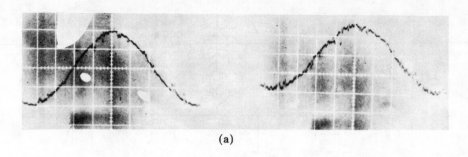

(a)

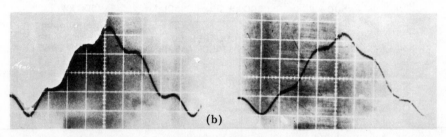

(b)

Fig 113
Oscillograms Showing Transient Voltages and Harmonics of Line Voltage on 480 V Side
of 2000 kVA Transformer Loaded Mostly with Thyristor Drives
Having a Wide Range of Control Settings
(a) Without Capacitors. (b) With 350 kvar of Capacitors.
High-Frequency Effects Have Been Practically Eliminated but 7th Harmonic Accentuated.
Left Oscillograms — at Transformer 200 ft from Load; Right Oscillograms — at Load

The addition of capacitors to the power system can either reduce or increase the harmonic and transient voltages. Fig 113 illustrates an application where capacitors accentuated harmonics. This shows the difference in harmonic reinforcement effected by the intervening busway.

7.13 Resonances and Harmonics. In considering circuit natural frequencies and the effect of harmonic voltages, both the normal system configuration and unusual system operations must be considered. Single-phasing of capacitors due to blown fuses or single-phase switching and single-phase faults occassionally leads to resonances not observed during balanced system operation.

Substantial insight into the performance of a complete circuit comes from a determination of the frequencies and impedances of individual circuit loops.

7.13.1 *Generation of Harmonic Voltages and Currents.* A sinusoidal voltage across a nonlinear impedance will result in a nonsinusoidal current through the impedance. A sinusoidal current through a nonlinear impedance will result in a nonsinusoidal voltage across it. The nonsinusoidal voltages and currents associated with transformers operating in saturation are a familiar example of the generation of harmonics by a nonlinear impedance. The nonlinear magnetization curve, especially noticeable in saturation, along with hysteresis in this characteristic, create a nonlinear impedance. This effect is present in all iron-core devices, and the magnitude depends upon the design of the device and the voltage under which it is operated.

A rectifier is another common example of a nonlinear impedance. A sinusoidal voltage across the rectifier does not result in a sinusoidal current through the rectifier. Many other circuit components have varying degrees of nonlinearity inherent in their impedance. Arc furnaces, for example, impose a high harmonic duty on the system. Any device which has a nonsinusoidal voltage across it with a sinusoidal current through it, or a nonsinusoidal current through it with a sinusoidal voltage across it, is a nonlinear impedance. These nonlinear impedances create harmonic voltages in power systems, which can create resonant circuits. The more nonsinusoidal the voltage or current of a component, the more likely it is to create problems.

Although capacitors in themselves do not generate harmonics, the effects of a capacitor on the circuit impedance may cause the harmonic voltages to either decrease or increase. However, since the reactance of a capacitor is inversely proportional to the frequency, the current through a shunt capacitor per volt of impressed voltage is proportional to the order of the harmonic.

7.13.2 *Harmonic Generators.* Rectifiers generate harmonics of the $(pn \pm 1)$th order, where p is the number of rectifier pulses and n is 1, 2, 3, $\cdots$. Thus a six-pulse rectifier will generate harmonics of the 5th, 7th, 11th, 13th, 17th, $\cdots$ orders. A 12-pulse rectifier will generate harmonics of the 11th, 13th, 23rd, 25th, $\cdots$ orders.

Other components which generate harmonics include inverters; overexcited transformers, generators, or motors; salient-pole motors; and solid-state speed controls.

In a symmetrical three-phase system, evenmultiple harmonics of the fundamental are absent. The 3rd, 9th, 15th, 21st, and 27th are zero-sequence harmonics. They can flow in a three-phase grounded-wye capacitor installation having a ground neutral — one reason why industrial capacitor banks should not be grounded.

The $(pn + 1)$th order of harmonics, for example, the 7th, 13th, 19th, and 25th harmonics, are positive-sequence harmonics. These are present in the balanced-system three-phase line-to-line loads.

The $(pn - 1)$th order of harmonics, for example, the 5th, 11th, 17th, and 23rd harmonics, are negative-sequence harmonics. These will cause additional heating in motors. In rectifier circuits the theoretical maximum magnitude of the harmonic current generated is the reciprocal of the harmonic number. For example, the 5th harmonic current is 20 percent of the fundamental.

7.13.3 *Harmonic Resonance.* When shunt capacitors are used on power systems having rectifiers, there is the possibility of resonance between the capacitor and the power system reactance. (It is convenient to consider the rectifier as a harmonic current generator, thus the capacitor and power system are in parallel as seen by the rectifier. If reactances of the capacitor and the power system are nearly equal at one of the harmonic frequencies generated by the rectifier, the parallel combination approaches resonance and results in a high impedance to the flow of that harmonic current. As a result, a relatively high harmonic voltage will exist by virtue of the harmonic current flowing through the apparent high impedance.

As resonance is approached, the magnitude of harmonic current in the system and capacitor becomes much larger than the harmonic current generated by the rectifier. The current may be high enough to blow capacitor fuses, one indication of the possibility of resonance.

The following expression may be used to determine the potential of harmonic resonance with a capacitor bank for radial systems:

$$h \cong \sqrt{\frac{\text{kVA}_{SC}}{\text{kvar}_C}} \qquad \text{(Eq 19)}$$

where

h = Order of the harmonic
kVA_{SC} = System short-circuit duty
kvar_C = Capacitor rating
This is graphically shown in Fig 114.

Eq 19 may also be approximated to

$$h \cong \sqrt{\frac{\text{kVA}_T \times 100}{\text{kvar}_C \times \%X}} \qquad \text{(Eq 20)}$$

where

kVA_T = Transformer rating
$\%X = X_T + X_{SYS}$

with X_T being the transformer reactance in percent and X_{SYS} the equivalent system reactance in percent of the transformer rating.

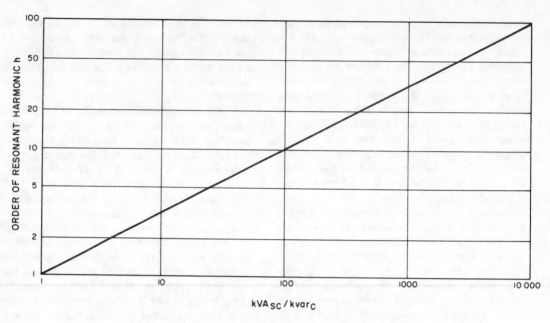

Fig 114
Order of Resonant Harmonic as a Function of kVA$_{SC}$/kvar$_C$, Based on Eq 19

Fig 115 is a plot Eq 20 and is useful for indicating the critical capacitor ratings to avoid harmonic resonance. The 5th and 7th harmonics are common with six-phase rectifiers, and resonance at these harmonics is to be especially avoided.

In the case of the unit substation transformer used in industry (500—2000 kVA) the utility system reactance %X_{SYS} in Eq 20 may be assumed as 1/2-1 percent if the actual system short-circuit-current capacity is not known. Thus values of X = 6—7 percent can be used as an approximation for %X in Eq 20. In applications involving large power transformers, the actual system short-circuit-current capacity should be obtained. An alternate approach is included in [8].

Fig 115 is limited to applications on radial systems. For more complex arrangements, which utilize capacitors in a number of scattered locations, the analytical determination of harmonic distribution is much more difficult and recourse to a computer study of the harmonic current flow may be necessary. Added capacitor locations multiply the combinations by which resonances may occur. It may be difficult to define accurately in the planning stage the necessary harmonic impedances upon which to base a meaningful harmonic analysis. Sometimes a wait-and-see attitude is adopted in such cases. If trouble becomes apparent, such as blown capacitor fuses or cell failures, then corrective steps must be taken.

7.13.4 *Application Guide Lines.* If excessive harmonic currents or voltages are suspected, remedies fall into several general categories.

(1) Detuning consists of changing the capacitance or inductance of the circuit so that the circuit natural frequency will not fall near an expected or known integral multiple of the fundamental frequency. This usually takes the form of removing or adding capacitor units. If that does not solve the problem, then the addition of tuning reactors in series with the capacitor bank can be used to shift the parallel harmonic resonant point out of the range of predominant orders of harmonic current.

(2) Wye-connected capacitor banks should be ungrounded, eliminating a path for the zero-sequence harmonics that should flow through a grounded neutral. Another important reason

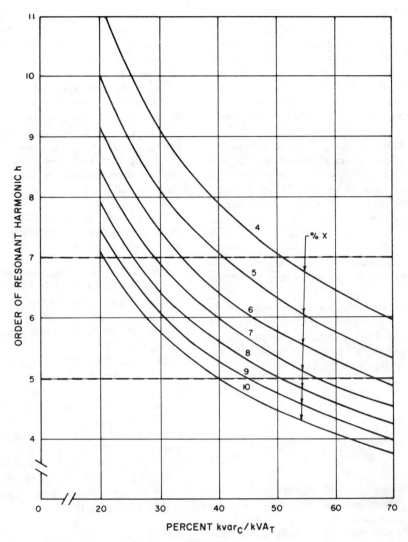

Fig 115
Order of Resonant Harmonic Versus Capacitor Size for Selected
Short-Circuit Impedance

NOTES: (1) Both short-circuit impedance and capacitor size are in
percent on transformer kVA base.
(2) $\%X = (\%X_T + \%X_{SYS})$ on transformer base.

for operating capacitors ungrounded is that grounding them could interfere with the performance of the plant ground-relaying system.

(3) Harmonic input to the system can be reduced by operating at a lower level on the saturation curve of transformers and motors.

(4) Increasing the number of phases of a rectifier or converter will reduce the harmonic input.

7.14 Capacitor Switching. Switching capacitors on an energized system will generally result in a transient voltage. The magnitude of the transient voltage depends upon the available short-circuit current, the amount of capacitance switched, and the point on the voltage wave where switching occurs. Except for switching of very large capacitor banks or the presence of extremely sensitive loads (semiconductor controls, computers, etc), these transients can generally be ignored.

The approximately maximum peak inrush current on energizing a discharged capacitor from normal line voltage, in the absence of other capacitors that may be charged, is in per unit of capacitor peak rated current:

$$I_{inrush} = \frac{I_{inrush\,(peak)}}{I_{rated\,(peak)}}$$

$$\cong 1 + \sqrt{\frac{kVA_{SC}}{kvar_C}} \qquad (Eq\ 21)$$

where kVA_{SC} is the short-circuit duty of the system in kilovolt-amperes and $kvar_C$ is the rating of the capacitor.

Generally this peak current value is less than the system momentary short-circuit current. Thus the switching device selected for the system short-circuit-current capacity should be adequate. An exception may be a contactor having a limited inrush withstand ability. When a capacitor is energized in parallel with an existing capacitor, the inrush current will be much higher, as the inherent inductive reactance between banks is usually small in the case of bank-to-bank switching. If the inrush current is excessive, a small inductance or resistance can be inserted in series with the capacitor banks.

When a capacitor is energized alone, the capacitor acts initially as a short circuit, causing an initial inrush of current to build up a back voltage in the capacitor. This creates an oscillation at the natural frequency of the circuit until it is damped out by the circuit resistance. The amount of oscillation depends on where in the voltage wave the switch is closed.

The current and voltage transients occurring during capacitor energizaiton vary from operation to operation, depending upon the exact voltage conditions across the contacts for each operation. Difficulties occur only for the most severe voltage conditions, usually for applications over 600 V.

7.15 Standards References. The following standards publications were used as references in preparing this chapter.

ANSI C37.0732-1972, Schedule of Preferred Ratings for Capacitance Current Switching for AC High-Voltage Circuit Breakers Rated on a Symmetrical Current Basis

IEEE Std 18-1968, Shunt Power Capacitors (ANSI C55.1-1968)

IEEE Std 341-1972, Requirements for Capacitance Current Switching for AC High-Voltage Circuit Breakers Rated on a Symmetrical Current Basis (ANSI C37.073-1972)

IEEE Std 342-1973, Application Guide for Capacitance Current Switching for AC High-Voltage Circuit Breakers Rated on a Symmetrical Current Basis (ANSI C37.0731-1973)

NEMA CP1-1976, Shunt Capacitors

NFPA No 70, National Electrical Code (1975), (ANSI C1-1975)

7.16 References and Bibliography

7.16.1 *References*

[1] JACOBS, A.P., and WALSH, G.W. Application Considerations for SCR DC Drives and Associated Power Systems. *IEEE Transactions on Industry and General Applications*, vol IGA-4, Jul/Aug 1968, pp 396-404.

[2] BLOOMQUIST, W.C., CRAIG, C.R., PARTINGTON, R.M., and WILSON, R.C. *Capacitors for Industry*. New York: Wiley, 1950. (Out of print. Now available in paperback from University Microfilm, 300 N. Zeeb Road, Ann Arbor, MI 38106.)

[3] STANGLAND, G. The Economic Limit of Capacitor Application for Load Relief. *Power Engineering*, Nov 1950, pp 78-80.

[4] MARBURY, R.E. *Power Capacitors.* New York: McGraw-Hill, 1949.

[5] MOORE, R.C., and SCHWARTZBURG, W.E. Applying Capacitors at Motor Terminals. *Allis-Chalmers Electrical Review,* 4th Quarter 1958, pp 20-22.

[6] DEMELLO, F.P., and WALSH, G.W. Reclosing Transients in Induction Motors with Terminal Capacitors. *AIEE Transactions (Power Apparatus and Systems),* pt III, vol 79, Feb 1961, pp 1206-1213.

[7] BECK, C.D., and RHUDY, R.G. Plugging an Induction Motor. *IEEE Transactions on Industry and General Applications,* vol IGA-6, Jan/Feb 1970, pp 10-18.

[8] STRATFORD, R.P. Capacitors on AC System Having Large Rectifier Loads. *Industrial Power Systems Magazine,* vol 4, Mar 1961, pp 3-6.

7.16.2 *Bibliography*

[9] *Electric Utility Engineering Reference Book;* vol 3, *Distribution Systems.* Trafford, PA: Westinghouse Electric Corporation, 1965.

[10] ERLICKI, M.S., SCHIEBER, D., and BEN URI, J. Power-Measurement Errors in Controlled Rectifier Circuits. *IEEE Transactions on Industry and General Applications,* vol IGA-2, Jul/Aug 1966, pp 309-311.

[11] Greenwood, A. *Electrical Transients in Power Systems.* New York: Wiley, 1971.

[12] IEEE COMMITTEE REPORT. Bibliography on Switching of Capacitive Circuits Exclusive of Series Capacitors. *IEEE Transactions on Power Apparatus and Systems,* vol PAS-89, Jul/Aug 1970, pp 1203-1207.

[13] NAILON, R.L. Use Capacitors with Large Motors. *Power,* Jul 1971, pp 81-83.

[14] OSCARSON, G.L. Telephone Influence Factor and what to Do about It. *Power Generation,* Dec 1949, pp 68-71.

[15] PALKO, E. Cutting Costs with Power Capacitors. *Plant Engineering,* Feb 14, 1972, pp 55-59.

[16] STEEPER, D.E., and STRATFORD, R.P. Reactive Compensation and Harmonic Suppression for Industrial Power Systems Using Thyristor Converters. *IEEE Transactions on Industry Applications,* vol IA-12, May/Jun 1976, pp 232-254.

8. Power Switching, Transformation, and Motor-Control Apparatus

8.1 Introduction. This chapter provides information on the requirements for and application of major apparatus utilized in an industrial electric distribution system More detailed information on this apparatus is available in the standards of the American National Standards Institute (ANSI), National Electrical Manufacturers Association (NEMA), and Underwriters' Laboratories, Inc (UL), as well as in manufacturers' publications.

The engineer must make basic decisions in his choice of equipment for his particular electric system. He should include all facets of the actual project such as protection, coordination, initial cost including installation, costs of maintenance, space, and operations, and the procurement time to meet schedules.

8.1.1 *Equipment Installation.* Electric equipment must be installed to be safe and accessible to persons frequently in the area. Sufficient access and working space should be provided and maintained about all electric apparatus to permit ready and safe operation and maintenance of such equipment. NFPA No 70, the National Electrical Code (1975), (ANSI C1-1975), (NEC) lists in Section 110-16 minimum working clearances for electric equipment operating at 600 V or less, whereas Section 110-34 lists these minimum requirements for voltages over 600 V. Table 55 combines these requirements.

Installations in industrial plants require that adequate aisles, hatchways, wall openings, etc, be provided for easy removal and replacement of all electric equipment. For metal-enclosed switchgear extreme care should be exercised in setting and aligning leveling channels flush with the floor in order to prevent stressing of porcelain and bus structures and to provide easy insertion and removal of circuit breakers.

The NEC, Article 450, outlines the installation requirements for transformers of all types. In some industrial plants, the unit substation transformer is located outside a pressure-ventilated switchgear room with a secondary throat connection for buswork through the wall of the room, and with a primary connection to an air-filled terminal box with stress-cone termination for 15 kV and lower voltage cable. This pressure-ventilated switchgear room may also house motor-control centers, panelboards, and other electric equipment in addition to the switchgear. This installation method is used

(1) To protect electric equipment from accumulation of dirt, dust, and other foreign material

(2) To enable the use of less expensive and more readily maintainable general-purpose electric equipment enclosures in lieu of costly explosion-proof enclosures in areas classified as hazardous in accordance with Article 500 of the NEC

(3) To prevent access by unauthorized people

8.1.2 *Maintenance, Testing, and Safety.* The use of sound engineering principles in the de-

Table 55
Minimum Clear Working Space in Front
of Electric Equipment

Voltage to Ground (volts)	Working Space (feet) for Conditions*		
	(1)	(2)	(3)
0—150	2½	2½	3
151—600	2½	3½	4
601—2500	3	4	5
2501—9000	4	5	6
9001—25 000	5	6	9
25 001—75 000	6	8	10

Based on the NEC (1975).

*Conditions:

(1) Exposed live parts on one side and no live or grounded parts on the other side of the working space, or exposed live parts on both sides effectively guarded by suitable wood or other insulating materials. Insulated wire or insulated bus bars operating at not over 300 V shall not be considered live parts.

(2) Exposed live parts on one side and grounded parts on the other side. Concrete, brick, or tile walls will be considered as grounded surfaces.

(3) Exposed live parts on both sides of the work space [not guarded as provided in Condition (1)] with the operator between.

Exception: Working space is not required in back of equipment such as deadfront switchboards or control assemblies where there are no renewable or adjustable parts such as fuses or switches and when all connections are accessible from other locations than the back.

sign of electric systems dictates consideration of maintenance, testing, and safety as factors always present during the life of the plant.

An effective maintenance program must sustain production at required levels, protect valuable investment, and reduce down time and maintenance costs. Planned maintenance must provide a series of tasks to be performed on each unit of equipment on a regular schedule basis in order to detect areas of potential failure and to provide a warning to plan for necessary repairs. Record keeping is an important part of the planned maintenance program. Inherent with any maintenance program is the testing of electric equipment and the appropriate safety precautions for personnel and equipment. Chapter 4 discusses methods of testing protective equipment such as circuit breakers, relays, etc.

Testing procedures for other equipment are found in appropriate standards of the Institute of Electrical and Electronics Engineers (IEEE), ANSI, and NEMA, as well as in manufacturers' literature.

8.1.3 *Heat Losses.* The heat generated by power losses in electrical distribution equipment, particularly transformers, switchgear, rectifiers, and motor-control centers, must be considered in the initial design of a project. A realistic summation of the electric losses, which show up as heat, must be developed and added to the cooling requirements of the building and unit substation areas in manufacturing plants. Simply exhausting the air from the equipment location and allowing air to enter through screens or filters is not necessarily the best solution. Drawing in large quantities of cooler outside air can cause condensation in enclosed electric equipment, increase cost of power for ventilation and filter replacement, and possibly create a hazardous atmosphere in the building. Alternate methods include heat removal by ventilation, air-conditioning, or outdoor location of the main heat-producing equipment. Data on the heat losses for the electrical distribution apparatus can be obtained from the manufacturer.

8.2 Switching Apparatus for Power Circuits

8.2.1 *Definition.* Switching apparatus can be defined as a device for opening and closing or for changing the connections of a circuit. The general classification of switching apparatus as used in this chapter is switches, fuses, circuit breakers, and contactors.

8.2.2 *Switches.* The types of switches normally applied for power circuits include (1) disconnecting, (2) load interrupter, and (3) safety switches for 600 V and lower power applications, including bolted-pressure switches, and (4) transfer switches for emergency power.

(1) A disconnecting switch is used for changing the connections in a circuit or for isolating a circuit or equipment from the source of power for all voltage classes. It has no interrupting rating and is intended to be operated only after the circuit has been opened by other means. Interlocking is generally provided to prevent operation when the switch is carrying current. Latches may be required to prevent the switch

from being opened by magnetic forces under heavy fault currents.

(2) An interrupter or load-break switch, generally associated with unit substations supplied from the primary distribution system, is a switch combining the functions of a disconnecting switch and a load interrupter for interrupting, at rated voltage, currents not exceeding the continuous-current rating of the switch. Load-break switches are of the air or fluid-immersed type. The interrupter switch is usually manually operated and has a quick-make, quick-break mechanism independent of the speed of handle operation. Load-break switches may have a close and latch rating; these provide maximum safety in the event of closing in on a faulted circuit.

With a current-limiting fuse and load-break switch combination, fast fault clearing and circuit isolation can be provided. This application, if properly coordinated to protect the transformer and to interrupt transformer magnetizing currents and load currents within the switch rating, may be more economical than a circuit breaker. Rollout fuse—switch assemblies, supplied by several manufacturers, have advantages such as (a) replacement of a fuse without the necessity of de-energizing the primary circuit, (b) ability to fuse on the line side of the switch, and (c) complete access to mechanical parts for checkout and maintenance.

It is desirable, from a safety standpoint, to interlock the operation of an interrupter switch with the secondary circuit breaker in the case of a transformer application to minimize the chance of operating the interrupter switch at a time when the current exceeds the rating of the switch. Many interrupter switches, however, have interrupting ratings in excess of the transformer full-load current so that interlocking is not required.

(3) For service of 600 V and below safety switches are commonly used. These are enclosed and may be fused or unfused. This type of switch is operable by a handle from outside the enclosure and is so interlocked that the enclosure cannot be opened unless the switch is open or the interlock defeater is operated. Many safety switches have quick-make and quick-break contact features.

A safety switch for motors is rated in horse-power and voltage and is capable of interrupting the maximum operating overload current of a motor, that is, the stalled-rotor current of the same horsepower and the switch rating at the rated voltage. The NEC recognizes six times full-load motor current as the stalled-rotor current.

The application of fused enclosed switches is limited by rules of the NEC to a current not in excess of 80 percent of the current rating of the same switch without fuses.

Although the safety switch with the NEC fuse has an interrupting ability no higher than the fuse, switches with current-limiting fuses are available with interrupting ratings of the switch—fuse combination up to 200 000 A, symmetrical rms. This entire assembly must be labeled by the manufacturer or by UL for the particular fuse to be applied. The interruptible rating of the switch—fuse combination must be utilized. It should never be assumed that the application of a high-interrupting-capacity fuse with a lower rated switch will result in an elevation of the switch rating.

A bolted-pressure switch consists of movable blades and stationary contacts with arcing contacts and a simple toggle mechanism for applying bolted pressure to both the hinge and jaw contacts in a manner similar to a bolted bus joint. The operation mechanism consists of a spring which is compressed by the operating handle and released at the end of the operating stroke to provide quick-make and -break switching action.

The electrical-trip bolted pressure switch is basically the same as the manually operated bolted-pressure switch, except that a stored-energy latch mechanism and a solenoid trip release are added to provide automatic electrical opening. These switches are designed specifically for use with ground-fault protection equipment which may be required by the NEC and have a contact-interrupting rating of 12 times continuous rating. These switches are available in ratings of 800, 1200, 1600, 2000, 2500, 3000, and 4000 A, 480 V, alternating current, and are suitable for use on circuits having available fault currents of 200 000 A, symmetrical rms.

(4) Automatic transfer switches of double-throw construction are primarily used for

emergency and standby power generation systems rated 600 V and less. These transfer switches do not normally incorporate overcurrent protection and are designed and applied in accordance with the NEC, particularly Articles 517, 700, and 750. They are available in ratings from 30 to 3000 A. For reliability, most automatic transfer switches rated above 100 A are mechanically held and are electrically operated from the power source to which the load is to be transferred.

These switches provide protection against failure of the utility service. In addition to utility failures, continuity of power to critical loads can also be disrupted by (1) an open circuit within the building area on the load side of the incoming service, (2) overload or fault conditions, or (3) electrical or mechanical failure of the electric power distribution system within the building. Therefore many engineers advocate the use of multiple transfer switches of lower current rating located near the load rather than one large transfer switch at the point of incoming service. For additional information see IEEE Std 446-1974, Emergency and Standby Power Systems.

8.2.3 Fuses

(1) *Types and Rating Basis.* A fuse is an overcurrent protective device with a circuit-opening fusible part that is heated and severed by the passage of overcurrent through it. Fuses may be used instead of circuit breakers for most applications. They are available in a wide range of voltage, current, and interrupting ratings, current-limiting and non-current-limiting types, and for indoor and outdoor applications.

Fuses over 600 V have an interrupting capability based on asymmetrical current, although their published ratings are expressed in symmetrical amperes. Current-limiting fuses interrupt a short circuit within the first half-cycle, and their equivalent asymmetrical rating includes a 1.6 multiplier to provide for the maximum expected current asymmetry. Fuse ratings for 600 V and below are also published as symmetrical current values.

Current-limiting fuses are extremely fast in operation at very high values of fault current, and act to limit the current in less than one quarter-cycle to a value well below the available peak short-circuit current. Non-current-limiting

fuses may operate in one or two cycles depending on the level of fault current. Several types of current-limiting fuses for 600 V and below are now available for alternating-current service with interrupting ratings as high as 200 000 A, symmetrical rms, in accordance with NEMA FU 1-1972, Low-Voltage Cartridge Fuses. For more information on fuses see Chapter 4.

(2) *Application Considerations.* There is no general rule whether fuses or a circuit breaker should be used. The designer must evaluate the total performance in terms of what his particular application demands. The following application considerations may be of assistance to the designer.

(a) *Interrupting Ratings.* For some applications, 200 000 A symmetrical rms interrupting ratings of 600 V and below, current-limiting fuses are available.

(b) *Component Protection*

(i) Current-limiting fuses will permit lower momentary and interrupting ratings by limiting let-through current to within equipment rating.

(ii) In a polyphase circuit, the opening of one fuse, in response to a fault condition, may severely reduce the magnitude of current continuing to flow to the fault and make impossible the operation of the remaining fuses in the circuit. Thus the fault current will not be totally cleared. For general protection against motor single-phasing damage, three properly sized overcurrent running devices must be used.

(iii) Dual-element fuses can provide more time delay, allowing closer sizing and better overload protection on high current inrush loads.

(iv) Fuses cannot provide sensitive ground-fault protection.

(c) *Selective Coordination*

(i) It may be more difficult to apply circuit breakers with instantaneous elements in a coordinated scheme on 600 V and below systems. Circuit breakers can be equipped with short-time-delay trips to improve coordination, if care is exercised to ensure that equipment ratings are not exceeded.

(ii) Fuse time—current clearing characteristics are less accurate than relay-controlled circuit breaker tripping characteristics with the result that coordinated circuit protection is less reliable.

(iii) Coordination of power circuit breakers without short-time delay is difficult, and tolerances of time—current curves in the overload and short-circuit range can vary.

(d) *Space Requirements.* Fusible switching devices can take more space; however, fuses alone are generally smaller than mechanical protective devices if switching is not required.

(e) *Economics*

(i) First cost, life cycle, and maintenance cost are lower for fusible equipments.

(ii) The simpler mechanical operation of fusible equipments results in lower maintenance costs. Circuit breakers require periodic maintenance in accordance with NEMA and manufacturers' recommendations.

(f) *Automatic Switching.* Circuit breakers can be automatically switched for a variety of needs as remote control or ground-fault protection. Fuses alone are not capable of automatic switching, but can be installed in suitable shunt-trip equipped switches to provide this service.

8.2.4 *Circuit Breakers.* A circuit breaker is a device designed to open and close a circuit by nonautomatic means, and to open the circuit automatically on a predetermined overload of current without injury to itself when properly applied within its rating. Ordinarily, circuit breakers are required to operate only infrequently, although some classes of circuit breakers are suitable for frequent operation. The interrupting rating of a circuit breaker shall be equal to or greater than the available system short-circuit current.

Power circuit breakers are used extensively on utility and industrial (largely over 1000 V) and industrial (predominantly 1000 V and below) power distribution systems to provide essential switching flexibility and circuit protection.

Circuit breakers are available for the entire voltage range and may be furnished single-, double-, or triple-pole, and arranged for indoor or outdoor use. Circuit breakers for 34.5 kV service and above are generally not enclosed or available for indoor location.

(1) *Circuit Breakers Over 1000 V.* The rated close and latch and interrupting-current capabilities are very important factors for use in the application of circuit breakers over 1000 V. The close and latch capability is a measure of

the equipment's ability to withstand the mechanical stresses produced by the asymmetrical short-circuit current during the first few cycles without severe mechanical damage, and is normally expressed in total rms amperes. This asymmetrical current consists of a direct-current component superimposed on an alternating-current component. The direct-current component decays with time, depending upon the resistance and reactance or the X/R of the circuit. The initial value of the direct-current component of the short-circuit current depends on the point of the normal voltage wave at which the fault occurs. The procedure to be used for short-circuit selection of power circuit breakers in the over 1000 V class is covered in Chapter 5. Application data can be found in IEEE Std 320-1972, Application Guide for AC High-Voltage Circuit Breakers Rated on a Symmetrical Current Basis (ANSI 37.010-1972).

(2) *Ratings.* For the rating of power circuit breakers in the over 1000 V class refer to ANSI C37.06-1971, Schedules of Preferred Ratings and Related Required Capabilities for AC High-Voltage Circuit Breakers Rated on a Symmetrical Current Basis. Present circuit breakers being manufactured are rated on the symmetrical basis. In specifying these circuit breakers consideration must be given to the related values and required capabilities listed as headings in Table 56. This table lists preferred ratings for indoor oilless circuit breakers. These ratings are applicable for service at altitudes up to 3300 ft. For service beyond 3300 ft, derating factors must be applied in accordance with ANSI C37.04-1964 (R 1969), Rating Structure for AC High-Voltage Circuit Breakers.

Power circuit breakers used for applications up through 15 kV are predominantly of the air-magnetic circuit breaker type, with limited usage of vacuum-type interrupters. For voltages above 15 kV, the available types of circuit breakers include oil, compressed air or gas, and vacuum interrupters.

(3) *Circuit Breakers 1000 V and Below.* Circuit breaker 1000 V and below are divided into (1) power circuit breakers and (2) molded-case circuit breakers. The basis for their comparison can be found in ANSI and NEMA standards for power circuit breakers, and UL and NEMA standards for molded-case circuit breakers.

Table 56
Preferred Ratings for Indoor Oilless Circuit Breakers
(Symmetrical Current Basis of Rating)

Identification		Rated Values									Related Required Capabilities		
		Voltage		Insulation Level Rated Withstand Test Voltage		Current					Current Values		
Nominal Voltage Class (1)* kV, rms	Nominal 3 Phase MVA Class (1)	Rated Max Voltage (2) kV, rms	Rated Voltage Range Factor, K (3)	Low Frequency kV, rms	Impulse (4) kV, Crest	Rated Continuous Current at 60 Hz (5) Amperes, rms	Rated Short-Circuit Current (at Rated Max kV) (6)(7) kA, rms	Rated Interrupting Time (8) Cycles	Rated Permissible Tripping Delay, Y Seconds	Rated Max Voltage Divided by K kV, rms	Max Symmetrical Interrupting Capability (9) / K Times Rated Short-Circuit Current kA, rms	3 Second Short-Time Current Carrying Capability (10) kA, rms	Closing and Latching Capability 1.6 K Times Rated Short-Circuit Current (10)(11) kA, rms
Col 1	Col 2	Col 3	Col 4	Col 5	Col 6	Col 7	Col 8	Col 9	Col 10	Col 11	Col 12	Col 13	Col 14
4.16	75	4.76	1.36	19	60	1200	8.8	5	2	3.5	12	12	19
4.16	150	4.76	1.36	19	60	1200	18	5	2	3.5	24	24	39
4.16	250	4.76	1.24	19	60	1200	29	5	2	3.85	36	36	58
4.16	250	4.76	1.24	19	60	2000	29	5	2	3.85	36	36	58
4.16	350	4.76	1.19	19	60	1200	41	5	2	4.0	49	49	78
4.16	350	4.76	1.19	19	60	3000	41	5	2	4.0	49	49	78
7.2	250	8.25	1.79	36	95	1200	17	5	2	4.6	30	30	49
7.2	500	8.25	1.25	36	95	1200	33	5	2	6.6	41	41	66
7.2	500	8.25	1.25	36	95	2000	33	5	2	6.6	41	41	66
13.8	250	15	2.27	36	95	1200	9.3	5	2	6.6	21	21	34
13.8	500	15	1.30	36	95	1200	18	5	2	11.5	23	23	37
13.8	500	15	1.30	36	95	2000	18	5	2	11.5	23	23	37
13.8	750	15	1.30	36	95	1200	28	5	2	11.5	36	36	58
13.8	750	15	1.30	36	95	2000	28	5	2	11.5	36	36	58
13.8	1000	15	1.30	36	95	1200	37	5	2	11.5	48	48	77
13.8	1000	15	1.30	36	95	3000	37	5	2	11.5	48	48	77

(Line Nos. 1–16)

From ANSI C37.06-1971.

NOTES to Table 56

*Numbers in parentheses refer to the notes below.

NOTES: These ratings were prepared by the EEI-AEIC-NEMA Joint Committee on Power Circuit Breakers.

For service conditions, definitions, and interpretation of ratings, tests, and qualifying terms, see ANSI C37.03-1964 (R 1969), C37.04-1964 (R 1969) C37.04a-1964 (R 1969), C37.04b-1970, C37.09-1964 (R 1969), and C37.09a-1970.

The interrupting ratings are for 60 Hz systems. Applications on 25 Hz systems should receive special consideration.

Current values have been rounded off to the nearest kiloampere except below 10 kA where two significant figures are used.

(1) For reference only. Figures in Col 2 must not be used for evaluation of circuit breaker in any specific application. Actual application must be based on rated short-circuit current at rated maximum voltage and in accordance with Notes (6) and (7).

(2) The voltage rating is based on ANSI C84.1-1970, Voltage Ratings for Electric Power Systems and Equipment (60 Hz), where applicable and is the maximum voltage for which the circuit breaker is designed and the upper limit for operation.

(3) The rated voltage range for factor K is the ratio of rated maximum voltage to the lower limit of the range of operating voltage in which the required symmetrical and asymmetrical current interrupting capabilities vary in inverse proportion to the operating voltage.

(4) 1.2×50 μs positive and negative wave. All impulse values are phase to phase and phase to ground and across the open contacts.

(5) The 25 Hz continuous-current ratings in amperes are given herewith following the respective 60 Hz rating: 1200—1400; 2000—2250; 3000—3500.

(6) To obtain the required symmetrical current interrupting capability of a circuit breaker at an operating voltage between $1/K$ times rated maximum voltage and rated maximum voltage, the following formula shall be used:

required symmetrical current
interrupting capability

$$= \text{rated short-circuit current}$$

$$\times \frac{\text{rated maximum voltage}}{\text{operating voltage}}$$

For operating voltages below $1/K$ times rated maximum voltage, the required symmetrical current interrupting capability of the circuit breaker shall be equal to K times rated short-circuit current.

(7) With the limitation stated in 04-4.5 of ANSI C37.04-1964 (R 1969), all values apply for polyphase and line-to-line faults. For single-phase-to-ground faults, the specific conditions stated in 04-4.5.2.3 of ANSI C37.04-1964 (R 1969) apply.

(8) The ratings in this column are on a 60 Hz basis and are the maximum time interval to be expected during a circuit breaker opening operation between the instant of energizing the trip circuit and interruption of the main circuit on the primary arcing contacts under certain specified conditions. The values may be exceeded under certain conditions as specified in 04-4.8 of ANSI C37.04-1964 (R 1969).

(9) Current values in this column are not to be exceeded even for operating voltages below $1/K$ rated maximum voltage. For voltages between rated maximum voltage and $1/K$ times rated maximum voltage, follow (6) above.

(10) Current values in this column are independent of operating voltage up to and including rated maximum voltage.

(11) If currents are to be expressed in peak amperes, multiply values in this column by a factor of 1.69, which is a ratio of 2.7/1.6.

Power circuit breakers 1000 V and below are open-construction assemblies on metal frames with all parts designed for accessible maintenance, repair, and ease of replacement. They are intended for service in switchgear compartments or other enclosures of dead front construction. Tripping units are field adjustable over a wide range and are completely interchangeable within their frame sizes. The tripping units used have been the electromagnetic overcurrent direct-acting type; however, static-type tripping units are now available from most manufacturers.

Power circuit breakers 1000 V and below can be used with integral current-limiting fuses in drawout construction to meet interrupting current requirements up to 200 000 A, symmetrical rms, of the system to which they are applied. When part of the circuit breaker, the fuses are combined with an integral mounted anti-single-phasing device to eliminate the possibility of single-phasing.

A molded-case circuit breaker (NEMA AB 1-1975) is a switching device and an automatic protective device assembled in an integral housing of insulating material. They are available in the following general types.

(a) *Thermal Magnetic.* This type employs thermal tripping for overloads and instantaneous magnetic tripping for short circuits. These are the most widely applicable molded-case circuit breakers.

(b) *Magnetic.* This type employs only instantaneous magnetic tripping where short-circuit protection only is required.

(c) *Fused Molded Case.* This type combines regular thermal magnetic protection against normal short-circuit and overcurrent faults with current-limiting fuse protection against higher short-circuit faults.

(d) *High Interrupting Capacity.* This type provides protection for higher short-circuit currents than do standard constructed thermal magnetic circuit breakers without the use of fuses and without increasing the amount of space. This line incorporates sturdier construction of contacts and mechanism plus a special high-impact molded casing.

Circuit breaker interrupting devices for 1000 V and below are capable of operating to clear a fault current more rapidly than over 1000 V power circuit breakers. In present designs, 1000 V and below air circuit breaker contacts often begin to part during the first cycle of short-circuit current. Consequently, such equipment must be rated to interrupt the maximum available first-cycle asymmetrical current. However, air circuit breakers 1000 V and below are rated on a symmetrical current basis, eliminating the need for applying direct-current offset multipliers in determining equipment ratings. Molded-case circuit breakers fall into the same classification; thus momentary withstand and interrupting ratings are the same.

Circuit breakers 1000 V and below can be applied on a symmetrical basis if the system X/R ratio does not exceed 6.6. If the system X/R ratio is higher, then the asymmetrical capability in the pole of the circuit breaker having maximum offset must be checked against the maximum phase asymmetrical duty available at the circuit breaker location.

Power circuit breakers are designed for periodic planned maintenance. This design permits their higher endurance ratings and repetitive duty capabilities and forms one basis for their broader range of applications. It is important to recognize that difference in test criteria between power circuit breakers and molded-case circuit breakers can be significant in the application of circuit breakers at or near their interrupting rating. A combination of circumstances can reduce the performance of molded-case circuit breakers to less than adequate if they are applied at or near their interrupting rating on a par with power circuit breakers.

Molded-case circuit breakers generally are not designed to be maintained in the field as are 1000 V and below power circuit breakers. Molded cases are sealed to prevent tampering, thereby precluding inspection of the contacts. In addition, replacement parts are not generally available. Manufacturers recommend total replacement of the molded-case circuit breaker if a defect appears, or if the unit begins to overheat. Molded-case circuit breakers, particularly the larger sizes, are not suitable for repetitive switching.

The circuit breaker must be capable of closing, carrying, and interrupting the highest fault current possible at that location. It is essential to select a circuit breaker whose short-circuit (in-

Table 57
Preferred Ratings for 1000 V and Below AC Power Circuit Breakers With
Instantaneous Trip Devices (Releases)

Line No.	System Nominal Voltage (volts)	Rated Maximum Voltage (volts)	Insulation (Dielectric) Withstand (volts)	Three-Phase Short-Circuit Current Rating, Symmetrical (amperes)*	Frame Size (amperes)	Range of Trip-Device Current Ratings (amperes)
	Col 1	Col 2	Col 3	Col 4	Col 5	Col 6
1	600	635	2200	14 000	225	40–225
2	600	635	2200	22 000	600	40–600
3	600	635	2200	42 000	1600	200–1600
4	600	635	2200	42 000	2000	200–2000
5	600	635	2200	65 000	3000	2000–3000
6	600	635	2200	85 000	4000	4000
7	480	508	2200	22 000	225	40–225
8	480	508	2200	30 000	600	100–600
9	480	508	2200	50 000	1600	400–1600
10	480	508	2200	50 000	2000	400–2000
11	480	508	2200	65 000	3000	2000–3000
12	480	508	2200	85 000	4000	4000
13	240	254	2200	25 000	225	40–225
14	240	254	2200	42 000	600	150–600
15	240	254	2200	65 000	1600	600–1600
16	240	254	2200	65 000	2000	600–2000
17	240	254	2200	85 000	3000	2000–3000
18	240	254	2200	130 000	4000	4000

From ANSI C37.16-1973.

*Single-phase short-circuit current ratings are 87% of these values.

Table 58
Preferred Ratings for 1000 V and Below AC Power Circuit Breakers Without
Instantaneous Trip Devices (Releases)

Line No.	System Nominal Voltage (volts)	Rated Maximum Voltage (volts)	Insulation (Dielectric) Withstand (volts)	Three-Phase Short-Circuit Current Rating* or Short-Time Current Rating, Symmetrical (amperes)†	Frame Size (amperes)	Range of Trip-Device Current Ratings (amperes) Setting of Short-Time-Delay Trip Element		
						Minimum Time Band	Intermediate Time Band	Maximum Time Band
	Col 1	Col 2	Col 3	Col 4	Col 5	Col 6	Col 7	Col 8
1	600	635	2200	14 000	225	100–225	125–225	150–225
2	600	635	2200	22 000	600	175–600	200–600	250–600
3	600	635	2200	42 000	1600	350–1600	400–1600	500–1600
4	600	635	2200	42 000	2000	350–2000	400–2000	500–2000
5	600	635	2200	65 000	3000	2000–3000	2000–3000	2000–3000
6	600	635	2200	85 000	4000	4000	4000	4000
7	480	508	2200	14 000	225	100–225	125–225	150–225
8	480	508	2200	22 000	600	175–600	200–600	250–600
9	480	508	2200	42 000	1600	350–1600	400–1600	500–1600
10	480	508	2200	50 000	2000	350–2000	400–2000	500–2000
11	480	508	2200	65 000	3000	2000–3000	2000–3000	2000–3000
12	480	508	2200	85 000	4000	4000	4000	4000
13	240	254	2200	14 000	225	100–225	125–225	150–225
14	240	254	2200	22 000	600	175–600	200–600	250–600
15	240	254	2200	42 000	1600	350–1600	400–1600	500–1600
16	240	254	2200	50 000	2000	350–2000	400–2000	500–2000
17	240	254	2200	65 000	3000	2000–3000	2000–3000	2000–3000
18	240	254	2200	85 000	4000	4000	4000	4000

From ANSI C37.16-1973.

*Short-circuit current ratings for circuit breakers without direct-acting trip devices, opened by a remote relay, are the same as those listed here.

†Single-phase short-circuit current ratings are 87 percent of these values.

Table 59
Standard Ratings of Molded-Case Circuit Breakers

Frame size, amperes	50, 100, 125, 150, 200, 225, 400, 600, 800, 1000, 1200, 1600, 2000, 2500
Frequency, hertz	60
Rated voltage, volts	
alternating current	120, 120/240, 240, 277, 277/480, 480, 600
direct current	125, 125/250, 250
Rated interrupting currents,	7500, 10 000, 14 000, 18 000, 22 000, 25 000,
symmetrical rms amperes	30 000, 35 000, 42 000, 50 000, 65 000, 85 000,
(60 Hz alternating and direct current)	100 000, 125 000, 150 000, 200 000

terrupting) rating at the circuit voltage is equal to or greater than the available short-circuit current at the point of installation. The procedure to be used for short-circuit selection of 1000 V and below power circuit breakers is covered in Chapter 5.

Manufacturers' publications give specific information on mechanical and electrical features of circuit breakers 1000 V and below. Refer to Tables 57—59 for lists of standard ratings for 1000 V and below power circuit breakers and molded-case circuit breakers. For service at altitudes above 3300 ft above sea level, derating factors must be applied in accordance with IEEE Std 20-1973, Low-Voltage AC Power Circuit Breakers Used in Enclosures (ANSI C37.13-1973).

NOTE (applicable to Tables 57 and 58): Solid-state trip devices for both overcurrent and ground-fault protection are readily available from most manufacturers although the trip ranges given here are based on electromechanical direct or indirect acting types. Standards for solid-state trip devices should be forthcoming in later editions of this publication. In the meantime it is recommended that the manufacturer be consulted relative to solid-state trip characteristics. Solid-state trip devices provide many advantages over conventional types and should not be overlooked. A major one is the inherent provision for sensitive ground-fault protection. For further discussion and trip characteristic curves, see Chapter 4. Particular attention must be given to coordination of load-side fuse devices with instantaneous trip devices of circuit breakers.

(4) *Vacuum Power Circuit Breakers.* In general, vacuum power circuit breakers are applied in accordance with the specific continuous and short-circuit current requirements in the same manner as air-magnetic circuit breakers. It must be recognized, however, that under certain conditions, vacuum interrupters have characteristics which are different from air-magnetic power circuit breakers. Vacuum interrupters will sometimes, in special applications, force a premature current zero by opening the circuit in an unusually short time. When this occurs, a higher than normal transient recovery voltage occurs that can be of a magnitude that will impose excessive dielectric stress on the connected equipment. There are times when this magnitude may be greater than the basic impulse insulation level of any connected device and failure may result.

When applying vacuum power circuit breakers, the following precautions should be taken.

(a) *Switching Unloaded Transformers.* When switching power transformers that are unloaded, that is, interrupting just the small magnetizing current on an infrequent basis (less than 50 operations per year), and where the basic impulse level is 95 kV or higher, no special attention is required. However, should a dry-type transformer be involved with a less than 95 kV basic impulse level rating, or else if all switching is highly repetitive, then the application should be checked with the transformer manufacturer.

(b) *Switching Loaded Transformers.* When a power transformer has a permanently connected load in kilovolt-amperes of 5 percent or more, no special consideration is needed.

(c) *Switching Motors.* When vacuum power circuit breakers are utilized to switch motors, the standard rotating-machine protection package of capacitors and surge arresters must always be used.

Unusual service conditions as defined in ANSI C37.04-1964 (R 1969) and IEEE Std 20-1973 must be considered when applying power circuit breakers. Such conditions should be brought to the attention of the circuit breaker manufacturer at the earliest possible time.

(5) *Service Protectors.* A service protector is a quick-make, quick-break current-limiting fuse and nonautomatic circuit breaker type switching and protective device. Stored energy operation provides for manual or electrical closing. The service protector, utilizing basic circuit breaker principles, permits frequent repetitive operation under normal and abnormal current conditions up to 12 times the device's continuous-current rating. In combination with current-limiting fuses it is capable of closing and latching against fault currents up to 200 000 A, symmetrical rms. During fault interruption, the service protector will withstand the stresses created by the let-through current of the fuses.

Service protectors are available at continuous-current ratings of 800, 1200, 1600, 2000, 3000, and 4000 A for use on 240 and 480 V alternating-current systems, in two-pole and three-pole construction. Single-phase protection is included in the design of the service protector.

8.3 Switchgear

8.3.1 *General Discussion.* Switchgear is a general term covering switching and interrupting devices alone or their combination with other associated control, metering, protective, and regulating equipment.

A power switchgear assembly consists of a complete assembly of one or more of the above noted devices and main bus conductors, interconnecting wiring, accessories, supporting structures, and enclosure. Power switchgear is applied throughout the electric power system of an industrial plant but is principally used for incoming line service and to control and protect load centers, motors, transformers, motor control centers, panelboards, and other secondary distribution equipment.

Outdoor switchgear assemblies can be of the non-walk-in (without enclosed maintenance aisle) or walk-in (with an enclosed maintenance aisle) variety. Switchgear for industrial plants is generally located indoors; such location is

preferred because of easier maintenance, avoidance of weather problems, and shorter runs of feeder cable or bus duct. In outdoor applications the effect of external influences, principally the sun, wind, moisture, and local ambient temperatures, must be considered in determining the suitability and current-carrying capacity of the switchgear. Further information on this evaluation is contained in IEEE Std 144-1971, Guide for Evaluating the Effect of Solar Radiation on Outdoor Metal-Clad Switchgear (ANSI C37.24-1971).

In many locations, the use of lighter colored (nonmetallic) paints will minimize the effect of solar energy loading and increase the rating of the equipment in outdoor locations. Future editions of IEEE Std 144 will note this effect.

8.3.2 *Classifications.* An open switchgear assembly is one that does not have an enclosure as part of the supporting structure. An enclosed switchgear assembly generally consists of a metal-enclosed supporting structure with the switchgear enclosed on the top and all sides with sheet metal (except for ventilating openings and inspection windows). Access within the enclosure is provided by doors or removable panels.

Metal-enclosed switchgear is universally used throughout industry for utilization and primary distribution voltage service, for alternating- and direct-current applications, and for indoor and outdoor locations.

8.3.3 *Types.* Specific types of metal-enclosed power switchgear are used in industrial plants. These types are defined as (1) metal-clad switchgear, (2) low-voltage power circuit breaker switchgear, and (3) interrupter switchgear. Metal-enclosed bus will also be discussed because it is frequently used in conjunction with power switchgear in modern industrial power systems.

8.3.4 *Definitions.* Metal-Clad switchgear is metal-enclosed power switchgear characterized by the following necessary features.

(1) The main circuit switching and interrupting device is of the removable type arranged with a mechanism for moving it physically between connected and disconnected positions and equipped with self-aligning and self-coupling primary and secondary disconnecting devices. There are two basic designs. In one the circuit

breaker is withdrawn horizontally to achieve connected, test, disconnected, and fully withdrawn positions. In the other the circuit breaker is disconnected by being lowered vertically.

(2) Major parts of the primary circuit, such as the circuit switching or interrupting devices, buses, potential transformers, and control power transformers, are enclosed by grounded metal barriers. Specifically included is an inner barrier in front of or a part of the circuit interrupting device to ensure that no energized primary circuit components are exposed when the unit door is opened.

(3) All live parts are enclosed within grounded metal compartments. Automatic shutters prevent exposure of primary circuit elements when the removable element is in the test, disconnected, or fully withdrawn position.

(4) Primary bus conductors and connections are covered with insulating material where practicable. This constitutes only a small portion of the effective bus insulation.

(5) Mechanical interlocks are provided to ensure a proper and safe operating sequence.

(6) Instruments, meters, relays, secondary control devices, and their wiring are isolated by grounded metal barriers from all primary circuit elements with the exception of short lengths of wire such as at instrument transformer terminals.

(7) The door through which the circuit-interrupting device is inserted into the housing may serve as an instrument or relay panel and may also provide access to a secondary or control compartment within the housing.

Auxiliary frames may be required for mounting associated auxiliary equipment such as potential transformers, control power transformers, etc.

The term metal-clad switchgear can be properly used only if metal-enclosed switchgear conforms to the foregoing definition. All metal-clad switchgear is metal-enclosed, but not all metal-enclosed switchgear can be correctly designated as metal-clad. The most prevalent type of switching and interrupting device used in metal-clad switchgear is the air-magnetic power circuit breaker over 1000 V.

Metal-enclosed 1000 V and below power circuit breaker switchgear is metal-enclosed power switchgear, including the following equipment

as required:

(1) 1000 V and below power circuit breakers (fused or unfused)

(2) Bare bus and connections

(3) Instrument and control power transformers

(4) Instruments, meters, and relays

(5) Control wiring and accessory devices

(6) Cable and busway termination facilities

The 1000 V and below power circuit breakers are contained in individual grounded metal compartments and controlled either remotely or from the front of the panels. The circuit breakers are usually of the drawout type, but may be nondrawout. When drawout type circuit breakers are used, mechanical interlocks must be provided to ensure a proper and safe operating sequence.

Metal-enclosed interrupter switchgear is metal-enclosed power switchgear including the following equipment as required:

(1) Interrupter switches

(2) Power fuses

(3) Bare bus and connections

(4) Instrument and control power transformers

(5) Control wiring and accessory devices

The interrupter switches and power fuses may be of the stationary or removable type. For the removable type, mechanical interlocks are provided to ensure a proper and safe operating sequence.

Metal-enclosed bus is an assembly of rigid electrical buses with associated connections, joints, and insulating supports, all housed within a grounded metal enclosure. Three basic types of metal-enclosed bus construction are recognized: nonsegregated phase, segregated phase, and isolated phase. The most prevalent type used in industrial power systems is the nonsegregated phase which is defined as "one in which all phase conductors are in a common metal enclosure without barriers between the phases." When metal-enclosed bus over 1000 V is used with metal-clad switchgear, the bus commonly has the same continuous-current rating as the switchgear main bus, and its primary bus conductors and connections are covered with insulating material throughout. When metal-enclosed bus is associated with metal-enclosed 1000 V and below power circuit breaker switch-

Table 60

Rated Voltages and Insulation Levels for AC Switchgear Assemblies

Rated Voltage (rms)		Insulation Levels (kV)		
Rated Nominal Voltage	Rated Maximum Voltage	Power Frequency Withstand (rms)	DC Withstand*	Impulse Withstand
Metal-Enclosed Low-Voltage Power Circuit Breaker Switchgear				
Volts	Volts			
240	250	2.2	3.1	—
480	500	2.2	3.1	—
600	630	2.2	3.1	—
Metal-Clad Switchgear				
kV	kV			
4.16	4.76	19	27	60
7.2	8.25	36	50	95
13.8	15.0	36	50	95
34.5	38.0	80	†	150
Metal-Enclosed Interrupter Switchgear				
kV	kV			
4.16	4.76	19	27	60
7.2	8.25	26	37	75
13.8	15.0	36	50	95
14.4	15.5	50	70	110
23.0	25.8	60	†	125
34.5	38.0	80	†	150
Station-Type Cubicle Switchgear				
kV	kV			
14.4	15.5	50	†	110
34.5	38.0	80	†	150
69.0	72.5	160	†	350

From IEEE Std 27-1974.

*The column headed "DC Withstand" is given as a reference only for those using direct-current tests and represents values believed to be appropriate and approximately equivalent to the corresponding power frequency withstand test values specified for each voltage class of switchgear. The presence of this column in no way implies any requirement for a direct-current withstand test on alternating-current equipment. When making direct-current tests, the voltage should be rasied to the test value in discrete steps and held for a period of 1 min.

†Because of the variable voltage distribution encountered when making direct-current withstand tests, the manufacturer should be contacted for recommendations before applying direct-current withstand tests to the switchgear. Potential transformers above 34.5 kV should be disconnected when testing with direct current. Refer to 6.8 of ANSI C57.13-1968, and in particular to 6.8.2 which reads "Periodic kenotron tests should not be applied to transformers of higher than 34.5 kV voltage ratings."

Table 61
Voltage Ratings for Metal-Enclosed Bus

Rated AC Voltage (kV rms)		Insulation Level (kV)			
		Power Frequency Withstand (rms)		DC Withstand (Dry)†	Impulse Withstand
Nominal	Rated Maximum	(Dry 1 Minute)	(Dew 10 Seconds)*		
0.6	0.63	2.2	—	3.1	—
4.16	4.76	19.0	15	27.0	60
13.8	15.00	36.0	24 (36)	50.0	95
14.4	15.50	50.0	30 (50)	70.0	110
23.0	25.80	60.0	40 (60)	85.0	150
34.5	38.00	80.0	70 (80)	‡	200
69.0	72.50	160.0	140 (160)	‡	350

For applications of isolated phase bus to generators, the following voltage ratings apply:§

Rated kV of Generator (rms)	Power Frequency Withstand (rms)		DC Withstand (Dry)	Impulse Withstand
	(Dry 1 Minute)	(Dew 10 Seconds)		
14.4 to 24	50	50	70	110

From IEEE Std 27-1974.

*Applied to porcelain insulation only. Values in parentheses apply to "high creepage" designs.

†The column headed "DC Withstand" is given as a reference only for those using direct-current tests and represents equivalent to the corresponding power frequency withstand test values specified for each voltage class of bus. The presence of this column in no way implies any requirement for a direct-current withstand test on alternating-current equipment. When making direct-current tests the voltage should be raised to the test value in discrete steps and held for a period of 1 min.

‡Because of the variable voltage and distribution encountered when making direct-current withstand tests, the manufacturer should be contacted for recommendations before applying direct-current withstand tests to these voltage ratings. Potential transformers above 34 5 kV should be disconnected when testing with direct current. Refer to 6.8 of ANSI C57.13-1968, and in particular to 6.8.2 which reads, "Periodic kenotron tests should not be applied to transformers of higher than 34.5 kV voltage rating."

§ These ratings are applicable to generators rated 14.4 to 24 kV which are directly connected to transformers without intermediate circuit breakers and where adequate surge protection is provided. These bus withstand ratings are compatible with or in excess of required withstand values of the generators.

gear or metal-enclosed interrupter switchgear, the primary bus conductors and connections are usually bare.

8.3.5 *Ratings.* The ratings of switchgear assemblies and metal-enclosed bus are designations of the operational limits of the particular equipment under specific conditions of ambient temperature, altitude, frequency, duty cycle, etc. Table 60 lists the rated voltages and insulation levels for alternating-current switchgear assemblies discussed in this section. Table 61 lists similar ratings for metal-enclosed bus. Rated voltages and insulations levels for direct-current switchgear assemblies can be found by referring to IEEE Std 27-1974, Switchgear Assemblies Including Metal-Enclosed Bus [ANSI C37.20 (1974 ed)]. The definition of the ratings listed in Tables 60 and 61, and other subsequently discussed, can be found in ANSI C37.100-1972, Definitions for Power Switchgear.

Standard self-cooled continuous-current ratings of the main bus in metal-enclosed power switchgear are listed in Table 62. Metal-enclosed bus standard self-cooled continuous-current ratings are shown in Table 63.

The momentary and short-time short-circuit current ratings of power switchgear assemblies shall correspond to the equivalent ratings of the switching or interrupting devices used.

The limiting temperature for a power switchgear assembly or metal-enclosed bus (where applicable) is the maximum temperature permitted:

(1) For any component such as insulation, buses, instrument transformers, and switching and interrupting devices; or

(2) For air in cable termination compartments; or

(3) For any non-current-carrying structural parts

The hot-spot temperature rise and hot-spot total temperature to which insulating materials are subjected shall not exceed established values for the various classes of insulating materials. The hot-spot temperature rise and hot-spot total temperature of buses and connections for metal-enclosed power switchgear assemblies and bus shall not exceed the limits as established in Table 64. These temperature limitations apply for all bus continuous-current ratings shown in Tables 62 and 63. Additional information re-

Table 62
Continuous-Current Ratings of Buses for Metal-Enclosed Power Switchgear

Type of Assembly	Rated Continuous Current of Buses (amperes)
Metal-clad switchgear	1200, 2000, 3000
Metal-enclosed interrupter switchgear	600, 1200, 2000
Metal-enclosed bus	See Table 63
Station-type cubicle switchgear	2000, 3000, 4000, 5000

From IEEE Std 27-1974.

NOTE: The numerous combinations of circuit-breaker sizes and ratings in similar housings make it impractical to provide current ratings of buses for metal-enclosed low-voltage power circuit breaker switchgear.

garding limits of temperature rise for air surrounding enclosed power switchgear devices, power cable, parts handled by operating personnel, exposed surfaces, etc, can be obtained by referring to IEEE Std 27-1974.

8.3.6 *Application Guides.* Metal-enclosed switchgear is available for application at voltages up through 34.5 kV. Metal-clad switchgear with vacuum power circuit breakers is more available for use on industrial power systems having nominal voltage ratings of 13.8 and 34.5 kV. Rating information on circuit breakers for use in metal-enclosed switchgear is shown in Tables 56—58. Rating data for metal-clad switchgear with vacuum power circuit breakers should be obtained directly from the equipment manufacturer. Because of rapid advances in the state-of-the-art, new application rules are being generated as field experience with vacuum interrupters is increased.

Metal-enclosed switchgear with air-magnetic or vacuum power circuit breakers when properly applied as protective equipment on power systems meets the following requirements: (1) personnel safety, (2) system reliability, (3) adaptability, (4) minimal maintenance, and (5) low total cost. Personnel safety and equipment reliability are two of the prime reasons for user insistence on metal-enclosed switchgear to perform the industrial power

Table 63
Current Ratings for Metal-Enclosed Bus, in Amperes

| 0.6 AC and All DC | Voltage Ratings (kV) | | | | | |
	2.4, 4.16	13.8	14.4	23.0	34.5	69.0
600	—	—	—	—	—	—
1200	1200	1200	1200	1200	1200	1200
1600	—	—	—	—	—	—
2000	2000	2000	2000	2000	2000	2000
—	—	—	2500	2500	2500	2500
3000	3000	3000	3000	3000	3000	3000
—	—	—	3500	3500	—	—
4000	—	—	4000	4000	—	—
—	—	—	4500	4500	—	—
5000	—	—	5000	5000	—	—
—	—	—	5500	5500	—	—
6000	—	—	6000	6000	—	—
—	—	—	6500	—	—	—
—	—	—	7000	—	—	—
—	—	—	7500	—	—	—
—	—	—	8000	—	—	—
—	—	—	9000	—	—	—
—	—	—	10 000	—	—	—
—	—	—	11 000	—	—	—
—	—	—	12 000	—	—	—

Table 64
Temperature Limits for Buses and Connections in Switchgear Assemblies

Type of Bus or Connection	Limit of Hottest Spot Temperature Rise (°C)	Limit of Hottest Spot Total Temperature (°C)
Buses and connections with copper-to-copper connecting joints	30	70
Buses and connections with silver-surfaced (or equivalent) connecting joints	65	105
Terminations to insulated cables (copper to copper)	30	70
Terminations to insulated cables (silver surfaced or equivalent)	45	85

From IEEE Std 27-1974.

NOTE: The temperature of the air surrounding all devices within an enclosed assembly, considered in conjunction with their rating and loading as used, shall not cause these devices to operate outside their normal temperature range when the enclosure is surrounded by air within the range of −30° C to +40° C.

system protective function. Metal-enclosed switchgear enhances system reliability because of its basic construction features and the flexibility derived from the multitude of main bus configurations available to the user. It is adaptable to many applications because it is easily expanded and can be specified and designed with load location and load characteristics in mind. Reduced maintenace cost results if metal-enclosed switchgear with drawout interrupting devices are applied. Ease of maintenance is facilitated because of the accessibility of most components. On the average, metal-enclosed switchgear represents a small percentage of total plant cost. Metal-enclosed switchgear is generally shipped factory assembled and reduces the amount of expensive field assembly.

The following steps are normally taken in applying switchgear equipment.

(1) Develop a single-line diagram.

(2) Determine proper power circuit breaker based on required service voltage, continuous and momentary current, and interrupting capability (see Tables 56—58).

(3) Select main bus rating.

(4) Select current transformers.

(5) Select potential transformers.

(6) Select metering, relaying, and control power.

(7) Determine closing, tripping, and other control power requirements.

(8) Consider special applications.

Essentially all recognized basic bus arrangements, radial, double bus, circuit breaker and half, main and transfer bus, sectionalized bus, synchronizing bus, and ring bus, are available in metal-enclosed switchgear to ensure the desired system reliability and flexibility. A choice is made based on an evaluation of initial cost, installation cost, required operating procedures, and total system requirements.

The continuous-current rating of the switchgear main bus must be no less than that of the highest rated circuit breaker. Table 62 lists standard main bus continuous-current ratings for metal-enclosed power switchgear. The rated continuous current of a switchgear assembly is the maximum current in rms amperes, at rated frequency, which can be carried continuously by the primary circuit components

Table 65
Standard Accuracy Class Ratings* of Current Transformers in Metal-Enclosed Low-Voltage Power Circuit Breaker Switchgear

Ratio	B 0.1	B 0.2
100/5	1.2	2.4†
150/5	1.2	2.4†
200/5	1.2	1.2
300/5	0.6	0.6
400/5	0.6	0.6
600/5	0.6	0.6
800/5	0.3	0.3
1200/5	0.3	0.3
1500/5	0.3	0.3
2000/5	0.3	0.3
3000/5	0.3	0.3
4000/5	0.3	0.3

From IEEE Std 27-1974.

*See ANSI C57.13-1968.
†Not in ANSI C57.13-1968.

without causing temperatures in excess of limits specified in Table 64. The switchgear main bus will be designed and rated for the full current capacity specified and will not be tapered. As power system facilities are increased to serve larger loads, it is advisable to consider future expansion when selecting the bus continuous-current rating. The switchgear assembly should have momentary and short-time ratings equal respectively to the close and latch capability and short-time rating of the circuit breaker.

Current transformers are used to develop scale replica secondary currents, separated from the primary current and voltage, to provide a readily usable current for application to instruments, meters, and relays. For switchgear applications they are manufactured in single and double secondary types, also in the tapped multiratio type. The double secondary is suitable where two transformers of the same ratio would otherwise be required at the same location, with a resulting saving in space. The primary current rating should be no less than 125 percent of the ultimate full-load current of the circuit.

The metering and relaying accuracy must be adequate for the imposed burdens. The current transformer accuracy and excitation characteristics must be checked for proper relay applica-

Table 66
Standard Accuracy Class Ratings* of Current Transformers in Metal-Clad Switchgear

| Ratio | 60 Hz Standard Burden | | | Relaying Accuracy |
	B 0.1	B 0.5	B 2.0	
50/5†	1.2	2.4	—	10-H 10
75/5†	0.6	2.4	—	10-H 20
100/5	0.6	2.4	—	10-H 20
150/5	0.6	2.4	—	10-H 20
200/5	0.6	2.4	—	10-H 20
300/5	0.6	2.4	2.4	10-H 20
400/5	0.3	1.2	2.4	10-H 50
600/5	0.3	0.3	2.4	10-H 50
800/5	0.3	0.3	1.2	10-H 50
1200/5	0.3	0.3	0.3	10-H 100
1500/5	0.3	0.3	0.3	10-H 100
2000/5	0.3	0.3	0.3	10-H 100
3000/5	0.3	0.3	0.3	10-H 100
4000/5	0.3	0.3	0.3	10-H 100

From IEEE Std 27-1974.

*See ANSI C57.13-1968.

†These ratios and transformer accuracies do not apply for metal-clad switchgear assemblies having rated momentary currents above 60 000 A. Where such assemblies have a rated momentary current above 60 000 A, the minimum current transformer ratio shall be 100/5.

tion. Tables 65 and 66 list standard ratios and relaying and metering accuracies for current transformers.

Potential transformers are used to transform primary voltage to a nominal safe value, usually 120 V. The primary rating normally is that of the system voltage, though slightly higher ratings may be used, that is, a 14 400 V rating on a 13 800 V nominal system. These transformers are used to isolate the primary voltages from the instrumentation, metering, and relaying systems, yet provide replica scale values of the primary voltage. All ratings, such as impulse, dielectric, etc, must be adequate for the purpose. Table 67 lists standard potential transformer ratios.

8.3.7 *Control Power.* Successful operation of switchgear embodying electrically operated devices is dependent on a reliable source of control power which at all times will maintain voltage at the terminals of such devices within their rated operating voltage range. These ranges are established by appropriate ANSI standards for power circuit breakers (Table 68).

There are two primary uses for control power in switchgear: tripping power and closing power.

Table 67
Standard Potential Transformer Ratios

2400/4160Y—120
2400—120
4200—120
4800—120
7200—120
8400—120
12 000—120
14 400—120

Since an essential function of switchgear is to provide instant and unfailing protection in emergencies, the source of tripping power must always be available. The source of closing power must be independent of voltage conditions on the power system associated with the switchgear. However, other methods may be used, considering cost or maintenance a factor in selecting a source of closing power that is permissible with the essential tripping power. For example, for 1000 V and below circuit breakers, manual closing for circuit breakers up through 1600 A frame is not uncommon.

Table 68
Rated Control Voltages and Their Ranges for 1000 V and Below Power Circuit Breakers

| Rated Voltage (volts) | Control Voltage (volts) | Power Supply (volts) | | Tripping Voltage Range (volts) |
		Solenoid or Motor Operator	Stored Energy Operator†	
Direct Current*				
24‡	—	—	—	14—30‡
48	—	—	—	28—60
125	90—130	90—130§	90—130	70—140
250	180—260	180—260§	180—260	140—280
Alternating Current				
115	95—125**	—	95—125	95—125
230	190—250**	190—250§	190—250	190—250
460	380—500**	380—500§	380—500	380—500

From ANSI C37.15-1954 (R 1967) consolidated with ANSI C37.16-1973.

NOTE: It is recommended that trip, closing, relay coils, etc, normally connected continuously to one direct-current potential should be connected to the negative wire of the control circuit to minimize electrolytic deterioration.

*Control from exciter circuits is not recommended.
†For driving motor for air compressors and compressed spring mechanisms.
‡ Unless the circuit breaker is located close to the battery and relay and adequate electric conductors are provided between the battery and trip coil, 24 V tripping is not recommended.
**Includes heater circuits.
§ Some operating mechanisms will not meet all the closing requirements over the full control voltage range. In such cases it will be necessary to provide for two ranges of closing voltage. Where applicable, the preferred method of obtaining the double range of closing voltage is by the use of tapped coils. Otherwise it will be necessary for the user to designate one of the two closing voltage ranges listed below as representing the condition existing at the circuit breaker location due to battery or lead voltage drop or control power transformer regulation.

Rated Voltage	Closing Voltage Ranges for Power Supply
125 (dc)	90—115 or 105—130
250 (dc)	180—230 or 210—260
230 (dc)	190—230 or 210—250
460 (ac)	380—460 or 420—500

Four practical sources of tripping power are
(1) Direct current from a storage battery
(2) Direct current from a charged capacitor
(3) Alternating current from the secondaries of current transformers in the protected power circuit
(4) Direct or alternating current in the primary circuit passing through direct-acting trip devices

Where a storage battery has been chosen as a source of tripping power, it can also supply closing power. The battery ampere-hour and in-rush requirements have been reduced considerably with the advent of the stored-energy spring mechanism closing on power circuit breakers up through 34.5 kV. General distribution systems, whether alternating current or direct current, cannot be relied upon to supply tripping power because outages are always possible. These are most likely to occur in times of emergency when the switchgear is required to perform its protective functions.

Other factors influencing the choice of control power are

(1) Availability of adequate maintenance for a battery and its charger

(2) Availability of suitable housing for a battery and its charger

(3) Advantages of having removable circuit breaker units interchangeable with those in other installations

(4) Necessity for closing circuit breakers with the power system de-energized

The importance of periodic maintenance and testing of the tripping power source cannot be overemphasized. The most elaborate protective relaying system is useless if tripping power is not available to open the circuit breaker under abnormal conditions. Alarm monitoring for abnormal conditions of the tripping source and for circuits is a general requirement.

Space heaters are supplied as standard on outdoor metal-enclosed switchgear. Often ambient temperature or other environmental conditions dictate the use of space heaters in indoor switchgear as well. When space heaters are furnished, they must be continuously energized from an alternating-current power source. If alternating-current closing power is available, this source can also be used for the heaters, provided it is of sufficient capacity to supply the maximum current requirements of the space heaters and the inrush current of circuit breaker closing.

Standard air-magnetic or vacuum power circuit breakers are rated at 60 Hz but can be applied as low as 50 Hz without derating. For 25 Hz application, however, there is a derating factor which must be applied to the circuit breaker interrupting rating. Equipment manufacturers should be consulted to determine the proper derating factor for low-frequency power switchgear applications.

The application of metal-enclosed switchgear in contaminated atmospheres may create many problems if adequate precautions are not taken. Typical precautions include, but are not limited to, (1) location of equipment away from localized sources of contamination, (2) isolation of equipment through use of air-conditioning or pressurization equipment, (3) development of an appropriate supplemental maintenance program, and (4) maintenance of adequate spare part replacements.

8.4 Transformers

8.4.1 *Classifications.* Transformers have many classifications which are useful in the industry to distinguish or define certain characteristics of design and application. Some of these classifications are the following.

(1) *Distribution and Power.* A classification according to the rating in kilovolt-amperes. The distribution type covers the range of 3 kVA through 500 kVA, and the power type all ratings above 500 kVA.

(2) *Insulation.* This grouping includes liquid and dry types. Liquid insulated can be further defined as to the types of liquid: mineral oil, askarel, or other synthetic liquids. The dry-type grouping includes the ventilated and sealed gas-filled types.

(3) *Substation or Unit Substation.* The title "substation" transformer usually denotes a power transformer with direct cable or overhead line termination facilities to distinguish it from a unit substation transformer designed for integral connection to primary or secondary switchgear, or both, through enclosed bus connections. The substation classification is further defined by the terms "primary" and "secondary." The primary substation transformer has a secondary or load-side voltage rating of 1000 V or higher, whereas the secondary substation transformer has a load-side voltage rating of less than 1000 V.

Most transformer ratings and design features have been standardized by ANSI and NEMA, and these are listed as such in manufacturers' publications. The selection of other than standard ratings will usually result in higher costs.

8.4.2 *Specifications.* In specifying a transformer for a particular application, the following items comprising the rating structure must be included:

(1) Rating in kilovolt-amperes or megavolt-amperes

(2) Single phase or three phase

(3) Frequency

(4) Voltage ratings

(5) Voltage taps

(6) Winding connections, delta or wye

(7) Impedance (base rating)

(8) Basic impulse level (BIL)

(9) Temperature rise

The desired construction details to be specified must include

(10) Insulation medium, dry or liquid type [see Section 8.4.1(2)]

(11) Indoor or outdoor service

(12) Accessories

(13) Type and location of termination facilities

(14) Sound level limitations if the installation site requires this consideration

(15) Manual or automatic load tap changing

(16) Grounding requirements

(17) Provisions for future cooling of the specified type

8.4.3 *Power and Voltage Ratings.* Ratings in kilovolt-amperes or megavolt-amperes will include the self-cooled rating at a specified temperature rise, as well as the forced cooled rating if the transformer is to be so equipped. The standard self-cooled ratings, and self-cooled/forced-cooled relationships, are listed in Tables 69—71. As a minimum consideration, the self-cooled rating should be at least equal to the expected peak demand, and an allowance is recommended for load growth.

The standard allowable average winding temperature rise (by resistance test) for the modern liquid-filled power transformer is either $55°C/65°C$ or $65°C$, based on an average ambient of $30°C$ ($40°C$ maximum) for any 24 hour period. A transformer specified with a $55°C/65°C$ rise will permit 100 percent loading with a $55°C$ rise, and 112 percent loading at the $65°C$ rise with no additional loss of life.

Transformers have certain overload capabilities, varying with ambient temperature, preloading, and overload duration. These capabilities are defined in ANSI C57.92 (1962 ed), Guide for Loading Oil-Immersed Distribution and Power Transformers, and ANSI C57.96 (1959 ed), Guide for Loading Dry-Type Distribution and Power Transformers, for both the liquid-insulated and dry types.

The modern substation transformer for industrial plant service is an integral three-phase unit, as contrasted to three single-phase units. The advantages of the three-phase unit, such as lower cost, higher efficiency, less space, and elimination of exposed interconnections, have contributed to its widespread acceptance. The one advantage of single-phase units is that they can be operated at a reduced three-phase equivalent capacity by connecting two units in open delta if one unit should fail. However, the excellent reliability record of the three-phase unit has tended to lessen the importance of this advantage.

The transformer voltage ratings will include the primary and secondary continuous-duty levels at the specified frequency, as well as the basic impulse level for each winding. The continuous rating specified for the primary winding will be the nominal line voltage of the system to which the transformer is to be applied, and preferably within ± 5 percent of the normally sustained voltage. The secondary or transformed voltage rating will be the value under no-load conditions. The change in secondary voltage experienced under load conditions is termed regulation, and is a function of the impedance of the system and the transformer and the power factor of the load. Table 72 illustrates the proper designation of voltage ratings for three-phase transformers.

The basic impulse level rating for a transformer winding signifies the design and tested capability of its insulation to withstand transient overvoltages from lightning and other surges. Standard values of basic impulse level established for each nominal voltage class are listed in Table 73. A description of test requirements for these values is given in IEEE Std 462-1973, General Requirements for Distribution, Power, and Regulating Transformers (ANSI C57.12.00-1973). Transformer bushings may be specified with extra creepage distance and higher than standard basic impulse level ratings, if required by local conditions.

A cost advantage can be realized in a high-voltage transformer if it is protected by surge arresters of reduced rating which have been applied in accordance with the requirements for "effectively grounded" systems. This cost advantage is obtained by the use of a lower level of insulation dielectric. Refer to Chapter 4 for a discussion of arrester application for the protection of transformers.

8.4.4 *Voltage Taps.* Voltage taps are usually found necessary to compensate for small changes in the primary supply to the transformer, or to vary the secondary voltage level with changes in load requirements. The most

Table 69
Transformer Standard Base kVA Ratings

Single-Phase				Three-Phase			
3	75	1250	10 000	15	300	3750	25 000
5	100	1667	12 500	30	500	5000	30 000
10	167	2500	16 667	45	750	7500	37 500
15	250	3333	20 000	75	1000	10 000	50 000
25	333	5000	25 000	112½	1500	12 000	60 000
37½	500	6667	33 333	150	2000	15 000	75 000
50	833	8333		225	2500	20 000	100 000

From IEEE Std 462-1973.

Table 70
Transformer Capabilities with Forced Cooling

Class	Self-Cooled Ratings* (kVA)		Percent of Self-Cooled Ratings with Auxiliary Cooling	
	Single-Phase	Three-Phase	First Stage	Second Stage
OA/FA	501—2499	501—2499	115	—
OA/FA	2500—9999	2500—11 999	125	—
OA/FA	10 000 and above	12 000 and above	133 1/3	—
OA/FA/FA	10 000 and above	12 000 and above	133 1/3	166 2/3
OA/FA/FOA	10 000 and above	12 000 and above	133 1/3	166 2/3
OA/FOA/FOA	10 000 and above	12 000 and above	133 1/3	166 2/3
AA/FA†	501 and above	501 and above	133 1/3	—

From NEMA TR1-1974.

*In the case of multiwinding transformers or autotransformers, the ratings given are the equivalent two-winding ratings.
†Not applicable to sealed dry-type transformers.

Table 71
Classes of Transformer Cooling Systems

Type Letters	Method of Cooling
OA	Oil-immersed, self-cooled
OW	Oil-immersed, water-cooled
OW/A	Oil-immersed, water-cooled/self-cooled
OA/FA	Oil-immersed, self-cooled/forced-air-cooled
OA/FA/FA	Oil-immersed, self-cooled/forced-air-cooled/forced-air-cooled
OA/FA/FOA	Oil-immersed, self-cooled/forced-air-cooled/forced air—forced-oil-cooled
OA/FOA/FOA	Oil-immersed, self-cooled/forced-air-cooled/forced air—forced-oil-cooled
FOA	Oil-immersed, forced-oil-cooled with forced-air cooler
FOW	Oil-immersed, forced-oil-cooled with forced-water cooler
AA	Dry-type, self-cooled
AFA	Dry-type, forced-air-cooled
AA/FA	Dry-type, self-cooled/forced-air-cooled

From IEEE Std 462-1973.

Table 72
Designation of Voltage Ratings of Three-Phase Windings (Schematic Representation)

Identi-cation	Nomenclature	Nameplate Marking	Typical Winding Diagram	Condensed Usage Guide
(2) (a)	E	2400		E shall indicate a winding which is permanently Δ connected for operation on an E volt system.
(2) (b)	E_1Y	4160Y		E_1Y shall indicate a winding which is permanently Y connected without a neutral brought out (isolated) for operation on an E_1 volt system.
(2) (c)	E_1Y/E	4160Y/2400		E_1Y/E shall indicate a winding which is permanently Y connected with a fully insulated neutral brought out for operation on an E_1 volt system, with E volts available from line to neutral.
(2) (d)	E/E_1Y	2400/4160Y		E/E_1Y shall indicate a winding which may be Δ connected for operation on an E volt system, or may be Y connected without a neutral brought out (isolated) for operation on an E_1 volt system.
(2) (e)	$E/E_1Y/E$	2400/4160Y/2400		$E/E_1Y/E$ shall indicate a winding which may be Δ connected for operation on an E volt system or may be Y connected with a fully insulated neutral brought out for operation on an E_1 volt system with E volts available from line to neutral.
(2) (f)	E_1GrdY/E	67 000GrdY/38 700		E_1GrdY/E shall indicate a winding with reduced insulation and permanently Y connected, with a neutral brought out and effectively grounded for operation on an E_1 volt system with E volts available from line to neutral.
(2) (g)	$E/E_1GrdY/E$	38 700/67 000GrdY/38 700		$E/E_1GrdY/E$ shall indicate a winding, having reduced insulation, which may be Δ connected for operation on an E volt system or may be connected Y with a neutral brought out and effectively grounded for operation on an E_1 volt system with E volts available from line to neutral.
(2) (h)	$V \times V_1$	7200 × 14 400 4160Y/2400 × 12 470Y/7200		$V \times V_1$ shall indicate a winding, the sections of which may be connected in parallel to obtain one of the voltage ratings (as defined in a, b, c, d, e, f, and g) of V, or may be connected in series to obtain one of the voltage ratings (as defined in a, b, c, d, e, f, and g) of V_1. Windings are permanently Δ or Y connected.

Key: $E_1 = \sqrt{3}$ E.

From IEEE Std 462-1973.

Table 73
Transformer Standard Basic Impulse Insulation Levels Usually
Associated with Nominal System Voltages

Nominal System Line-to-Line Voltage (volts)	Insulation Class (kV)	Basic Impulse Insulation Level (kV)			
		Liquid Insulated		Dry-Type	
		Power	Distribution	Ventilated*	Gas-Filled and Sealed
120—600	1.2	45	30†	10	30
2400	2.5	60	45†	20	45
4160	5.0	75	60†	25	60
4800	5.0	75	60†	25	60
6900	8.7	95	75†	35	75
7200	8.7	95	75†	35	75
12 470	15.0	110	95†	50	95
13 200	15.0	110	95†	50	95
13 800	15.0	110	95†	50	95
14 400	15.0	110	95†	50	95
22 900	25.0	150	150	—	—
23 000	25.0	150	150	—	—
26 400	34.5	200	200	—	—
34 500	34.5	200	200	—	—
43 800	46.0	250	250	—	—
46 000	46.0	250	250	—	—
67 000	69.0	350	350	—	—
69 000	69.0	350	350	—	—
92 000	92.0	450	—	—	—
115 000	115.0	550	—	—	—
		350‡	—	—	—
		450‡	—	—	—
138 000	138.0	650	—	—	—
		550‡	—	—	—
		450‡	—	—	—
161 000	161.0	750	—	—	—
		650‡	—	—	—
		550‡	—	—	—

From IEEE Std 462-1973 and NEMA 201-1970 and 210-1970.

*Special ratings for ventilated dry-type transformers, equivalent to the gas-filled type, can be obtained.

†Rating are also applicable to primary and secondary unit substation transformers.

‡Optional reduced levels applicable if equivalent reduced rating arresters are properly applied on the system.

commonly selected tap arrangement is the manually adjustable no-load type, consisting of four 2½ percent steps or variations from the nominal primary voltage rating. These tap positions are usually numbered 1 through 5, with the number 1 position providing the greatest number of effective turns. Based on a specific incoming voltage, selection of a higher voltage tap (lower tap number) will result in a lowering of the output voltage. The changing of tap positions is performed manually only with the transformer de-energized. In addition to the no-load taps, and considered desirable when load swings are larger and more frequent, or voltage levels more critical, the "automatic tap changing under load" feature is available. This provides an additional ± 10 percent voltage adjustment automatically in incremental steps, with continuous monitoring of the secondary terminal voltage or of a voltage level remote from the transformer.

8.4.5 *Connections.* Connections for the standard two-winding power transformers are preferably delta primary and wye secondary. The wye secondary, specified with external neutral bushing, provides a convenient neutral point for establishing a system ground, or run as a phase conductor for phase-to-neutral load. The delta-connected primary isolates the two systems with respect to the flow of zero-sequence currents resulting from third-harmonic exciting current or a secondary ground fault, and may be used without regard to whether the system to which the primary is connected is three wire or four wire.

For those applications which require a wye—wye transformation, a third winding, delta connected and designated as the tertiary, is recommended to provide a low-impedance path for zero-sequence currents. Since the third winding usually has a minimum of 35 percent of the rating of the primary winding in kilovolt-amperes, and its voltage rating is optional, it can be used to serve an auxiliary load. If the auxiliary loading feature is desired, the tertiary winding must be specified to include external connection facilities.

8.4.6 *Impedance.* Impedance voltage of a transformer is the voltage required to circulate rated current through one of two specified windings of a transformer when the other wind-

ing is short-circuited, with the windings connected as for rated voltage operation.

Impedance voltage is normally expressed as a percent value of the rated voltage of the winding in which the voltage is measured on the transformer self-cooled rating in kilovolt-amperes. The percent impedance voltage levels considered as standard for two-winding transformers rated up through 10 000 kVA are listed in Table 74, and a value specified above or below those listed will usually result in higher costs. For transformer ratings above 10 000 kVA or 67 kV, a percent impedance voltage may be selected and considered standard if it lies within a published minimum and maximum range. The percent impedance voltage of a two-winding transformer shall have a tolerance of 7.5 percent of the specified value. For three-winding or autotransformers, the manufacturing tolerance is ± 10 percent. When considering a low-impedance voltage level, it should be remembered that the standard transformer is designed with a limited ability to withstand the stresses imposed by external faults. Each winding is capable of withstanding the thermal and magnetic stresses resulting from current flow not exceeding 25 times the full-load rating of any winding for a period of 2 s. A combined primary system and transformer impedance voltage permitting symmetrical rms fault magnitudes in excess of this should be avoided.

With respect to impedance, transformers are generally considered suitable for parallel operation if their impedances match within 10 percent. The importance of minimizing the mismatch becomes greater as the total load approaches the combined capacity of the paralleled transformers, since load division is in direct proportion to the impedance of each transformer. The impedance mismatch should be checked throughout the entire range of taps (both no-load and load).

8.4.7 *Insulation Medium.* The selection of the insulation medium is dictated mainly by the installation site and cost. For outdoor installations, the oil-insulated transformer has widespread use due to its lowest cost and inherent weatherproof construction. When located close to combustible buildings, safeguards are required as specified in the NEC. Although oil-

Table 74
Standard Impedance Values for Three-Phase Transformers

High-Voltage Rating (volts)	kVA Rating	Percent Impedance Voltage	
Secondary Unit Substation Transformers*			
2400—13 800	112.5—225	Not less than 2.0	
2400—13 800	300—500	Not less than 4.5	
2400—13 800	750—2500	5.75	
22 900	All	5.75	
34 400	All	6.25	
Liquid-Immersed Transformers, 501—30 000 kVA†			
		Low Voltage, 480 V	Low Voltage, 2400 V and Above
2400—22 900		5.75	5.5
26 400, 34 400		6.25	6.0
43 800		6.75	6.5
67 000			7.0
115 000			7.5
138 000			8.0

NOTES: (1) Ratings separated by hyphens indicate that all intervening standard ratings are included. Ratings separated by a comma indicate that only those listed are included.

(2) Percent impedance voltages are at self-cooled rating and as measured on rated voltage connection.

*From NEMA 210-1970.
†From ANSI C57.12.10-1969.

insulated transformers can be located indoors, the construction costs for the required vault and venting facilities usually exceed the price premium for the askarel (nonflammable liquid insulated) transformer. However, askarel-insulated transformers rated over 25 kVA must be furnished with a pressure-relief vent and those over 35 kV must be installed in a vault. Due to possible environmental pollution, the users of askarel-insulated transformers should consult the manufacturer of the transformer or the manufacturer of the liquid for selection as well as proper safeguard in the disposal of used liquid (see ANSI C107.1-1974, Guidelines for Handling and Disposal of Capacitor- and Transformer-Grade Askarels, Containing Polychlorinated Biphenyls, and IEEE Std 76-1974, Guide for Acceptance and Maintenance of Transformer Askarel in Equipment).

The ventilated dry-type transformer has application in industrial plants for indoor installation where floor space, weight, and regard for liquid maintenance and safeguards would be important factors. Since the basic impulse level for the ventilated dry-type transformer winding is usually less than that of the liquid or gas-filled dry type, surge arresters must be included for the primary winding in order to obtain suitable protection.

The sealed or gas-filled dry-type transformer has very limited use, due principally to the higher price. A preference factor over the ventilated dry type is less maintenance, and over the liquid type the elimination of necessary safeguards against a ruptured tank.

On oil-insulated transformers a sealed-tank construction with welded cover is standard with manufacturers. Optional oil-preservation systems may be specified as follows.

(1) A gas-oil seal which consists of an auxiliary tank mounted on the transformer. This seal provides for the safe expansion and contraction of the transformer gas and oil without exposing the transformer oil to the atmosphere.

Table 75
Audible Sound Levels for Two-Winding Substation Transformers

Self-Cooled Rating (kVA)	Average Sound Level (decibels)*		
	Liquid Type (69 kV and Below)	Dry Type (15 kV and Below) Ventilated	Sealed
151—300	55	58 (67)	57
301—500	56	60 (67)	59
501—700	57 (67)	62 (67)	61
701—1000	58 (67)	64 (67)	63
1001—1500	60 (67)	65 (68)	64
1501—2000	61 (67)	66 (69)	65
2001—3000	62, 63 (67)	68 (71)	66
3001—4000	64 (67)	70 (73)	68
4001—5000	65 (67)	71 (74)	69
5001—6000	66 (67)	72 (75)	70
6001—7500	67 (68)	73 (76)	71
10 000	68 (70)	—	—
12 500	69 (70)	—	—
15 000	70 (71)	—	—
20 000	71 (72)	—	—
25 000	72 (73)	—	—
30 000	73 (74)	—	—
40 000	74 (75)	—	—
50 000	75 (76)	—	—
60 000	76 (77)	—	—
80 000	77 (78)	—	—
100 000	78 (79)	—	—

From NEMA TR 1-1974.

NOTE: The tabulated values represent the average sound level which will not be exceeded on the base transformer, and is exclusive of sound emitted by integral load tap-changing mechanisms, disconnecting switches, or close-coupled switchgear. These values are not necessarily applicable to rectifier transformers or furnace transformers. Refer to NEMA TR 1-1974 for measurement and test conditions. (Noise in dBA at 1 ft from external surface.)

*The values in parentheses represent the sound level with one stage of forced-cooling equipment in operation. (Noise in dBA at 6 ft from external surface.)

(2) An automatic gas seal which maintains a constant positive nitrogen pressure with the tank. A combination regulating valve and pressure relief operates with a cylinder of high-pressure nitrogen to control the proper functioning of this seal.

8.4.8 *Accessories.* Accessories furnished with the transformer include those identified as standard and optional in manufacturers' publications. The standard items will vary with different types of transformers. Some of the optional items which offer protective features include the following.

(1) Winding temperature equipment in addi-

tion to the standard top-oil temperature indicator. This device is calibrated for use with specific transformers and automatically takes into account the hottest spot temperature of the windings, ambient temperature, and load cycling. For this reason it provides a more accurate, continuous, and automatic measure of the transformer loading and overloading capacity. It may have contacts which can be set to alarm and even subsequently trip a circuit breaker.

(2) The pressure relay for sensitive high-speed indication of liquid-filled transformer internal faults. Since the device is designed to operate on the rate of change of internal pressure, it is

sensitive only to that resulting from internal faults and not to pressure changes due to temperature and loading.

(3) Alarm contacts can be included on the standard devices such as temperature indicators, liquid-level and pressure vacuum gauges, and pressure relief devices for more effective utilization of such devices.

(4) Surge arresters mounted directly on the transformer tank provide maximum surge protection for the transformer. The type of arrester specified and its voltage rating must be coordinated with the voltage parameters of the system on which it is applied. Refer to Chapter 4 for a detailed discussion of surge arrester application.

8.4.9 *Termination Facilities.* Termination facilities are available to accommodate most types of installation. For the unit substation arrangement, indoors or outdoors, the incoming and outgoing bushings are usually side-wall mounted and enclosed in a throat or transition section for connection to adjacent switchgear assemblies. Tank-wall mounted enclosures, oil or air insulated, with or without potheads or cable clamps, are available for direct cable termination. The size and number of conductors must be specified, along with minimum space for stress cone termination, if required. For the non-unit-substation type of outdoor installation, cover-mounted bushings provide the simplest facility for overhead lines.

8.4.10 *Sound Levels.* The transformer sound level is of importance in certain installations, and the maximum standard levels are listed in Table 75. These can be reduced to some extent by special design, so the transformer manufacturer should be consulted regarding the possible reduction for a particular type and rating.

8.5 Unit Substations

8.5.1 *General Discussion.* A unit substation consists of the following sections.

(1) A primary section which provides for the connection of one or more incoming high-voltage circuits, each of which may or may not be provided with a switching device or a switching and interrupting device.

(2) A transformer section with includes one or more transformers with or without automatic load-tap-changing equipment. The use of automatic load-tap-changing equipment is not common in unit substations.

(3) A secondary section which provides for the connection of one or more secondary feeders, each of which is provided with a switching and interrupting device.

8.5.2 *Types.* Unit substation sections are normally subassemblies for connection in the field and are usually one of the following types. The application of these to industrial power systems with the appropriate single-line diagrams is described in Chapter 2.

(1) *Radial.* One primary feeder to a single stepdown transformer with a secondary section for the connection of one or more outgoing radial feeders (see Figs 1 and 2).

(2) *Primary Selective and Primary Loop.* Each stepdown transformer connected to two separate primary sources through switching equipment to provide a normal and alternate source. Upon failure of the normal source, the transformer is switched to the alternate source (see Figs 3 and 4).

(3) *Secondary Selective.* Two stepdown transformers, each connected to a separate primary source. The secondary of each transformer is connected to a separate bus through a suitable switching and protective device. The two sections of bus are connected by a normally open switching and protective device. Each bus has provisions for one or more secondary radial feeders (see Fig 5).

(4) *Secondary Spot Network.* Two stepdown transformers, each connected to a separate primary source. The secondary side of each transformer is connected to a common bus through a special type of circuit breaker called a network protector, which is equipped with relays to trip the circuit breaker on reverse power flow to the transformer and reclose the circuit breaker upon restoration of the correct voltage, phase angle, and phase sequence at the transformer secondary. The bus has provisions for one or more secondary radial feeders (see Fig 6).

(5) *Distributed Network.* A single stepdown transformer having its secondary side connected to a bus through a special type of circuit breaker called a network protector, which is equipped with relays to trip the circuit breaker upon restoration of the correct voltage,

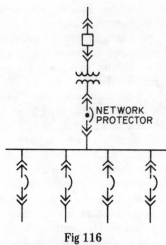

Fig 116
Distributed Network

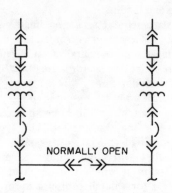

Fig 117
Duplex (Circuit-Breaker-and-a-Half Scheme)

phase angle, and phase sequence at the transformer secondary. The bus has provisions for one or more secondary radial feeders and one or more tie connections to a similar unit substation (Fig 116).

(6) *Duplex (Breaker-and-a-Half Scheme).* Two stepdown transformers, each connected to a separate primary source. The secondary side of each transformer is connected to a radial feeder. These feeders are joined on the feeder side of the power circuit breakers by a normally open tie circuit breaker. This type is used mostly on electric utility primary distribution systems (Fig 117).

8.5.3 *Selection and Location.* Considerations in the selection and location of unit substations are apparent power and voltage ratings, allowance for future growth, appearance, atmospheric conditions, and outdoor versus indoor location. NEMA 201-1970, Primary Unit Substations, and NEMA 210-1970, Secondary Unit Substations, contain useful information, which is also applicable for general substation design when the components are purchased separately.

8.5.4 *Advantages of Unit Substations.* The engineering of the components is coordinated by the manufacturer, the costs of field labor and installation time are reduced, substation appearance is improved, and the substation is safer to operate. The operating costs are reduced due to the reduced power losses from shorter secondary feeders, and a power system

using unit substations is flexible and easy to expand.

Unit substations are available for either indoor or outdoor location. In some applications, the heat-producing transformer is located outdoors and connected by metal-enclosed bus duct to indoor switchgear.

Primary unit substations may be located outdoors, particularly when the primary supply is above 34.5 kV. The high-voltage equipment including the design, all components, the supporting structures, and installation drawings may be obtained as a package. There is a trend toward metal-enclosed equipment above 34.5 kV in a unit substation arrangement because of the increasing requirements for safety, compactness, appearance, and reduction of installation labor and time.

Most secondary unit substations are located indoors to reduce costs and improve voltage reduction by placing the transformer as close as possible to the center of the load in the area being supplied.

8.5.5 *Application Guides*

(1) The transformer main secondary circuit breakers and connections should have a continuous-current rating which is approximately 25 percent greater than the continuous-current rating of the transformer. This is necessary since transformers are often required to carry short-time overloads above their nameplate ratings for a short time such as during plant

startup. When selecting the continuous-current rating of the transformer main secondary circuit breaker and connections, consideration should also be given to whether or not the transformer has, or will have in the future, a continuous forced-air-cooled rating, dual temperature rating, or other extension of the basic continuous rating.

(2) The effects of solar radiation and atmospheric conditions should be considered in the selection of outdoor equipment. IEEE Std 144-1971 gives guidance for evaluating the effect of solar radiation. The lighter the color of the exterior paint, the lower the effect of solar radiation on the equipment. This will be included in future editions of IEEE Std 144.

(3) When connected to circuits which are subject to lightning or switching surges, substations should be equipped with surge-protective equipment, selected to limit voltage surges to values below the impulse withstand ratings of the transformers and the switching equipment.

NEMA 201-1970 and 210-1970 contain sections which give additional information on unit substation application.

8.6 Motor-Control Equipment

8.6.1 *General Discussion.* The majority of motors utilized by industrial firms are integral horsepower of induction squirrel-cage design powered from three-phase alternating-current 600 V and below distribution systems. The choise of an integral horsepower controller depends on a number of factors.

(1) *Power System.* Direct or alternating-current; single phase or three phase? What is the voltage and frequency? Will the system permit the large inrush currents during the full-voltage starting without excessive voltage drop?

(2) *Motor.* Is the controller to be used with a direct-current, squirrel-cage induction, wound-rotor induction, or synchronous motor? What is the horsepower? Will the motor be "plugged" or reversed frequently? What is the acceleration time from start to full speed?

(3) *Load.* Is the load geared, belt-driven, or direct coupled? Loaded or unloaded start?

(4) *Operation.* Is operation to be manual or automatic?

(5) *Protection.* Are fuses or circuit breakers to be used for short-circuit protection? To size the thermal overload elements, the full-load current of the motor and the ambient temperature at the motor and controller must be known.

(6) *Environment.* Will the motor and controller be subjected to excessive vibration, dirt, dust, oil, water? Will either be located in a hazardous or corrosive area?

(7) *Cable Connections and Space.* Will there be the required space for cable entrance, bending radius, and terminations, and for reliable connections to line and load buses?

To answer these questions for proper application of motor controllers, the specifying engineer should seek the assistance of the application engineers from the utility and manufacturers. In addition, process engineers and operating personnel associated with the installation should be consulted.

8.6.2 *Starters Over 600 V.* Starters for motors from 2300 through 13 200 V are designed as integrated complete units based on maximum horsepower ratings for use with squirrel-cage, wound-rotor, synchronous, and multispeed motors for full- or reduced-voltage starting. Alternating-current magnetic fused type starters, NEMA class E-2, employ current-limiting power fuses and magnetic air-break contactors. Each starter will be completely self-contained, prewired, and with all components in place. Air-break contactors will be current rated based on motor horsepower requirements. Combination starters will provide an interrupting fault capacity of 200 MVA symmetrical on a 2300 V system, and 350 MVA symmetrical on a 4160 or 4800 V system. This starter will conform to NEMA ICS-1970, Industrial Controls and Systems, class E-2 controllers, and applicable IEEE and ANSI standards. There is also a UL certification standard on this equipment. Combinations of motor controllers and switchgear are available as assemblies.

8.6.3 *Starters 600 V and Below.* Table 76 summarizes the NEMA standard for magnetic controller ratings of 115 through 575 V, NEMA ICS-1970. In alternating-current motor starters, contactors are generally used for controlling the circuit to the motor. Starters must be carefully applied on circuits and in combination with associated short-circuit protective devices

Table 76
NEMA Standard Magnetic Controller Ratings for 600 V and Below

NEMA Size	Continuous Current Rating, General Purpose (amperes)		Voltage (volts)	Maximum Horsepower Rating		Maximum Horsepower Rating for Plugging Service*	
	Open	Enclosed		Single-Phase	Three-Phase	Single-Phase	Three-Phase
00	10	9	115	1/3	—	—	—
			200	—	1 1/2	—	—
			230	1	1 1/2	—	—
			460	—	2	—	—
			575	—	2	—	—
0	20	18	115	1	—	1/2	—
			200	—	3	—	1 1/2
			230	2	3	1	1 1/2
			460	—	5	—	2
			575	—	5	—	2
1	30	27	115	2	—	1	—
			200	—	7 1/2	—	3
			230	3	7 1/2	2	3
			460	—	10	—	5
			575	—	10	—	5
2	50	45	115	3	—	2	—
			200	—	10	—	7 1/2
			230	7 1/2	15	5	10
			460	—	25	—	15
			575	—	25	—	15
3	100	90	115	—	—	—	—
			200	—	25	—	15
			230	—	30	—	20
			460	—	50	—	30
			575	—	50	—	30
4	150	135	115	—	—	—	—
			200	—	40	—	25
			230	—	50	—	30
			460	—	100	—	60
			575	—	100	—	60
5	300	270	115	—	—	—	—
			200	—	75	—	60
			230	—	100	—	75
			460	—	200	—	150
			575	—	200	—	150
6	600	540	115	—	—	—	—
			200	—	150	—	125
			230	—	200	—	150
			460	—	400	—	300
			575	—	400	—	300

From NEMA ICS-1970.

*An example is plug stop or jogging (inching duty) which requires continuous operation with more than five openings per minute.

Table 76 (Cont'd)

NEMA Size	Continuous Current Rating, General Purpose (amperes) Open	Enclosed	Voltage (volts)	Maximum Horsepower Rating Single-Phase	Three-Phase	Maximum Horsepower Rating for Plugging Service* Single-Phase	Three-Phase
7	900	810	115	—	—	—	—
			200	—	—	—	—
			230	—	300	—	—
			460	—	600	—	—
			575	—	600	—	—
8	1350	1215	—	—	—	—	—
			200	—	—	—	—
			230	—	450	—	—
			460	—	900	—	—
			575	—	900	—	—
9	2500	2250	—	—	—	—	—
			200	—	—	—	—
			230	—	800	—	—
			460	—	1600	—	—
			575	—	1600	—	—

From NEMA ICS-1970.

*An example is plug stop or jogging (inching duty) which requires continuous operation with more than five openings per minute.

(circuit breakers, fusible disconnects) which will limit the available fault current and the let-through energy to a level the starter can safely withstand. Some of the common motor starting devices used in industry are as follows.

(1) *Across-the-Line Starter*

(a) *Manual.* The manual starter provides overload protection but not undervoltage protection. One- or two-pole single phase for motor ratings to 3 hp. Single- or polyphase motor control for motor ratings up to 5 hp at 230 V single phase, 7½ hp at 230 V three phase, and 10 hp at 460 V three phase. Operating control available in toggle, rocker, or push-button design.

(b) *Magnetic, Nonreversing.* For full-voltage frequent starting of alternating-current motors, suitable for remote control with push-button station, control switch, or with automatic pilot devices. Undervoltage protection is obtained by using momentary contact-starting push-button in parallel with interlock contact in starter and series-connected stop push-button. Available in single-phase construction up to 15 hp at 230 V, and three-phase ratings up to 1600 hp at 460 V.

(c) *Magnetic, Reversing.* For full-voltage starting of single-phase and polyphase motors where application requires frequent starting and reversing or plugging operation. It consists of two contactors wired to provide phase reversal, mechanically and electrically interlocked to prevent both contactors from being closed at the same time.

(2) *Combination Across-the-Line Starter*

(a) *Magnetic, Nonreversing.* For full-voltage starting of polyphase motors. It provides motor overcurrent protection with thermal overload relays. Available with a nonfuse disconnect; provides branch circuit protection when specified with a fusible disconnect or circuit breaker. Check available fault current before deciding which fuses or circuit breakers will be used. It also provides undervoltage protection and is suitable for remote control. Readily available up to NEMA size 5 from most manufacturers and size 9 from several.

(b) *Magnetic, Reversing.* Same as reversing starter, except equipped with nonfusible disconnect, fusible disconnect, or circuit breaker.

(3) *Reduced Voltage Starter*

(a) *Autotransformer, Manual.* For limiting starting current and torque on polyphase induction motors to comply with power supply regulations or to avoid excessive shock to the driven machine. Overload and undervoltage protection are provided. Equipped with mechanical interlock to assure proper starting sequence. Taps are provided on the autotransformer for adjusting starting torque and current.

(b) *Autotransformers, Magnetic.* Same as manual, but suitable for remote control. It has timing relay for adjustment of time at which full voltage is applied.

To overcome the objection of the open-circuit transition associated with an autotransformer starter, a circuit known as the Korndorfer connection is in common use. This type starter requires a two-pole and a three-pole start contactor. The two-pole contactor opens first on the transition from start to run, opening the connections to the neutral of the autotransformer. The windings of the transformer are then momentarily used as series reactors during the transfer, allowing a closed-circuit transition.

(c) *Primary Resistor Type.* Automatic reduced voltage starter designed for geared or belted drive where sudden applicaton of full-voltage torque must be avoided. Inrush current is limited by the value of the resistor; starting torque is a function of the square of the applied voltage. Therefore if the initial voltage is reduced to 50 percent, the starting torque of the motor will be 25 percent of its full-voltage starting torque. A compromise must be made between the required starting torque and the inrush current allowed on the system. It provides both overload and undervoltage protection and is suitable for remote control.

(d) *Part Winding Type.* Used on light or low-inertia loads where the power system requires limitations on the increments of current inrush. It consists of two magnetic starters, each selected for one of the two motor windings, and a time-delay relay controlling the time at which the second winding is energized. It provides overload and undervoltage protection and is suitable for remote control.

(e) *Wye—Delta Type.* This type of starter is most applicable to starting motors which drive high-inertia loads with resulting long acceleration times. When the motor has accelerated on the wye connection, it is automatically reconnected by contactors for normal delta operation.

In selecting the type of reduced-voltage starter, consideration must be given to the motor-control transition from starting to running. In a closed-circuit transition, power to the motor is not interrupted during the starting sequence, whereas on open-circuit transition it is. Closed-circuit transition is recommended for all applications to minimize inrush voltage disturbances.

A comparison of starting currents and torques produced by various kinds of reduced-voltage starters is shown in Table 77.

(4) *Slip-Ring Motor Controller.* The wound-rotor or slip-ring motor functions in the same manner as the squirrel-cage motor, except that the rotor windings are connected through slip rings and brushes to external circuits with resistance to vary motor speed. Increasing the resistance in the rotor circuit reduces motor speed and decreasing the resistance increases motor speed. Some variation of this type controller employs silicon-controlled rectifiers in place of contactors and resistors. Some even rectify the secondary current and invert it to line frequency to supply back into the input, raising the efficiency appreciably.

(5) *Controller for Direct-Current Motors.* These motors have favorable speed—torque characteristics, and their speed is easily controlled. Large direct-current motors are started with resistance in the armature circuit, which is reduced step by step until the motor reaches its base speed. Higher speeds are provided by weakening the motor field.

(6) *Multispeed Controller.* These controllers are designed for the automatic control of two-, three-, or four-speed squirrel-cage motors of either the consequent-pole or separate-winding types. They are available for constant-horsepower, constant-torque, or variable-torque three-phase motors used on fans, blowers, refrigeration compressors, and similar machinery.

8.6.4 *Motor-Control Center.* Most control centers are tailor-made assemblies of conveniently grouped control equipment primarily used

Table 77
Comparison of Different Reduced-Voltage Starters

	Autotransformer*			Primary Resistance		Part Winding‡		Wye Delta
	50% Tap	65% Tap	80% Tap	65% Tap	80% Tap	2-Step	3-Step	
Starting current drawn from line as percentage of that which would be drawn upon full-voltage starting†	28%	45%	67%	65%	80%	60%‡	25%	33 1/3%
Starting torque developed as percentage of that which would be developed on full-voltage starting	25% Increases slightly with speed	42%	64%	42% Increases greatly with speed	64%	50%	12½%	33 1/3%
Smoothness of acceleration	Second in order of smoothness			Smoothness of reduced-voltage types. As motor gain speed, current decreases. Voltage drop across resistor decreases and motor terminal voltage increases		Fourth in order of of smoothness		Third in order of smoothness
Starting current and torque adjustment	Adjustable within limits of various taps			Adjustable within limits of various taps				Fixed

*Closed transition.
†Full-voltage start usually draws between 500 and 600 percent of full-load current.
‡Approximate values only. Exact values can be obtained from motor manufacturer.

for power distribution and associated control of motors. They contain all necessary bus, incoming line facilities, and safety features to provide the utmost in convenience for space and labor saving and adaptability to ever-changing conditions with a minimum of effort and a maximum of safety. NEMA standards govern the type of enclosure and wiring; NEMA type 1, 2, 3, and 12 enclosures are generally available. Wiring of motor-control centers conforms to two NEMA classes and three types. Class I provides for no wiring by the manufacturer between compartments of the center. Class II requires prewiring by the manufacturer with interlocking and other control wiring completed between compartments of the center. Type A, no terminal blocks are provided; type B, all connections within individual compartments are made to terminal blocks; type C, all connections are made to a master terminal block located in the horizontal wiring trough at the top or bottom of the center. The ideal wiring specification for minimum field installation time and labor is NEMA class II, type C wiring. The wiring specifications most frequently used by industrial contractors are NEMA class I, type B wiring. Refer to NEMA ICS-1970 for definitions of wiring class and type.

The NEMA standard specifies that a control center shall carry a short-circuit rating defined as the maximum available symmetrical rms current in amperes permissible at line terminals. The available short-circuit current at the line terminals of the motor-control center is computed as the sum of the maximum available current of the system at the point of connection and the short-circuit current contribution of the motors connected to the control center. It is common practice by many manufacturers to show only the short-circuit rating of the buswork on the nameplate. As a result, it is very important to establish the actual rating of the entire unit and, in particular, the plug-in units (that is, circuit breakers, disconnects, starters, etc), especially for applications where available fault currents exceed 10 000 A.

8.6.5 *Control Circuits.* Conventional starters for 600 V and below are factory wired with coils of the same voltage rating as the phase voltage to the motor. However, there are many cases in which it is desirable or necessary to use control circuits and devices of lower voltage rating than the motor. In such cases, control transformers are used to step the voltage down to permit the use of lower voltage coil circuits. Control transformers can be supplied by manufacturers as separate units with provisions for mounting external to the controller, or they can be incorporated in the controller enclosure, wired in with an operating coil of proper voltage rating. Such transformers can be obtained with fused or otherwise protected secondaries to meet code requirements on control-circuit overcurrent protection. Selection of the proper control transformer for a controller is a simple matter of matching the characteristics of the control circuit to the specifications of the transformer. The line voltage of the supply to the motor determines the required primary rating of the transformer. The secondary must be rated to provide the desired control-circuit voltage to match the voltage of the contactor operating coil. The continuous secondary current rating of the transformer must be sufficient for the magnetizing current of the operating coil and must also be able to handle the inrush current. In addition the control transformer must be of sufficient capacity to supply power requirements of control devices associated with the particular control circuit, that is, indicating lamps, solenoids, etc.

Two forms of control, undervoltage release and undervoltage protection, can be provided in the motor starters. In the first, if the voltage drops below a set minimum, or if the control-circuit voltage fails, the contactor will drop out but will reclose as soon as the voltage is restored. With undervoltage protection, low voltage or failure of the control-circuit voltage will cause the contactor to drop out, but the contactor will not reclose upon restoration of voltage.

8.6.6 *Overload Protection.* Motor starters are equipped with overload relays. These relays may be of the magnetic or the thermal type. If of the magnetic type, a dash pot provides the necessary delay time for the flow of starting current. If of the thermal type, this time delay is derived from the behavior of certain subcomponents of the overload in response to internally generated heat corresponding to line current. The magnetic type requires a particular coil to adapt the relay for a given size motor. In the thermal type, adaption is accomplished by a heater of a particular size, corresponding to the motor rating. Motor load current flowing through a compensated heater element simulates the motor overtemperature characteristics.

The types of overload relays to be used on a particular application depend on required reliability, type of load, ambient conditions, motor type and size, safety factors, acceleration time, and probability of an overload.

Table 430-37 in the 1975 NEC lists the minimum number of motor overcurrent units required to protect single- and polyphase motors connected to various supply systems. The minimum number of overload relays required for any three-phase motor application is three.

8.6.7 *Solid-State Control.* Solid-state controls are frequently applied with variable-speed systems, such as pump motor drives. Most solid-state systems consist of these basic sections, the sensor, the programmer, and the adjustable-speed unit.

(1) The sensor generates an electric signal proportional to the changing system conditions (pressure, flow, etc). This direct-current signal is applied to the programmer.

(2) The programmer determines the automatic starting and stopping sequence of each motor and sets the speed range of each adjustable-speed drive. One type of solid-state programmer consists of plug-in printed-circuit cards and a mother board. The input signal is applied to the plug-in cards through printed circuits on the mother board. The card functions are power supply, speed programming control, sequencing control, and metering.

(3) The adjustable-speed unit controls the electric energy to an alternating-current motor to change speed by means of a frequency-control system, a voltage-control system, or an impedance-control system. This unit contains the thyristors (silicon-controlled rectifiers) which provide the power control function to change motor speed.

8.7 Standards References. The following standards publications were used as references in preparing this chapter.

ANSI C2-1973, National Electrical Safety Code

ANSI C37.04-1964 (R 1969), Rating Structure for AC High-Voltage Circuit Breakers, including Supplements C37.04a-1964 (R 1969) and C37.04b-1970

ANSI C37.06-1971, Schedules of Preferred Ratings and Related Required Capabilities for AC High-Voltage Circuit Breakers Rated on a Symmetrical Current Basis

ANSI C37.09-1964 (R 1969), Test Procedure for AC High-Voltage Circuit Breakers, including Supplement C37.09a-1970

ANSI C37.16-1973, Preferred Ratings, Related Requirements, and Application Recommendations for Low-Voltage Power Circuit Breakers and AC Power Circuit Protectors

ANSI C37.100-1972, Definitions for Power Switchgear

ANSI C57.12.10-1969, Requirements for Transformers, 138 000 Volts and Below, 501 Through 10 000/13 333/16 667 kVA, Single-Phase, 501 Through 30 000/40 000/50 000 kVA, Three-Phase

ANSI C57.13-1968, Requirements for Instrument Transformers

ANSI C57.92 (1962 ed), Guide for Loading Oil-Immersed Distribution and Power Transformers

ANSI C57.96 (1959 ed), Guide for Loading Dry-Type Distribution and Power Transformers

ANSI C84.1-1970, Voltage Ratings for Electric Power Systems and Equipment (60 Hz)

ANSI C107.1-1974, Guidelines for Handling and Disposal of Capacitor- and Transformer-Grade Askarels Containing Polychlorinated Biphenyls

IEEE Std 20-1973, Low-voltage AC Power Circuit Breakers Used in Enclosures (ANSI C37.13-1973)

IEEE Std 27-1974, Switchgear Assemblies Including Metal-Enclosed Bus [ANSI C37.20 (1974 ed)]

IEEE Std 76-1974, Guide for Acceptance and Maintenance of Askarel in Equipment

IEEE Std 100-1972, Dictionary of Electrical and Electronics Terms (ANSI C42.100-1972)

IEEE Std 142-1972, Grounding of Industrial and Commercial Power Systems (ANSI C114.1-1973)

IEEE Std 144-1971, Guide for Evaluating the Effect of Solar Radiation on Outdoor Metal-Clad Switchgear (ANSI C37.24-1971)

IEEE Std 241-1974, Electric Power Systems in Commercial Buildings

IEEE Std 242-1975, Protection and Coordination of Industrial and Commercial Power Systems

IEEE Std 320-1972, Application Guide for AC High-Voltage Circuit Breakers Rated on a Symmetrical Current Basis (ANSI C37.010-1972)

IEEE Std 446-1974, Emergency and Standby Power Systems

IEEE Std 462-1973, General Requirements for Distribution, Power and Regulating Transformers (ANSI C57.12.00-1973)

NEMA AB 1-1975, Molded-Case Circuit Breakers

NEMA FU 1-1972, Low-Voltage Cartridge Fuses

NEMA ICS-1970, Industrial Controls and Systems

NEMA SG 2-1975, High-Voltage Fuses

NEMA SG 4-1975, AC High-Voltage Circuit Breakers

NEMA SG 5-1975, Power Switchgear Assemblies

NEMA SG 6-1974, Power Switching Equipment

NEMA TR 1-1974, Transformers, Regulators, and Reactors

NEMA 201-1970, Primary Unit Substations

NEMA 210-1970, Secondary Unit Substations

NFPA No 70, National Electrical Code (1975), (ANSI C1-1975), Articles 110-16, 110-34, 430-37, 450, 500, 517, 700, and 750

9. Instruments and Meters

9.1 Introduction. This chapter covers instruments and meters utilized in industrial power distribution systems. Metering and instrumentation are essential to satisfactory plant operation, the amount required depending upon the size and complexity of the plant, as well as on economic factors. Instruments and meters are needed for monitoring plant operating conditions, for power billing purposes, and for the determination of production costs.

An instrument is defined as a device for measuring the value of the quantity under observation. Instruments may be either indicating or recording type.

A meter is defined as a device that measures and registers the integral of a quantity with respect to time. The term meter is also commonly used in a general sense as a suffix or as part of a compound word (such as voltmeter, frequency meter), even though these devices are classed as instruments.

Alternating-current, voltage, etc, instruments and meters are designed to measure the root mean square value of the input wave of the frequency specified and are calibrated for a pure sine wave. Any deviation in frequency or wave shape will result in a decrease in the accuracy of the instrument or meter.

9.2 Basic Objectives. The basic objective of instrumention and metering is to assist operators in the operation of the plant. For proper operation, information relative to the magnitude of loads, energy consumption, load characteristics, load factor, power factor, voltage, etc, are required. Certain checks are required on plant electric equipment prior to placing it in service to be sure that the insulation is in proper condition for an application of voltage, and that connections have been properly made. After the equipment is in service, certain periodic checks are necessary to be sure that the equipment remains in proper operating condition. Instruments and meters are used to perform these and many other important functions.

9.3 Means Available. A large variety of instruments and meters are available to measure alternating current, voltage, power consumption, etc. In most cases their current coils are rated 5 A, and their potential coils are rated 120 V. Whenever the current and voltage of the circuit exceed the rating of the instruments, instrument current and potential transformers are required.

Current transformers serve to insulate the instrument circuit from the primary circuit and reduce the current through the instrument to values within the rating of the instrument element. The current transformer ratio selected should be as low as possible without exceeding rated current in the secondary winding. If possible, a ratio should be used which will give a normal current reading at about one-half to three-quarter scale on the instrument. On three-phase three-wire circuits, two current transformers are sufficient for metering, although a third transformer is sometimes used for checking the ratio of the others. On a three-phase four-wire grounded system, three current transformers are required. Usually the turns ratio of a current transformer is such that dangerously high potentials result when the secondary circuit is opened. Hence a test switch or current jack must be provided to short-circuit the transformer secondary while testing the instrument, or for use of plug-in portable meters. Current transformers must have their secondary circuits grounded.

Potential transformers serve to reduce the voltage of higher voltage circuits to values within the rating of the instrument potential coils. Single-phase transformers are usually employed with two connected in open delta for three-phase three-wire circuits. For three-phase four-wire systems, three potential transformers are required. Switches should be provided in the potential transformer secondary circuit to disconnect the instrument for testing. Potential transformer secondaries must also be grounded.

Direct-current measurements of current or energy utilize shunts to carry the main current to be measured. A shunt is made of metal with a low-temperature coefficient of resistance and low thermoelectric effect with respect to copper. Strips of resistance metal are brazed into heavy copper blocks which become the terminals for the line and the leads to the instrument. Ordinarily the leads must be calibrated with the shunt with which they are to be used. The direct-current ammeter actually measures the millivolt drop across its shunt and is calibrated in terms of the current rating of its associated shunt. Meters reading up to 50 A or less may have the shunt within the meter case. External shunts are available in ratings up to many thousands of amperes.

Another method of direct-current measurement uses a direct-current transformer which is a form of magnetic amplifier or saturable core reactor. Two double-circuit transformers are used, with one winding of each connected in the direct-current circuit and the secondary windings excited by an alternating voltage. An alternating-current instrument calibrated to read direct current is connected in the sec-

ondary circuit of the transformer. This method has two definite advantages over the shunt method. The secondary circuit is insulated from the measured source, and the user may add one or more instruments, relays, or other current-operated devices to the secondary of the transformer. For very high current values the lower cost and greater reliability of the direct-current transformer give it a decided advantage over the shunt. Direct-current transformers are especially useful where remote metering of large direct currents is involved since calibrated leads are not required.

Electrical quantities on alternating-current systems can be measured by the use of a transducer which utilizes the Hall effect, referred to as a Hall generator. The Hall generator acts as a multiplying device, outputting a direct-current millivolt signal which is proportional to the product of the current and a magnetic field input. The magnetic field may be developed from a current or a voltage source. The output signal may be used to operate a direct-current voltmeter calibrated to the product units or as an input to telemetering or recording devices.

The Hall generator principle has been applied to devices for the measurement of voltage, current, power, reactive power, power factor, and frequency.

Some of the advantages of the Hall generator type of transducer are its small all-solid-state construction, its relatively high output signal level, and its high speed. Also its output is isolated from the measured quantity, that is, the transducer inputs.

9.4 Instruments

9.4.1 *Ammeters.* Ammeters are used to measure the current which flows in a circuit. An ammeter is directly connected in series or has its associated current transformer primary in series with the circuit being measured.

9.4.2 *Voltmeters.* Voltmeters are used for measuring the potential difference between conductors or terminals. A voltmeter is connected directly, or through a potential transformer, across the points between which the potential difference is to be measured. For voltmeters operating on direct-current circuits above 300 V external series resistors are commonly required.

9.4.3 *Wattmeters.* A wattmeter is an instrument which measures the magnitude of the electric power being delivered to a load or group of loads. Indicating, in effect, the product of voltage and in-phase current, the wattmeter has both potential coils and current coils.

9.4.4 *Varmeters.* A varmeter is an instrument which measures reactive power. It is essentially a wattmeter which will indicate reactive power with the current coils connected in series with the circuit and the voltage element shifted 90 electrical degrees from the voltage across the circuit. Varmeters usually have the zero point at the center of the scale, since reactive power may be leading or lagging. The varmeter has an advantage over a power-factor meter in that the scale is linear, and small variations in reactive power can be read. A power-factor meter may be difficult to read near unity power factor.

9.4.5 *Power-Factor Meters.* A power-factor meter is an instrument that indicates the power factor of the load. It is a direct-reading instrument which will indicate the power factor of a three-phase load if the voltage and load are balanced on the three phases. The meter consists of both current and voltage elements, utilizing instrument transformers where necessary. The meter indicates unity power factor on scale center and lead or lag for any power factor other than unity. It is possible to obtain the average power factor over a definite period, like a day, week, or month, by use of the readings of an integrating kilowatt-hour meter and a kilovar-hour meter. However, a power-factor meter provides a convenient method for obtaining the power factor directly.

9.4.6 *Frequency Meters.* The frequency of an alternating-current power supply can be measured directly by frequency meters. Two commonly used types are the pointer-indicating type and the vibrating-reed type.

The pointer-indicating type is connected directly across the line or the secondary of the potential transformer, only single-phase connections being required. This instrument may have a scale range such as 55 to 65 Hz or 58 to 62 Hz for use on a 60 Hz system, and the moving pointer indicates the exact value.

The vibrating-reed frequency meter is based upon the principle of mechanical resonance. The instrument has a number of reeds in a line, which are free to vibrate. It is connected directly across the line or the secondary of the potential transformer, the same as the pointer-indicating type. The reed which is most nearly in tune with the line frequency will respond with the greatest amplitude of motion to indicate frequency.

9.4.7 *Synchroscopes.* Synchroscopes, which are synchronism indicators, are utilized whenever two generators or systems are to be connected in parallel. Lamps were used in the early days to indicate synchronism, but such lamps have given place to the more accurate synchroscope. A synchroscope has the appearance of a switchboard instrument, except that the pointer is free to revolve through 360 degrees. When the frequency to be synchronized is low, the pointer revolves in one direction; and conversely when high, the pointer rotates in the opposite direction. When the frequency is the same, the point stands still; and when the circuits are "in phase," the pointer so indicates, and the systems may be safely paralleled. A single-phase synchroscope may be used on a three-phase system, provided proper checks are made intially relative to phasing, and thereafter neither the direction of rotation nor the connections are changed.

9.4.8 *Elapsed-Time Meters.* Elapsed-time meters are small synchronous-motor-driven cyclometer dials for registering the amount of time a circuit or electrically driven piece of machinery is in operation. They provide important data for efficiency and life studies.

9.4.9 *Portable Instruments.* Most instruments are permanently mounted on switchboards, but many can be obtained in portable forms. Portable instruments are useful for special tests or for augmenting those mounted on the switchboard. Provisions should be made on the secondaries of instrument transformers for connecting portable instruments into the circuit. Current jacks are particularly applicable for current circuits. Portable current and potential transformers also are available for cases where the self-contained range of the portable instrument is not sufficient for the values to be measured.

Portable ammeters, voltmeters, and wattmeters may be indicating or recording depending on the type of test or data required. A combination alternating-current portable instrument, sometimes called a circuit analyzer, is available which will read current, voltage, power, and power factor. Others are available for current, voltage, and resistance, and may operate on either alternating or direct current, providing flexible instrumentation for various conditions.

A split-core current transformer, with its associated leads and ammeters either recording or indicating, is convenient for load checks. The clip-on type instrument is easier to use than the split-core current transformer type for reading currents in cables.

Ammeters with back-up pointers can be used to measure short-duration loads that a normal instrument will not indicate, such as welder loads. The pointer-stop instrument contains a hand-operated stop which restrains the indicating hand from returning to zero. The stop point is raised to successively higher levels until the surge of current just causes the hand or point to flutter, thus giving the actual current reading. If the duration of the load swing is less than 10 cycles, the point-stop instrument will not indicate accurately. An oscillograph or oscilloscope is recommended for load swings of short duration.

9.4.10 *Recording Instruments.* Most of the instruments available as direct-reading indicating instruments are also available as recording or curve-drawing instruments. A continuous record of current, voltage, power, frequency, etc, may be required for economic, statistical, and engineering studies as well as for checking operator and machine performance. The record is traced automatically on either a strip or a circular chart by a pen fastened to the end of the pointer of the instrument. The chart is moved at a constant speed by a clock mechanism. Certain design problems are present in recording instruments that do not exist with indicating instruments. One is the necessity of providing sufficient torque for overcoming pen friction without impairing the accuracy of the instrument.

Some recording instruments have adjustable speeds for movement of the chart. Normal records are obtained at a speed consistent with chart-changing schedules, types, and load characteristics, etc. Special tests may require a more rapid chart speed in order to obtain proper data. Then more sophisticated equipment, such as magnetic tape recorders, may be needed.

9.4.11 *Miscellaneous Instruments*

(1) *Temperature Indicator.* Temperature-indicating and temperature-control devices include liquid, gas, or saturated-vapor thermometers, resistance thermometers, bimetal thermometers, radiation pyrometers, and thermoelectric pyrometers. These devices may be obtained as indicating instruments or as recorders. They are used for measuring temperatures of electric windings, bearings, oil, air, and conductors. Some of them can be obtained with electric contacts for use on an alarm device or relay circuit. Pyrometers are generally used for indication of furnace temperatures and control thereof.

(2) *Cycle Counter.* This instrument, generally consisting of a synchronous motor together with necessary clutch, brake, and indicator, is used to indicate the number of cycles for an operation. One use of the cycle counter is to determine relay and circuit breaker operating times.

(3) *Megohmmeter.* A megohmmeter is an instrument for measuring the insulation resistance of electric cables, insulators, buses, motors, and other electric equipment. It consists of either a hand- or motor-driven direct-current generator and resistance indicator. It is calibrated in megohms with a maximum indication of 10 000 MΩ or infinity and is available in different voltage ratings, usually 250, 500, or 2500 V. An insulation resistance test on electrical insulation prior to placing equipment in service or during routine maintenance will give a good indication of the condition of the insulation. Wet insulation can be detected very readily. However, a high reading does not necessarily mean that equipment can withstand the rated potential since the instrument does not apply rated potential. A high-potential test along with this test is desirable but may not be practicable. In lieu of a high-potential test a dielectric-absorption test using a megohmmeter may be made. Recording and plotting periodic resistance readings will show trends and may often indicate imminent insulation failure.

(4) *Ground Ohmmeter.* A ground ohmmeter measures the resistance to earth of ground elec-

trodes. It is calibrated in ohms, usually zero to 300 Ω. Some types also include provision for measurement of soil resistance.

(5) *Low-Resistance Ohmmeter.* Very-low-reading ohmmeters are available for low-resistance measurements, such as electric conductors, joints, contact surfaces, and electric windings. This instrument is not considered as a laboratory high-precision type, but it is useful in field testing for either original installations or trouble shooting. Instruments of this type contain a source of direct current and a meter calibrated to read directly the low resistance in microhms.

(6) *Ground Detector.* A ground detector is a device which indicates a ground path on an ungrounded system. Direct-reading types are available to match the ammeter, voltmeter, etc. An incandescent-lamp ground detector connected either directly to low-voltage systems or to potential transformers on high-voltage systems also may be used. The voltage rating of the lamp is selected so that the lamp glows on the normal system. Whenever a ground fault occurs on one phase of the system, the lamp connected from that phase to ground will go out or dim and the other two lamps will glow brightly. With a fault of sufficiently low resistance on one phase, the lamp glow differential can be detected by the eye. A voltage relay can be added to the circuit and an alarm, such as a bell, sounded whenever a ground fault occurs.

(7) *Oscillograph.* An oscillograph is an instrument for observing and recording rapidly changing values of short duration, such as the waveform of alternating voltage, current, or power. This instrument has many uses in engineering fields, such as determining load characteristics, wave shapes, and phenomena which occur too rapidly for measurement by indicating meters. They are available for frequencies up to about 10 000 Hz. Most magnetic-type oscillographs consist of a galvanometer (which gives deflections closely proportional to the instantaneous value of current or voltage), an optical system (using a light beam from a mirror rather than usual pointers), and the recording device (film or light-sensitive material which can be moved rapidly). Multielement oscillographs are available for recording several different values

simultaneously, such as three-phase current, voltage, power, etc.

Direct-writing oscillographs record the phenomena or transients directly on a paper chart, utilizing an inking pen. Due to the pen inertia, this type of instrument has a limited frequency range in the order of 0.5—100 Hz. There are many investigations or measurements which can be made on this type of instrument.

(8) *Oscilloscope.* Oscilloscopes are electronic instruments which are available to study very high frequencies or phenomena of short duration. They can be used to study transients which occur in power circuits. These instruments use electronic controls and an electron beam, thereby eliminating the inertia of mechanical instruments. A fine electron beam is made to impinge on a fluorescent screen. The electron beam is deflected by the electrostatic or magnetic field set up by the voltage or current to be investigated. Oscilloscopes can be used for any frequencies up to millions of hertz. A camera can be used in conjunction with the oscilloscope to record the wave shape permanently.

(9) *Computer.* A computer is a device that will perform mathematical operations upon information fed into it according to a predetermined plan. There are two basic types of computers, the analog and the digital. Analog computers use physical units to simulate conditions of the problem to be solved. In contrast, the digital computer accepts discrete information and by use of logic rules and counting, solves a mathematical equation of the problem.

The accuracy of the analog computer is limited by the noise in its amplifiers and by the linearity of its components, while the accuracy of the digital computer is limited only by the number of digits in the computer word. Accuracy, speed, and the development of peripheral and conversion equipment which enable digital computers to accept analog as well as digital inputs, have made the digital computer almost universally accepted as the control computer today.

The information fed to computers can be actual numerical information, or electric or mechanical energy of a desired value, and the solution can be numerical information, or electric or mechanical energy for the control of some component. The purpose here is to call attention to computer systems not only for the solution of problems where previously gathered data are fed into the device, but also for the real-time solution with information fed from operating systems, for monitoring the more complex industrial power systems, and for complete automatic control of systems for optimum performance. This is now being done in power utility systems with a saving in manpower, maintenance, and expense due to operating errors. Such systems are custom designed but can be divided into three basic sections consisting of

(1) The sensing devices (current or potential transformers, relays, thermocouples)

(2) The computer with input and output circuits

(3) The control equipment (motor or solenoid-operated valves, control relays, data logging equipment, etc)

With today's technology, the digital computer has become a more desirable tool than ever before because of the increasing complexity of industrial systems and processes, and because of the decrease in the hardware cost of the computer itself.

Input and output of information to and from the computer can be accomplished through various devices, some of the more common of which are

(1) High-speed paper tape punch, capable of speeds to 150 characters per second

(2) High-speed paper tape reader, capable of speeds to 400 characters per second

(3) Punched-card reader, which can handle 1200 cards per minute

(4) High-speed printers, capable of speeds of 2000 lines per minute

(5) Magnetic tape

(6) Cathode ray tube display

(7) Manual keyboard

For the most efficient use of the computer today, one should consider the possibility of multiprogramming, so that the computer may be used for computation as well as control. With these ideas in mind, it is obvious that the computer provides unlimited capabilities for monitoring, control, and computation on any type of industrial system.

9.5 Meters

9.5.1 *Watthour Meters.* Watthour meters measure the amount of energy consumed by a load. Watthour meters with totalizing registers may be used in assigning electric energy costs to segments of production. Alternating-current watthour meters employ the induction-disk type of mechanism; the disk revolves at a speed proportional to the rate at which energy is passing through the meter. The number of revolutions, through a gear train, is recorded on a dial in kilowatt-hours. The watthour meter may be used to calculate the power used by a load. Count the number of revolutions of the disk for any number of seconds and use this formula:

$$\text{power (kilowatts)} = \frac{3600 \times r \times kh}{1000 \times s}$$

where kh is the meter constant (marked on the meter disk or nameplate), r the number of revolutions, and s the number of seconds.

Instrument transformer ratios must be taken into account in this formula if utilized. Current and potential instrument transformers are used, when necessary, with these meters, just as they are used with a wattmeter. On three-phase three-wire circuits, two current element meters are used. On four-wire circuits, three current element meters are necessary.

Dynamometer type watthour meters are used for direct-current measurement.

9.5.2 *Demand Meters.* Because of appreciable variations in power level of many industrial plants and the effect of resulting low load factors on the power system investment, the maximum power demand during a specified period of time is used as a factor in determining power bills. This maximum demand is measured by a demand meter. Both indicating and recording meters are available for this purpose.

(1) Curve-drawing wattmeters, which record the load—time curve of the system, can be used to determine the maximum demand by averaging the load over the selected demand interval of time.

(2) Integrating demand meters totalize the energy used over the demand interval and either record the average demand for each interval or, by means of a maximum indicating pointer, indicate the maximum demand that has occurred since the meter was last read and reset.

(3) Lagged demand meters usually obtain their demand interval by some thermal time lag means. Such meters are usually adjusted so that the indicating element reaches 90 percent of the maximum value of a suddenly applied steady load at the end of the selected demand interval, and 99 percent at the end of the succeeding interval.

(4) Contact-operated demand meters. Contactors can be attached to watthour meters so that impulses are transmitted at a rate proportional to the load on the meter. These impulses operate the demand meter to drive a pointer on the indicator type or a pen on the graphic type. These are reset at the end of the demand interval by a synchronous or key-wound clock. These demand meters have the advantage of easy servicing as compared to the combined watthour meter and demand meters. Also the demands of several lines or loads can be combined through a totalizing relay to operate one demand meter to give total demand. Phase-shifting transformers and scale plates are available to make the meters read the demand in kilovolt-amperes. The demand in kilovolt-amperes may be more useful in determining the actual load in terms of equipment rating.

(5) Printing demand meter. Another type of contact-operated demand meter is the printing demand meter which records the demand for each interval by printing it on a tape together with the time of day.

(6) A variation of the printing demand meter is the magnetic-tape metering concept. Instead of printing the demand and time on paper tape, kilowatt-hour pulses and time pulses may also be simultaneously recorded on magnetic tape. This tape can be fed into a translator which will transfer the demand information directly to key punch cards or magnetic tape for computer input. The major advantage of this type of system is the elimination of manual chart or tape reading and manual computations.

9.6 Typical Installations. The following combinations of instruments and meters are used in various applications.

9.6.1 *Above 600 V*
(1) *Utility supply lines*
Voltmeters
Ammeters

Wattmeters
Varmeters or power-factor meters
Watthour meters
Demand meters
Frequency meters

(2) *Plant feeders*
Ammeters
Wattmeters
Varmeters or power-factor meters
Watthour meters (demand attachment optional)
Test blocks for portable instruments

(3) *Generators*
Voltmeters, alternating and direct current
Ammeters, alternating and direct current
Wattmeters
Varmeters or power-factor meters (optional)
Synchronoscopes (with two or more generators or utility tie)
Other optional equipment may include
Watthour meters
Frequency meters
Recording meters and instruments

(4) *Motors, synchronous*
Voltmeters (optional or direct current only)
Ammeters, alternating and direct current
Wattmeters (optional)
Varmeters or power-factor meters (optional)
Watthour meters (optional)
Elapsed-time meters (optional)

(5) *Motors, induction*
Voltmeters (optional)
Ammeters
Watthour meters (optional)
Elapsed-time meters (optional)

9.6.2 *Low Voltage, 600 V and Below*
(1) *Utility supply lines.* Same, in general, as in Section 9.6.1(1).
(2) *Feeders*
Voltmeters (optional)
Ammeters (optional)
Watthour meters (optional)
Because of the added cost, these instruments and others are only used in low-voltage feeders when justified by a potential savings in operation and maintenance of equipment. Without permanent instrumentation checks are made periodically with portable instruments.

(3) *Generators.* Same, in general, as in Section 9.6.1(3).

(4) *Motors, synchronous.* Same as in Section 9.6.1(4).

(5) *Motors, induction*
    Ammeters on large motors

Generally no instruments are installed on small motors.

9.7 Standards References. The following standards publication was used as reference in preparing this chapter.

ANSI C12-1975, Code for Electricity Metering

IEEE Std 316-1971, Requirements for Direct Current Instrument Shunts

10. Cable Systems

10.1 Introduction. The primary function of cables is to carry energy reliably between source and utilization equipment. In carrying this energy there are heat losses generated in the cable that must be dissipated. The ability to dissipate these losses depends on how the cables are installed, and this affects their ratings.

Cables may be installed in raceway, cable trays, underground in duct or direct buried, preassembled on messenger, and in cable bus.

Selection of conductor size requires consideration of load current to be carried and loading cycle, emergency overloading requirements and duration, fault clearing time and interrupting capacity of the cable overcurrent protection or source capacity, and voltage drop for the particular installation conditions.

Insulations can be classified in broad categories as solid insulations, taped insulations, and special-purpose insulations. Cables incorporating these insulations cover a range of maximum and normal operating temperatures and exhibit varying degrees of flexibility, fire resistance, and mechanical and environmental protection.

Installation of cables requires care to avoid excessive pulling tensions that could stretch the conductor or insulation shield, or rupture the cable finish when pulled around bends.

Provisions must be made for the proper terminating, splicing, and grounding of cables.

There are minimum clearances between phases and between phase and ground for the various voltage levels. The terminating compartments must be either well ventilated or, if fully enclosed, heated or kept dry to prevent condensation from forming. Condensation or contamination on high-voltage terminations could result in tracking over the terminal surface with possible flashover.

Many users test cables after installation and periodically test important circuits. Test voltages are usually direct current of a level recommended by the cable manufacturer for his particular cable. Usually this test level is well below the direct-current strength of the cable, but it is possible for accidental flashovers to weaken or rupture the cable insulation due to the higher transient overvoltages that can occur from reflections of the voltage wave.

The application and sizing of cables rated 600 V through 15 kV is governed by NFPA No 70, National Electrical Code (1975), (ANSI C1-1975), (NEC). Cable use may also be covered in state and local regulations recognized by the local electrical inspection authority having jurisdiction in a particular area.

The various tables in this chapter are intended to assist the electrical engineer in laying out and understanding, in general terms, his cable requirements.

10.2 Cable Construction

10.2.1 *Conductors.* The two conductor materials in common use are copper and aluminum. Copper has historically been used for conductors of insulated cables due primarily to its desirable electrical and mechanical properties. The use of aluminum is based mainly on its favorable conductivity-to-weight ratio, the highest of the electrical conductor materials, and its ready availability and the low stable cost of the primary metal.

The need for mechanical flexibility usually determines whether a solid or a stranded conductor is used, and the degree of flexibility is a function of the total number of strands. A wire is regarded as a slender rod or filament of drawn metal where length is great in relation to diameter. A cable is defined as a stranded conductor or an assembly of insulated conductors.

Stranded conductors are available in various configurations such as standard concentric, compressed, compact, rope, and bunched, with the latter two generally specified for flexing service. Bunched stranded conductors, not illustrated, consist of a number of individual strand members of the same size which are twisted together to make the required area in circular mils for the intended service. Unlike the individual strands in the concentric-lay strands illustrated in Fig 118, no attempt is made during manufacture of bunch strand to control the position of each strand with respect to another. This type of conductor is usually found in lamp cords, fixture wires, and appliance cords.

10.2.2 *Comparison Between Copper and Aluminum.* Aluminum requires larger conductor sizes to carry the same current as copper. The equivalent aluminum cable when compared to copper in terms of ampacity will be lighter in weight and larger in diameter. The properties of these two metals are given in Table 78.

The 36 percent difference in thermal coefficients of expansion and the different electrical nature of the oxide films of copper and aluminum require consideration in connector designs. An aluminum oxide film forms immediately on exposure of fresh aluminum surface to air. Under normal conditions it slowly builds up to a thickness in the range of 3–6 nanometers and stabilizes at this thickness. The oxide film is es-

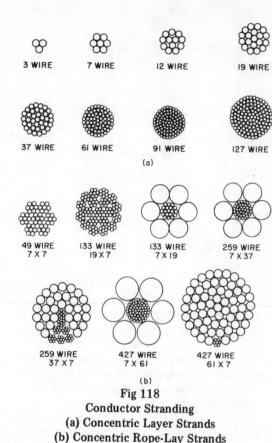

Fig 118
Conductor Stranding
(a) Concentric Layer Strands
(b) Concentric Rope-Lay Strands

sentially an insulating film or dieletric material and provides aluminum with its corrosion resistance. Copper produces its oxide rather slowly under normal conditions, and the film is relatively conducting, presenting no real problem at connections.

Approved connector designs for aluminum conductors essentially provide increased contact areas and lower unit stresses. These terminals possess adequate strength to ensure that the compression of the aluminum strands exceeds their yield strength and that a brushing action takes place which destroys the oxide film to form an intimate aluminum contact area yielding a low-resistance connection.

Water must be kept from entering the strand space in aluminum conductors at all times. Any moisture within a conductor, either copper or aluminum, is likely to cause corrosion of the

Table 78
Properties of Copper and Aluminum

	Copper Electrolytic	Aluminum EC Alloy
Conductivity, % IACS† at 20°C	100.0	61.0
Resistivity, $\Omega \cdot$ cmil/ft at 20°C	10.371	17.002
Specific gravity at 20°C	8.89	2.703
Melting point, °C	1083	660
Thermal conductivity at 20°C, $(cal \cdot cm)/(cm^2 \cdot °C \cdot s)$*	0.941	0.58
Specific heat, $cal/(g \cdot °C)$* for equal weights	0.092	0.23
for equal direct-current resistance	0.184	0.23
Thermal expansion, in; equal to constant $\times 10^{-6} \times$ length in inches $\times$ °F steel = 6.1 18-8 stainless = 10.2 brass = 10.5 bronze = 15	9.4	12.8
Relative weight for equal direct-current resistance and length	1.0	0.50
Modulus of elasticity, $(lb/in^2) \times 10^6$	16	10

*cal here denotes the gram calorie.
†International annealed copper standard.

conductor metal or impair insulation effectiveness.

10.2.3 *Insulation.* Basic insulating materials are either organic or inorganic, and there are a wide variety of insulations classed as organic. Mineral-insulated cable employs the one inorganic insulation (MgO) that is generally available.

Insulations in common use are

(1) Thermosetting compounds, solid dielectric

(2) Thermoplastic compounds, solid dielectric

(3) Paper-laminated tapes

(4) Varnished cloth, laminated tapes

(5) Mineral insulation, solid dielectric granular

Most of the basic materials listed in Table 79 must be modified by compounding or mixing with other materials to produce desirable and necessary properties for manufacturing, handling, and end use. The thermosetting or rubber-like materials are mixed with curing agents, accelerators, fillers, and anti-oxidants in varying proportions, and crosslinked polyethylene is in-

cluded in this class. Generally, smaller amounts of materials are added to the thermoplastics in the form of fillers, anti-oxidants, stabilizers, plasticizers, and pigments.

(1) *Insulation Comparison.* The aging factors of heat, moisture, and ozone are among the most destructive of organic based insulations, so the following comparisons are a gauge of the resistance and classification of these insulations.

(a) *Relative Heat Resistance.* The comparison in Fig 119 illustrates the effect of a relatively short period of exposure at various temperatures on the hardness characteristic of the material at that temperature. Basic differences between thermoplastic and thermosetting type insulation, excluding aging effect, are evident.

(b) *Heat Aging.* The effect on elongation of an insulation (or jacket) when subjected to aging in a circulating air oven is an acceptable measure of heat resistance. The air oven test at 121°C called for in some specifications is

Table 79
Commonly Used Insulating Materials

Common Name	Chemical Composition	Properties of Insulation Electrical	Physical
Thermosetting			
Crosslinked polyethylene	Polyethylene	Excellent	Excellent
EPR	Ethylene propylene rubber (copolymer and terpolymer)	Excellent	Excellent
Butyl	Isobutylene isoprene	Excellent	Good
SBR	Styrene butadiene rubber	Excellent	Good
Oil base	Complex rubber-like compound	Excellent	Good
Silicone	Methyl chlorosilane	Good	Good
TFE*	Tetrafluoroethylene	Excellent	Good
Natural rubber	Isoprene	Excellent	Good
Neoprene	Chloroprene	Fair	Good
Class CP rubber†	Chlorosulfonated polyethylene	Good	Good
Thermoplastic			
Polyethylene	Polyethylene	Excellent	Good
Polyvinyl chloride	Polyvinyl chloride	Good	Good
Nylon	Polyamide	Fair	Excellent

*For example, Teflon or Halon.
†For example, Hypalon.

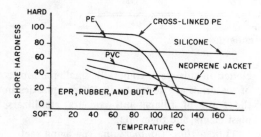

Fig 119
Typical Values for Hardness Versus
Temperature

severe, but provides a relatively quick method of grading materials for possible use at high conductor temperatures or in hot-spot areas. The 150°C oven aging is many times more severe and is used to compare materials with superior heat resistance. Temperature ratings of insulations in general use are shown in Table 80.

(c) *Ozone and Corona Resistance.* Exposure to accelerated conditions, such as higher concentrations of ozone [as standardized by the Insulated Power Cable Engineers Associa-

tion[12] (IPCEA), EM-60, for butyl, 0.03 percent ozone for 3 hours at room temperature], or air oven tests followed by exposure to ozone, or exposure to ozone at higher temperatures, aids in measuring the ultimate ozone resistance of the material. Insulations exhibiting superior ozone resistance under accelerated conditions are silicone, polyethylene, crosslinked polyethylene, EPR, and polyvinyl chloride. In fact, these materials are, for all practical purposes, inert in the presence of ozone. However, this is not the case with corona discharge.

The phenomenon of corona discharge produces concentrated and destructive thermal effects along with formation of ozone and other ionized gases. Although corona resistance is a property associated with cables over 600 V, in a properly designed and manufactured cable damaging corona is not expected to be present at operating voltage. Materials exhibiting less susceptibility than polyethylene and crosslinked polyethylene to such discharge activity are the ethylene/propylene rubbers.

---

[12] 192 Washington Street, Belmont, MA 02178.

Table 80
Rated Conductor Temperatures

Insulation Type	Maximum Voltage Class (kV)	Maximum Operating Temperature (°C)	Maximum Overload* Temperature (°C)	Maximum Short-Circuit Temperature (°C)
Paper (solid-type) multiconductor and single conductor, shielded	9	95	115	200
	29	90	110	200
	49	80	100	200
	69	65	80	200
Varnished cambric	5	85	100	200
	15	77	85	200
	28	70	72	200
Polyethylene (natural)†	5	75	95	150
	35	75	90	150
SBR rubber	2	75	95	200
Butyl rubber	5	90	105	200
	15	85	100	200
	35	80	95	200
Oil-base rubber	35	70	85	200
Polyethylene (cross-linked)†	35	90	130	250
Silicone rubber	5	125	150	250
EPR rubber†	35	90	130	250
Chlorosulfonated polyethylene‡	2	75	95	150
Polyvinyl chloride	2	60	85	150
	2	75	95	150
	2	90	105	150

*Operation at these overload temperatures shall not exceed 100 h/yr. Such 100-h overload periods shall not exceed five.
†Cables are available in 69 kV and higher ratings.
‡For example, Hypalon.

(d) *Moisture Resistance.* Insulations such as crosslinked polyethylene, polyethylene, and EPR exhibit excellent resistance to moisture as measured by standard industry tests such as IPCEA EM-60. The electrical stability of these insulations in water as measured by capacitance and power factor is impressive, and moisture is cause for little concern. The capacitance and percent power factor of the IPCEA EM-60 test of natural polyethylene and some crosslinked polyethylenes are lower than those of EPR or other elastomeric power cable insulations.

(2) *Insulations in General Use.* Insulations in general use for voltages above 2 kV are shown in Table 80. Solid dielectrics of both plastic and thermosetting types are being more and more commonly used while the laminated-type constructions, such as paper—lead cables, are declining in popularity in industrial service.

The generic names given for these insulations cover a broad spectrum of actual materials, and the history of performance on any one type may not properly be related to another in the same generic family.

10.2.4 *Cable Design.* The selection of power cable for particular circuits or feeders develops around the following considerations.

(1) *Electrical.* Dictates conductor size, type and thickness of insulation, correct materials for low- and medium-voltage designs, consideration of dielectric strength, insulation resistance, specific inductive capacitance (dielectric constant), and power factor

(2) *Thermal.* Compatible with ambient and overload conditions, expansion, and thermal resistance

(3) *Mechanical.* Involves toughness and flexibility, consideration of jacketing or armoring and resistance to impact, crushing, abrasion, and moisture

(4) *Chemical.* Stability of materials on exposure to oils, flame, ozone, sunlight, acids, and alkalies

The installation of cable in conformance with the NEC and state and local codes under the jurisdiction of a local electrical inspection authority requires evidence of approval for use in the intended service by a nationally recognized testing laboratory, such as Underwriters Laboratories Inc (UL) labeled cable. Some NEC types are discussed below.

(1) *600 V Cables.* Low-voltage power cables are generally rated at 600 V, regardless of the use voltage, whether 120, 240, 277, 480, or 600 V. The selection of 600 V power cable is oriented more to physical rather than to electrical service requirements. Resistance to forces such as crush, impact, and abrasion becomes a predominant factor, although good electrical properties for wet locations are also needed.

The 600 V compounds of crosslinked polyethylene are usually loaded (carbon black or mineral fillers) to further enhance the relatively good toughness of conventional polyethylene. The combination of crosslinking the polyethylene molecules through vulcanization plus fillers produces superior mechanical properties. Vulcanization eliminates polyethylene's main drawback of a relatively low melting point of 105°C. The 600 V cable consists simply of the conductor with a single extrusion of insulation in the specified thickness.

Rubber-like insulations such as EPR and SBR are provided with outer jackets for mechanical protection, usually of polyvinyl chloride, neoprene, or CP rubber.[13] A guide list of the more commonly used 600 V cables is outlined in the next section. Cables are classified by conductor operating temperatures and coverings with NEC insulation thicknesses.

(2) *600 V Cable Check List*

(a) EPR insulated, neoprene jacketed, type RHW/RHH, for 75°C maximum operating temperature in dry or wet locations and 90°C in dry locations only

(b) Polyvinyl chloride insulated, nylon jacketed, type THHN, for 90°C maximum operating temperature in dry locations

(c) Butyl insulated, neoprene jacketed, type RHW, for 75°C maximum operating temperature in dry or wet locations

(d) Crosslinked polyethylene insulated, without jacket, type XHHW, for 75°C maximum operating temperature in dry or wet locations and 90°C in dry locations only

(e) Polyvinyl chloride insulated, without jacket, type THW, for 75°C maximum operating temperature in dry or wet locations

---

[13] Such as Hypalon.

(f) Polyvinyl chloride insulated, nylon jacketed, type THWN, for 75°C maximum operating temperature in dry or wet locations

(g) SBR insulated, neoprene jacketed, type RHW, for 75°C maximum operating temperature in dry or wet locations

(h) Metal-clad or interlocked armor cable, type MC; individual conductors may be any of the above and cable takes rating of insulation selected; for use in partially protected areas, such as cable trays

(i) Tray cable, type TC; multiconductor with an overall flame-retardant nonmetallic jacket; individual conductors may be any of the above and cable takes rating of insulation selected; for use in cable trays, raceways, or where supported by a messenger wire

The preceding cables are suitable for applications in air, conduit, duct, direct burial, and tray, provided NEC requirements are satisfied. Cables (d) and (e) are usually restricted to conduit or duct. Single conductors may be furnished paralleled or multiplexed, as multiconductor cables with overall nonmetallic jacket (polyvinyl chloride or polyethylene), or as preassembled aerial cable on messenger.

Note that the temperatures listed are the maximum rated operating temperatures as specified in the NEC.

10.2.5 *Shielding of Higher Voltage Cable.* For operating voltages below 2 kV, nonshielded constructions are normally used, while above 8 kV, cables are required to be shielded to comply with the NEC. In the 2.1—8 kV range, both shielded and nonshielded cables are allowed, provided such cables meet NEC requirements.

As shielded cable is usually more costly than nonshielded cable, and the terminations require more care and space, the nonshielded cable has been used extensively in the 2.1—5 kV range with many applications at 7.2 kV. However, the following conditions may dictate the use of shielded cable:

(1) In direct earth burial

(2) Where the cable surface may collect unusual amounts of conducting materials (salt, soot, conductive pulling compounds)

NOTE: There are special constructions listed in the NEC for nonshielded cables for use in wet and dry locations.

Shielding of an electric power cable is defined as the practice of confining the electric field of the cable to the insulation surrounding the conductor by means of conducting or semiconducting layers, closely fitting or bonded to the inner and outer surfaces of the insulation. In other words, the outer shield confines the electric field to the space between conductor and shield. The inner or strand stress relief layer is at or near the conductor potential. The outer or insulation shield is designed to carry the charging currents and in many cases fault currents. The conductivity of the shield is determined by its cross-sectional area and the resistivity of the metal tapes or wires employed in conjunction with the semiconducting layer.

The stress control layer at the inner and outer insulation surfaces, by its close bonding to the insulation surface, presents a smooth surface to reduce the stress concentrations and minimize void formation. Ionization of the air in such voids can progressively damage certain insulating materials to eventual failure.

Insulation shields have several purposes.

(1) Confine the electric field within the cable.

(2) Equalize voltage stress within the insulation, minimizing surface discharges.

(3) Protect cable better from induced potentials.

(4) Limit electromagnetic or electrostatic interference (radio, TV).

(5) Reduce shock hazard (when properly grounded).

Fig 120 illustrates the electrostatic field of shielded cable.

The voltage distribution between an unshielded cable and a grounded plane is illustrated in Fig 121. There it is assumed that the air is the same, electrically, as the insulation, so that the cable is in a uniform dielectric above the ground plane to permit a simpler illustration of the voltage distribution and field associated with the cable.

In a shielded cable (Fig 120) the equipotential surfaces are concentric cylinders between conductor and shield. The voltage distribution follows a simple logarithmic variation, and the electrostatic field is confined entirely within the insulation. The lines of force and stress are

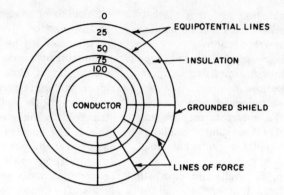

Fig 120
Electrostatic Field of Shielded Cable

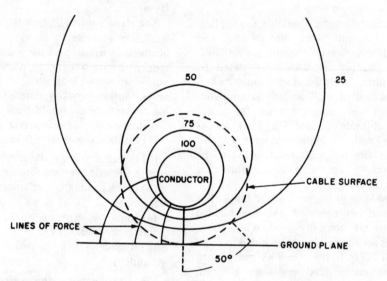

Fig 121
Field of Conductor on Ground Plane in Uniform Dielectric

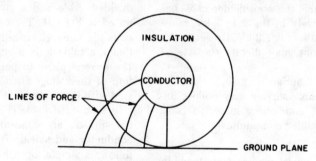

Fig 122
Unshielded Cable on Ground Plane

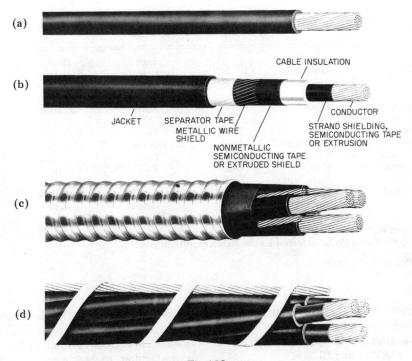

Fig 123
Commonly Used Shielded and Nonshielded Constructions
(a) 600 V Single-Conductor Cable. (b) 5.1—46 kV Single-Conductor Cable
(c) Three- or Four-Conductor 600 V Interlocked Armored Cable
(d) Preassembled Self-Supporting Aerial Cable

uniform and radial, and cross the equipotential surfaces at right angles eliminating any tangential or longitudinal stresses within the insulation or on its surface.

The equipotential surfaces for the unshielded system (Fig 122) are cylindrical but not concentric with the conductor, and cross the cable surface at many different potentials. The tangential creepage stress to ground at points along the cable may be several times the normal recommended for creepage distance at terminations in dry locations for unshielded cable operating on 4160 V systems.

Surface tracking, burning, and destructive discharges to ground could occur under these conditions. However, properly designed nonshielded cables as described in the NEC limit the surface energies available which could compromise the cable from any of these effects.

Typical cables supplied for shielded and nonshielded applications are illustrated in Fig 123.

10.3 Cable Outer Finishes. Cable outer finishes or outer coverings are used to protect the underlying cable components from the environmental and installation conditions associated with intended service. The choice of cable outer finishes for a particular application is based on the same performance categories as for insulations, specifically, electrical, thermal, mechanical, and chemical considerations. Except for highly protected and essentially room temperature operating environments, the usual condition associated with cable operation cannot be met with a single outer finishing material. Therefore, combinations of metallic and nonmetallic finishes are required to provide the total protection needed for the installation and opera-

Table 81
Properties of Jackets and Braids

Material	Abrasion Resistance	Flexibility	Low Temperature	Heat Resistance	Fire Resistance
Neoprene	Good	Good	Good	Good	Good
Class CP rubber*	Good	Good	Fair	Excellent	Good
Polyethylene					
low density	Fair	Poor	Poor	Fair	Poor
high density	Excellent	Poor	Poor	Good	Poor
cross linked	Good	Poor	Poor	Excellent	Poor
Polyvinyl chloride	Fair	Good	Fair	Good	Fair
Polyurethane	Excellent	Good	Good	Good	Poor
Asbestos braid	Fair	Good	Good	Excellent	Excellent
Nylon	Excellent	Fair	Good	Good	Fair

NOTE: Chemical resistance and barrier properties depend on the particular chemicals involved, and the question should be referred to the cable manufacturer.

*For example, Hypalon.

tion. Specific industry requirements for these materials are covered in standards of the IPCEA, the American Society for Testing and Materials[14] (ASTM), and UL.

10.3.1 Nonmetallic Finishes

(1) *Extruded Jackets.* There are outer coverings, either thermoplastic or vulcanized, which may be extruded directly over insulation or over electrical shielding systems of metal sheaths or tapes, copper braid, or semiconducting layers with copper drain wires or spiraled copper concentric wires, or over multiconductor constructions. Commonly used materials include polyvinyl chloride, NBR/PVC (nitrile butadiene/polyvinyl chloride), polyethylene, cross-linked polyethylene, polychloroprene (neoprene), chlorosulfonated polyethylene, and polyurethane. While the detailed characteristics may vary with individual manufacturers' compounding, these materials provide a high degree of moisture, chemical, and weathering protection, are reasonably flexible, provide some degree of electrical isolation, and are of sufficient mechanical strength to protect the insulating and shielding components from normal service and installation damage. Materials are available for service temperatures from -55°C to $+115^{\circ}$C (Table 81).

(2) *Fiber Braids.* This category includes braided, wrapped or served, and synthetic or natural fiber materials selected by the cable manufacturer to best meet the intended service. The most commonly used material is asbestos fiber, although special industrial applications may require synthetic or cotton fibers applied in braid form. All fiber braids require saturants or coating and impregnating materials to provide some degree of moisture and solvent resistance as well as abrasive and weathering resistance.

Asbestos braid is used on cables to minimize flame propagation, smoking, and other hazardous or damaging products of combustion which may be evolved by some extruded jacketing materials.

10.3.2 Metallic Finishes.
This category of materials is widely used where a high degree of mechanical, chemical, or short-time thermal protection of the underlying cable components may be required in the intended service. Commonly used materials are interlocked galvanized steel, aluminum, or bronze armor; extruded lead or aluminum; strip formed, welded, and corrugated steel and aluminum; and spirally laid round or flat armor wires. The use of any of these materials, alone or in combination with others, will reduce flexibility of the overall cable. This characteristic must be sacrificed to obtain the other benefits.

[14]1916 Race Street, Philadelphia, PA 19103.

Installation and operating conditions may involve localized compressive loadings, occasional impact from external sources, vibration and possible abrasion, heat shock from external sources, extended exposure to corrosive chemicals, and condensation.

(1) *Interlocked Armor.* The unprotected interlocked armor provides a high degree of mechanical protection without sacrificing flexibility significantly. While not entirely impervious to moisture or corrosive agents, interlocked armor does provide protection from thermal shock by acting as a heat sink for short-time localized exposure.

Where corrosion and moisture resistance are required in addition to mechanical protection, an overall jacket of extruded material may be used.

The use of interlocked galvanized steel armor must be avoided on single-conductor alternating-current power circuits due to its high hysteresis and eddy current losses. This effect, however, is minimized in three-conductor cables with armor overall and with aluminum armor on single-conductor cables.

Commonly used interlocked armor materials are galvanized steel, aluminum for less weight and general corrosion resistance, and marine bronze and other alloys for highly corrosive atmosphere.

(2) *Corrugated Metal Sheath.* Longitudinally corrugated metal sheaths (corrugations or bellows formed perpendicular to the cable axis) have been used for many years in direct-burial communications cables, but only recently has this method of cable core protection been applied to control and power cable. The sheath material may be of copper, aluminum, a corrosion resistant steel or copper alloy, or a bimetallic composition with the choice of material selected to best meet the intended service.

While some mechanical protection is sacrificed versus interlocked armor, its use is justified on the basis of lower cost, reduced weight, and ease of installation.

The sheath may be formed from thin strips or seamless or welded-seam tubing. The strip process results in longer use lengths than that obtained with the tube process, although mechanical connectors and splices are available

for the latter. The tube process offers maximum protection from moisture and liquid or gaseous contaminants, with the strip process depending upon an overall extruded jacket for protection from moisture penetration and to hold the metal sheath in position.

(3) *Lead.* Pure or lead alloy is used for industrial power cable sheaths for maximum cable protection in underground manhole and tunnel or underground duct distribution systems subject to flooding. While not as resistant to crushing loads as interlocked armor, its very high degree of corrosion and moisture resistance makes lead attractive in the above applications. Protection from installation damage can be provided by an outer jacket of extruded material.

Pure lead is subject to work hardening and should not be used in applications where flexing may be involved. Copper or antimony-bearing lead alloys are not as susceptible to work hardening as is pure lead, and may be used in applications involving limited flexing. Lead or its alloys must never be used for repeated flexing service.

One problem encountered today with the use of lead sheathed cable is in the area of splicing and terminating. Installing personnel experienced in the "art" of wiping lead sheath joints are not as numerous as they were many years ago, posing an installation problem for many potential users. However, many insulation systems do not require lead sleeves at splices and treat the lead like any other metallic sheath.

(4) *Aluminum or Copper.* Extruded aluminum or copper or die-drawn aluminum or copper sheaths are used in certain applications for weight reduction and moisture penetration protection. While more crush resistant than lead, aluminum sheaths are subject to electrolytic attack when installed underground. Under these conditions aluminum sheathed cable should be protected with an outer extruded jacket.

Mechanical splicing sleeves are available for use with aluminum sheathed cables, and sheath joints can be made by inert gas welding, provided that the underlying components can withstand the heat of welding without deterioration. Specifically designed hardware is available for terminating the sheath at junction boxes and enclosures.

(5) *Wire Armor.* A high degree of mechanical protection and longitudinal strength can be obtained with the use of spirally wrapped or braided round steel armor wire. This type of outer covering is frequently used in submarine cable and vertical riser cable for support. As noted for steel interlocked armor, this form of protection should be used only on three-conductor power cables to minimize sheath losses.

While not properly "armor," spirally laid tinned copper wires, round or flat, are used in direct-burial cables as concentrics or neutrals. In potentially corrosive environments, these should be protected with an extruded jacket.

10.3.3 *Single- and Multiconductor Constructions.* The single-conductor cables are usually easier to handle and can be furnished in longer lengths as compared to multiconductor cables. The multiconductor constructions give smaller overall dimensions in comparison with the equivalent number of single-conductor cables which can be an advantage where space is important.

Sometimes the outer finish can influence whether the cable should be supplied as a single- or multiconductor cable. For example, as mentioned previously, the use of steel interlock or steel wire armor on alternating-current cables is practical on multiconductor constructions, but must be avoided over single-conductor cables. It is also economical to apply the more rugged finishes over multiconductor constuctions rather than over each of the single-conductor cables.

10.3.4 *Physical Properties of Materials for Outer Coverings.* Depending on the environment and application, the selection of outer finishes can be complex as to the degree of protection needed. For a general appraisal, Table 81 lists the relative properties of various commonly used materials.

10.4 Cable Ratings

10.4.1 *Voltage Rating.* The selection of the cable insulation (voltage) rating is made on the basis of the phase-to-phase voltage of the system in which the cable is to be applied, and the general system category depending on whether the system is grounded or ungrounded and the time in which a ground fault on the system is cleared by protective equipment. It is

possible to operate cables on ungrounded systems for long periods of time with one phase grounded due to a fault. This results in line-to-line voltage stress across the insulation of the two ungrounded conductors. Therefore such cable must have greater insulation thickness than a cable used on a grounded system where it is impossible to impose full line-to-line potential on the other two unfaulted phases for an extended period of time.

Consequently 100 percent voltage rated cables are applicable to grounded systems provided with protection which will clear ground faults within 1 min. 133 percent rated cables are required on ungrounded systems where the clearing time of the 100 percent level category cannot be met, and yet there is adequate assurance that the faulted section will be cleared within 1 hour. 173 percent voltage level insulation is used on systems where the time required to de-energize a grounded section is indefinite.

10.4.2 *Conductor Selection.* The selection of conductor size is based on the following considerations:

(1) Load-current criteria as related to loadings, NEC requirements, thermal effects of the load current, mutual heating, losses produced by magnetic induction, and dielectric losses

(2) Emergency overload criteria

(3) Voltage-drop limitations

(4) Fault-current criteria

10.4.3 *Load-Current Criteria.* The NEC ampacity tables for low- and medium-voltage cables must be used where the code has been adopted. These are derived from IEEE S-135-1-1962, Power Cable Ampacities, Copper Conductors, and IEEE S-135-2-1962, Power Cable Ampacities, Aluminum Conductors.

All ampacity tables show the minimum size conductor required, but conservative engineering practice, future load growth considerations, voltage drop, and short-circuit heating may make the use of larger conductors necessary.

Large groups of cables must be carefully considered, as deratings due to mutual heating may be limiting. Conductor sizes over 500−750 kcmil require consideration of paralleling two or more smaller size cables because the current-carrying capacity per circular mil of conductor decreases for alternating-current circuits due to skin effect and proximity effect. The reduced ratio of surface to cross-sectional area of the larger size conductors is a factor in the reduced ability of the larger cable to dissipate heat. When cables are used in multiple, consideration must be given to the phase placement of the cable to minimize the effects of maldistribution of current in the cables, which will reduce ampacity. Although the material cost of cable may be less for two smaller conductors, this saving may be offset by higher installation costs.

The use of load factor in underground runs takes into account the heat capacity of the duct bank and surrounding soil which responds to the average heat losses. The temperatures in the underground section will follow the average loss, thus permitting higher short-period loadings. The load factor is the ratio of average load to peak load and is usually measured on a daily basis for the average load. The peak load is usually the average of a 1/2 to 1 hour period of the maximum loading that occurs in 24 hours.

For direct-buried cables, the average surface temperature is limited to 45—60°C, depending on soil condition, to prevent moisture migration and thermal runaway.

Cables must be derated when in proximity to other loaded cables or heat sources, or when the ambient temperature exceeds the ambient temperature on which the ampacity (current-carrying capacity) tables are based.

The normal ambient temperature of a cable installation is the temperature the cable would assume at the installed location with no load being carried on the cable. A thorough understanding of this temperature is required for a proper determination of the size cable required for a given load. For example, the ambient temperature for a cable exposed in the air isolated from other cables is the temperature of that cable before load is applied, assuming, of course, that this temperature is measured at the same time of day and with all other conditions exactly the same as they will be when the required load is being carried. It is assumed that for cables in air, the space around the cable is large enough so that the heat generated by the cable can be dissipated without raising the temperature of the room as a whole. Unless exact conditions are specified, the following ambients are commonly used for calculation of the current-carrying capacity.

(1) *Indoors.* The NEC ampacity tables are based upon an ambient temperature of 30°C for low-voltage cables. In most parts of the United States, 30°C is too low for summer months, at least for some part of the building. The NEC type MV cable ampacity tables use 40°C for air ambient temperature. In any specific case where the conditions are accurately known, the measured temperature should be used; otherwise, use 40°C.

Sources of heat adjacent to the cables under the most adverse condition must be taken into consideration in figuring the current-carrying capacity. This is usually done by correcting the ambient temperature for these localized hot spots. These may be caused by steam lines or heat sources adjacent to the cable or they may be due to sections of the cable running through boiler rooms or other hot locations. Rerouting may be necessary to avoid this problem.

(2) *Outdoors.* An ambient temperature of 40°C is commonly used as the maximum for cables installed in the shade and 50°C for cables installed in the sun. In using these ambient temperatures, it is assumed that the maximum load occurs during the time when the ambient will be as specified. Some circuits probably do not carry their full load during the hottest part of the day or when the sun is at its brightest, so that an ambient temperature of 40°C for outdoor cables is probably reasonably safe for such conditions. See NEC, notes to Tables 310-39 through 310-50, for the procedure to be used for outdoor installations.

(3) *Underground.* The ambient temperature used for underground cables varies in different sections of the country. For the northern section, an ambient of 20°C is commonly used. For the central part of the country, 25°C is commonly used, while for the extreme south and southwest, an ambient of 30°C may be necessary. The exact geological boundaries for these ambient temperatures cannot be set up, and the maximum ambient should be measured in the earth at a point away from any sources of heat at the depth at which the cable will be buried. Changes in the earth ambient will lag changes in the air ambient by several weeks.

The thermal characteristics of the medium surrounding the cable are of primary importance in determining the current-carrying capacity of cables. The type of soil in which the cable or duct bank is buried has a major effect on the current-carrying capacity of cables. Porous soils, such as gravel and cinder fill, usually result in higher temperatures and lower ampacities than normal sandy or clay soil. The type of soil and its thermal resistivity should be known before the size of the conductor is calculated.

The moisture content of the soil has a major effect on the current-carrying capacity of cables. In dry sections of the country, cables may have to be derated or other precautions taken to compensate for the increase in thermal resistance due to the lack of moisture. On the other hand, in ground which is continuously wet or under tidewater conditions, cables may carry higher than normal currents. Shielding for even 2400 V circuits is necessary for continuously wet or alternately wet and dry conditions, for where there is a change from dry to the "naturally shielded" wet cables, there will be an abrupt voltage gradient stress, just as at the end of shielded cables terminated without a stress cone, except where nonshielded cables are specifically designed for this service.

Ampacities in the 1975 NEC tables take into account the grouping of adjacent circuits. For ambient temperatures different from those shown in the tables, derating factors to be applied are shown in Article 310, Table 310-13, and notes to Tables 310-39 through 310-50.

10.4.4 *Emergency Overload Criteria.* Normal loading limits of insulated wire and cable are based on many years of practical experience and represent a rate of deterioration that results in the most economical and useful life of such cable systems. The rate of deterioration is expected to develop a useful life of about 20 to 30 years. The life of cable insulation is about halved and the average rate of thermally caused service failures about doubled for each $5-15°C$ increase in normal daily load temperature. Additionally, sustained operation over and above maximum rated operating temperatures or ampacities is not a very effective or economical expedient, because the temperature rise is directly proportional to the conductor loss which increases as the square of the current. The greater voltage drop might also increase the risks to equipment and service continuity.

As a practical guide, the IPCEA has established maximum emergency overload temperatures for various types of insulation. Operation at these emergency overload temperatures should not exceed 100 hours per year, and such 100 hour overload periods should not exceed five during the life of the cable. Table 82 gives uprating factors for short-time overloads for various types of insulated cables. The uprating factor, when multiplied by the nominal current rating for the cable in a particular installation, will give the emergency or overload current rating for the particular insulation type.

10.4.5 *Voltage-Drop Criteria.* The supply conductor, if not of sufficient size, will cause excessive voltage drop in the circuit, and the drop will be in direct proportion to the circuit length. Proper starting and running of motors, lighting equipment, and other loads having heavy inrush currents must be considered. The NEC recommends that the steady-state voltage drop in power, heating, or lighting feeders be no more than 3 percent, and the total drop including feeders and branch circuits be no more than 5 percent overall.

10.4.6 *Fault-Current Criteria.* Under short-circuit conditions, the temperature of the conductor rises rapidly. Then, due to thermal characteristics of the insulation, sheath, surrounding materials, etc, it cools off slowly after the short-circuit condition is removed. The IPCEA has recommended a transient temperature limit for each type of insulation for short-circuit duration times not in excess of 10 seconds.

Failure to check the conductor size for short-circuit heating could result in permanent damage to the cable insulation due to disintegration of insulation material, which may be accompanied by smoke and generation of combustible vapors. These vapors will, if sufficiently heated, ignite, possibly starting a serious fire. Less seriously, the insulation or sheath of the cable may be expanded to produce voids leading to subsequent failure. This becomes especially serious in 5 kV and higher voltage cables.

In addition to the thermal stresses, mechanical stresses are set up in the cable through expansion upon heating. As the heating is rapid, these stresses may result in undesirable cable movement. However, on modern cables, reinforcing binders and sheaths considerably re-

Table 82
Uprating for Short-Time Overloads*

Insulation Type	Voltage Class (kV)	Conductor Operating Temperature (°C)	Conductor Overload Temperature (°C)	Uprating Factors for Ambient Temperature							
				20°C		30°C		40°C		50°C	
				Cu	Al	Cu	Al	Cu	Al	Cu	Al
Paper (solid type)	9	95	115	1.09	1.09	1.11	1.11	1.13	1.13	1.17	1.17
	29	90	110	1.10	1.10	1.12	1.12	1.15	1.15	1.19	1.19
	49	80	100	1.12	1.12	1.15	1.15	1.19	1.19	1.25	1.25
	69	65	80	1.13	1.13	1.17	1.17	1.23	1.23	1.38	1.38
Varnished cambric	5	85	100	1.09	1.08	1.10	1.10	1.13	1.13	1.17	1.17
	15	77	85	1.05	1.05	1.07	1.07	1.09	1.09	1.13	1.13
	28	70	72								
Polyethylene (natural)†	35	75	95	1.13	1.13	1.17	1.17	1.22	1.22	1.30	1.30
SBR rubber	0.6	75	95	1.13	1.13	1.17	1.17	1.22	1.22	1.30	1.30
	5	90	105	1.08	1.08	1.09	1.09	1.11	1.11	1.14	1.14
Butyl RHH	15	85	100	1.09	1.08	1.10	1.10	1.13	1.13	1.17	1.17
	35	80	95	1.09	1.09	1.11	1.11	1.14	1.14	1.20	1.20
Oil-base rubber	35	70	85	1.11	1.11	1.14	1.14	1.20	1.20	1.29	1.29
Polyethylene (cross-linked)†	35	90	130	1.18	1.18	1.22	1.22	1.26	1.26	1.33	1.33
Silicone rubber	5	125	150	1.08	1.08	1.09	1.09	1.10	1.10	1.12	1.11
EPR rubber†	35	90	130	1.18	1.18	1.22	1.22	1.26	1.26	1.33	1.33
Chlorosulfonated polyethylene‡	0.6	75	95	1.13	1.13	1.17	1.17	1.22	1.22	1.30	1.30
Polyvinyl chloride	0.6	60	85	1.22	1.22	1.30	1.30	1.44	1.44	1.80	1.79
	0.6	75	95	1.13	1.13	1.17	1.17	1.22	1.22	1.30	1.30

*To be applied to normal rating determined for such installation conditions.
†Cables are available in 69 kV and higher ratings.
‡ For example, Hypalon.

Table 83
Minimum Conductor Sizes, in AWG or kcmil, for Indicated Fault Current and Clearing Times

Total RMS Current (amperes)	Polyethylene and Polyvinyl Chloride, 75–150°C				Oil Base and SBR, 75–200°C				Cross-Linked Polyethylene and EPR, 90–250°C			
	1/2 Cycle (0.0083 s)		10 Cycles (0.166 s)		1/2 Cycle (0.0083 s)		10 Cycles (0.166 s)		1/2 Cycle (0.0083 s)		10 Cycles (0.166 s)	
	Cu	Al	Cu	Al	Cu	Al	Cu	Al	Cu	Al	Cu	Al
5000	10	8	4	2	10	8	4	3	12	10	4	3
15 000	6	4	2/0	4/0	6	4	1/0	3/0	6	4	1	3/0
25 000	3	2	4/0	350	4	2	3/0	250	4	3	3/0	250
50 000	1/0	2/0	400	700	1	2/0	350	500	2	1/0	300	500
75 000	2/0	4/0	600	1000	1/0	3/0	500	750	1/0	3/0	500	700
100 000	4/0	300	800	1250	3/0	250	700	1000	2/0	4/0	600	1000

duce the effect of such stresses. Within the range of temperatures expected with coordinated selection and application, the mechanical aspects can normally be discounted except with very old or lead-sheathed cables.

During short-circuit or heavy pulsing currents, single-conductor cables will be subjected to forces tending to either attract or repel the individual conductors with respect to each other. Therefore such cables laid in trays, or racked, should be secured to prevent damage caused by such movements.

The minimum conductor size requirements for various rms short-circuit currents and clearing times are shown in Table 83. The IPCEA initial and final conductor temperatures are shown for the various insulations. Table 80 gives conductor temperatures (maximum operating, maximum overload, and maximum short-circuit current) for various insulated cables.

10.5 Installation. There are a variety of ways to install power distribution cables in industrial plants. The engineer's responsibility is to select the method most suitable for each particular application. Each mode has characterstics which make it more suitable for certain conditions than others, that is, each mode will transmit power with a unique degree of reliability, safety, economy, and quality for any specific set of conditions. These conditions include the quantity and characteristics of the power being transmitted, the distance of transmission, and the degree of exposure to adverse mechanical and environmental conditions.

10.5.1 *Layout.* The first consideration, of course, in laying out wiring systems is to keep the distance between the source and the load as short as possible. This consideration must be tempered by many other important factors to arrive at the lowest cost system that will operate within the reliability, safety, economy, and performance required. Some other factors that must be considered for various routings are the cost of additional cable and raceway versus the cost of additional supports, inherent mechanical protection provided in one alternative versus additional protection required in another, clearance for and from other facilities, and the need for future revision.

10.5.2 *Open Wire.* This mode was used extensively in the past. Although it has now been replaced in most applications, it is still quite often used for primary power distribution over large areas where conditions are suitable.

Open-wire construction consists of uninsulated conductors on insulators which are mounted on poles or structures. The conductor may be bare or it may have a covering for protection from corrosion or abrasion.

The attractive features of this method are its low initial cost and the fact that damage can be detected and repaired quickly. On the other hand, the uninsulated conductors are a safety hazard and are also highly susceptible to mechanical damage and electrical outage from birds, animals, lightning, etc. There is increased hazard where crane or boom truck use may be involved. In some areas contamination on insulators and conductor corrosion can result in high maintenance costs.

Due to the large conductor spacing, open-wire circuits have a higher reactance which results in a higher voltage drop. This problem is reduced with higher voltage and higher power-factor circuits.

Exposed open-wire circuits are more susceptible to outages from lightning than other modes. The effects may be minimized though by the use of overhead ground wires and lightning arresters.

10.5.3 *Aerial Cable.* Aerial cable is finding increased use in industrial wiring. The greatest gain is in replacing open wiring, where it provides greater safety and reliability and requires less space. Properly protected cables are not a safety hazard and are not easily damaged by casual contact. They are, however, open to the same objections as open wire so far as vertical clearance is concerned. Aerial cables are frequently used in place of the more expensive conduit systems, where the latter's high degree of mechanical protection is not required. They are also generally more economical for long runs of one or two cables than are cable tray installations. It is cautioned that aerial cable having a portion of the run in conduit must be derated to the in-conduit ampacity for this condition.

Aerial cables may be either self-supporting or messenger supported. They may be attached to pole lines or structures. Self-supporting aerial cables have high tensile strength for this application. Cables may be messenger supported either by spirally wrapping a steel band around the cables and the messenger or by pulling the cable into rings suspended from the messenger. The spiral wrap method is used for factory-assembled cable, while both methods are used for field assembly. A variety of spinning heads is available for application of the spiral wire banding in the field.

Self-supporting cable is suitable for only relatively short spans. Messenger-supported cable can span large distances, dependent on the weight of the cable and the tensile strength of the messenger. The supporting messenger provides high strength to withstand climatic rigors or mechanical shock. It may also serve as the grounding conductor of the power circuit.

A convenient feature available in one form of factory-assembled aerial cable makes it possible to form a slack loop to connect a circuit tap without cutting the cable conductors. This is done by reversing the direction of spiral of the conductor cabling every 10—20 ft.

Spacer cable is a type of electric supply-line construction that consists of an assembly of one or more covered conductors separated from each other and supported from a messenger by insulating spacers. This is another economical means of transmitting power overhead between buildings in an industrial facility. Available for use in three-phase grounded or ungrounded systems at utilization voltages of 5, 15, 25, or 35 kV, the insulated nonshielded phase conductors provide protection from accidental discharge through contact with ground-level equipment such as aerial ladders or crane booms. Uniform line electrical characteristics are obtained through the balanced geometric positioning of the conductors with respect to each other by the use of plastic or ceramic spacers located at regular intervals along the line. Low terminating costs are obtained because the conductors are unshielded.

10.5.4 *Direct Attachment.* This is a low-cost method where adequate support surfaces are available between the source and the load. It is most useful in combination with other methods such as branch runs from cable trays and when adding new circuits to existing installations. Its use in commercial buildings is usually limited to low-energy control and telephone circuits.

This method employs multiconductor cable attached to surfaces such as structural beams and columns. A cable with metallic covering should be used where exposed to adverse mechanical conditions. Otherwise, plastic or rubber jacketed cable is satisfactory, provided this is approved by local or regional codes. For architectural reasons it is usually limited to service areas, hung ceilings, and electric shafts.

10.5.5 *Cable Trays.* A cable tray is a unit or assembly of units or sections and associated fittings made of metal or other noncombustible material forming a continuous rigid structure used to support cables. These supports include ladders, troughs, channels, and are becoming increasingly popular in industrial electric systems for the following reasons: low installed cost, system flexibility, accessibility for repair or addition of cables, and space saving when compared with conduit where a larger number of circuits having common routing are involved.

Cable trays are available in a number of types and materials. Mechanical load carrying ability and careful consideration must be given to the selection of the best system suited to the intended installation to provide the lowest installed cost. For other uses, see NEC, Article 318.

Covers, either ventilated or nonventilated, may be used where additional mechanical protection is required or for additional electrical shielding where communication circuits are involved. Barrier strips for separation of voltages and special coatings or materials for corrosion protection are available.

Seals or fire stops may be required when passing through walls, partitions, or elsewhere to minimize flame propagation.

Initial planning of a cable tray system should consider occupancy requirements as given in the NEC and allow additional space for future system expansion.

In stacked tray installations it is good practice to separate voltages, locating the lowest voltage cables in the bottom tray and succeedingly higher voltage cables in ascending order of trays. In a multiphase system, all phase conductors must be installed closely grouped in the same tray.

10.5.6 *Cable Bus.* Cable bus is used for transmitting large amounts of power over relatively short distances. It is a more economical replacement of conduit or busway systems. It also offers higher reliability and safety and lower maintenance than open-wire or bus systems.

Cable bus is a cross between cable tray and busway. It uses insulated conductors in an enclosure which is similar to the cable tray with covers. The conductors are supported at maintained spacings by some form of nonmetallic spacer blocks. Cable buses are furnished either as components for field assembly or as completely assembled sections. The completely assembled sections are best if the run is short enough so that splices may be avoided. Multiple sections requiring joining may preferably employ the continuous conductors.

The spacing of the conductors is such that their maximum rating in air may be attained. This spacing is also close enough to provide low reactance, resulting in minimum voltage drop.

10.5.7 *Conduit.* Rigid-steel conduit systems afford the highest degree of mechanical protection available in above-ground conduit systems. Unfortunately, this is also a relatively high-cost system. For this reason their use is being superseded, where possible, by other types of conduit and wiring systems. Where applicable, rigid aluminum, intermediate grade steel conduit, thin-wall EMT, intermediate metal conduit, plastic, fiber, and asbestos-cement ducts may be used.

Conduit systems offer some degree of flexibility in permitting replacement of existing conductors with new ones. However, in case of fire or faults it may be impossible to remove the conductors. In this case it is necessary to replace both conduit and wire at great cost and delay. Also during fires, conduits may transmit corrosive fumes into equipment where these gases can do much damage. To keep flammable gases out of such areas, seals must be installed.

With magnetic conduits, an equal number of conductors of each phase must be installed; otherwise, losses and heating will be excessive. For example, a single-conductor cable should not be used in steel conduit.

NOTE: Refer to NEC for code regulations on conduit use.

Underground ducts are used where it is necessary to provide a high degree of mechanical protection. Two cases where they are used are when overhead conduits are subject to extreme severe abuse or when the cost of going underground is less than providing overhead supports. In the latter case, direct burial may be satisfactory under certain circumstances.

Underground ducts use rigid steel, plastic, fiber, and asbestos-cement conduits encased in concrete, or precast with multihole concrete with close fitting joints. Clay tile is also used to some extent. Where the added mechanical protection of concrete is not required, heavy wall versions of fiber and asbestos-cement conduits are direct buried as are rigid steel and plastic.

Cables used in underground conduits must be suitable for use in wet areas. Some cost savings can be realized by using flexible plastic conduits with conductors factory installed.

Where a relatively long distance between the point of service entrance into a building and the service entrance protective device is unavoidable, the conductors must be placed under at least 2 in of concrete or they must be placed in conduit or duct and enclosed by concrete or brick not less than 2 in thick and are thus considered outside the building by the NEC.

10.5.8 *Direct Burial.* Cables may be buried directly in the ground where permitted by codes when the need for future maintenance along the cable run is not anticipated nor the protection of conduit required. The cables used must be suitable for this purpose, that is, resistant to moisture, crushing, soil contaminants, and insect and rodent damage. The cost savings of this method over duct banks can vary from very little to a considerable amount. While this system cannot be readily added to or maintained, the current-carrying capacity is usually greater than that of cables in ducts. Buried cable must have selected back fill. It must be used only where chances of its being disturbed are small or it must be suitably protected if used where these chances exist. Relatively recent advances in the design and operating characteristics of cable fault location equipment

and subsequent repair methods and material have diminished the maintenance problem.

10.5.9 *Installation Procedures.* Care must be taken in the installation of raceways to make sure that no sharp edges exist to cut or abrade the cable as it is pulled in. Another important consideration is not to exceed the maximum allowable tensile strength or side-wall pressure of a cable. These forces are directly related to the force exerted on the cable when it is pulled in. This can be decreased by shortening the length of each pull and reducing the number of bends. The force required for pulling a given length can be reduced by the application of a pulling compound on cables in conduit and the use of rollers in cable trays.

If the cable is to be pulled by the conductors, the maximum tension in pounds is limited to 0.008 times the area of the conductors in circular mils, within the constructions. This tension may be further reduced if pulled by grips over the outside covering. A reasonable figure for most jacketed constructions would be 1000 lb per grip, but the calculated conductor tension should not be exceeded.

Side-wall pressures on most single-conductor constructions limit pulling tensions to approximately 450 lb times cable diameter, in inches, times radius of bend, in feet. Triplexed and paralleled cables would use their single-conductor diameters and a figure of 225 lb and 675 lb, respectively, instead of the 450 lb factor for single-conductor construction.

For duct installations involving many bends, it is preferable to feed the cable into the end closest to the majority of the bends (in that the friction through the longer duct portion without the bends is not yet a factor) and pull from the other end. Each bend gives a multiplying factor to the tension it sees; therefore the shorter runs to the bends will keep this increase in pulling tensions to a minimum. However, it is best to calculate pulling tensions for installation from either end of the run and install from the end offering the least tension.

The minimum bending radius for metal-taped cables is normally taken at twelve times the cable diameter, although cables with non-metallic covering can be bent to at least half this radius without disrupting the cable components.

When installing cables in wet underground locations, the cable ends must be sealed to prevent entry of moisture into the conductor strands. These seals should be left intact or remade after pulling if disrupted, until splicing, terminating, or testing is to be done. This practice is recommended to avoid unnecessary corrosion of the conductors and to safeguard against the generation of steam under overload, emergency loadings, or short-circuit conditions after the cable is placed in operation.

10.6 Connectors

10.6.1 *Types Available.* Connectors are classified as thermal or pressure, depending upon the method of attaching them to the conductor.

Thermal connnectors use heat to make soldered, silver-soldered, brazed, welded, or cast-on terminals. Soldered connections have been used with copper conductors for many years, and their use is well understood. Aluminum connections may also be soldered satisfactorily with the proper materials and technique. However, soldered joints are not commonly used with aluminum. Shielded arc welding of aluminum terminals to aluminum cable makes a satisfactory termination for cable sizes larger than no. 4/0. Torch brazing and silver soldering of copper cable connections are in use, particularly for underground connections with bare conductors such as are found in grounding mats. Thermite welding kits utilizing carbon molds are also in use for making connections with bare copper cable for ground mats and for junctions which will be below grade. The thermite welding process has also proved satisfactory for attaching connectors to insulated power cables.

Mechanical and compression pressure connectors are used for making joints in electric conductors. Mechanical type connectors obtain the pressure to attach the connector to the electric conductor from an integral screw, cone, or other mechanical parts. A mechanical connector thus applies force and distributes it suitably through the use of bolts or screws and properly designed sections. The bolt diameter and number of bolts are selected to produce the clamping and contact pressures required for the most satisfactory design. The sections are made heavy enough to carry rated current

and withstand the mechanical operating conditions.

Compression connnectors are those in which the pressure to attach the connector to the electric conductor is applied externally, changing the size and shape of the connector and conductor.

The compression connector is basically a tube with the inside diameter slightly larger than the outer diameter of the conductor. The wall thickness of the tube is designed to carry the current, withstand the installation stresses, and withstand the mechanical stresses resulting from thermal expansion of the conductor. A joint is made by compressing the conductor and tube into another shape by means of a specially designed die and tool. The final shape may be indented, cup, hexagon, circular, or oval. All methods have in common the reduction in cross-sectional area by an amount sufficient to assure intimate and lasting contact between the connector and the conductor. Small connectors can be applied with a small hand tool. Larger connectors are applied with a hydraulic compression tool.

A properly crimped joint deforms the conductor strands sufficiently to have good electrical conductivity and mechanical strength, but not so much that the crimping action overcompresses the strands, thus weakening the joint.

Mechanical and compression connectors are available as tap connectors. Many types have an independent insulating cover. After a connection is made, the cover is assembled over the joint to insulate, and in some cases to seal against the environment.

10.6.2 *Connectors for Aluminum.* Aluminum conductors are different from copper in several ways, and these differences should be considered in specifying and using connectors for aluminum conductors (see Table 78). The normal oxide coating on aluminum is of relatively high electrical resistance. Aluminum has a coefficient of thermal expansion higher than that of copper. The ultimate and the yield strength properties and the resistance to creep of aluminum are different from the corresponding properties of copper. Corrosion is possible under some conditions because aluminum is anodic to other commonly used metals, including copper, when electrolytes even from humid air are present.

(1) *Mechanical Properties and Resistance to Creep.* Creep has been defined as the continued deformation of the material under stress. The effect of excessive creep resulting from the use of an inadequate connector which applies excessive stress could be relaxation of contact pressure within the connector, and a resulting deterioration and failure of the electric connection. In mechanical connectors for aluminum, as for copper, proper design can limit residual unit-bearing loads to reasonable values, with a resulting minimum plastic deformation and creep subsequent to that initially experienced on installation. Connectors for aluminum wire can accommodate a range of conductor sizes, provided that the design takes into account the residual pressure on both minimum and maximum conductors.

(2) *Oxide Film.* The surface oxide film on aluminum, though very thin and quite brittle, has a high electrical resistance and therefore must be removed or penetrated to ensure a satisfactory electric joint. This film can be removed by abrading with a wire brush, steel wool, emery cloth, or similar abrasive tool or material. A plated surface, whether on the connector or bus, should never be abraded. It can be cleaned with a solvent or other means which will not remove the plating.

Some aluminum fittings are factory filled with a connection aid compound, usually containing particles which aid in obtaining low contact resistance. These compounds act to seal connections against oxidation and corrosion by preventing air and moisture from reaching contact surfaces. Connection to the inner strands of a conductor requires deformation of these strands in the presence of the sealing compound to prevent the formation of an oxide film.

(3) *Thermal Expansion.* The linear coefficient of thermal expansion of aluminum is greater than that of copper and is important in the design of connectors on aluminum conductors. Unless provided for in the design of the connector, the use of metals with coefficients of expansion less than that of aluminum can result in high stresses in the aluminum during heat cycles, causing additional plastic deformation and significant creep. Stresses can be quite high, not

only because of the differences of coefficients of expansion, but also because the connector may operate at an appreciably lower temperature than the conductor. This condition will be aggravated by the use of bolts which are of a dissimilar metal or have different thermal expansion characteristics from those of the terminal.

(4) *Corrosion.* Direct corrosion from chemical agents affects aluminum no more severely than it does copper and, in most cases, less. However, since aluminum is more anodic than other common conductor metals, the opportunity exists for galvanic corrosion in the presence of moisture and a more cathodic metal. For this to occur, a wetted path must exist between external surfaces of the two metals in contact to set up an electric cell through the electrolyte (moisture), resulting in erosion of the more anodic of the two, in this instance, the aluminum.

Galvanic corrosion can be minimized by the proper use of a joint compound to keep moisture away from the points of contact between dissimilar metals. The use of relatively large aluminum anodic areas and masses minimizes the effects of galvanic corrosion.

Plated aluminum connectors must be protected by taping or other sealing means.

(5) *Types of Connections Recommended for Aluminum.* UL has listed connectors approved for use on aluminum. Such connectors have successfully withstood UL performance tests which have recently been increased in severity. Both mechanical and compression-type connectors are available. The most satisfactory connectors are specifically designed for aluminum conductors to prevent any possible troubles from creep, the presence of oxide film, and the differences of coefficients of expansion between aluminum and other metals. These connectors are usually satisfactory for use on copper conductors in noncorrosive locations. The connection of an aluminum connector to a copper or aluminum pad is similar to the connection of bus bars. When both the pad and the connector are plated and the connection is made indoors, few precautions are necessary. The contact surfaces must be clean; if not, a solvent must be used. Abrasive means are undesirable since the plating may be removed. In

normal application, steel, aluminum, or copper alloy bolts, nuts, and flat washers may be used. A light film of a joint compound is acceptable, but not mandatory. When either of the contact surfaces is not plated, the bare surface should be cleaned by wire brushing and then coated with a joint compound. Belleville washers are suggested for heavy-duty applications where cold flow or creep may occur, or where bare contact surfaces are involved. Flat washers should be used wherever Belleville washers or other load-concentrating elements are employed. The flat washer must be located between the aluminum lug, pad, or bolt and the outside edge of the Belleville washer with the neck or crown of the Belleville against the bolting nut to obtain satisfactory operation. In outdoor or corrosive atmosphere, the above applies with the additional requirement that the joint be protected. An unplated aluminum-to-aluminum connection can be protected by the liberal use of compound.

In an aluminum-to-copper connection, a large aluminum volume compared to the copper is important as is the placement of the aluminum above the copper. Again, coating with a joint compound is the minimum protection; painting with a zinc chromate primer or thoroughly sealing with a mastic or tape is even more desirable. Plated aluminum should be completely sealed against the elements.

(6) *Welded Aluminum Terminals.* For aluminum cables in 250 kcmil and larger sizes carrying high currents, excellent terminations can be made by welding special terminals to the cable. This is best done by the inert-gas shielded metal arc method. The use of inert gas eliminates the need for any flux to be used in making the weld. The welding type terminal is shorter than a compression terminal because the barrel for holding the cable can be very short. It has the advantage of requiring less room in junction or terminal boxes of equipment. Another advantage is the reduced resistance of the connection. Each strand of the cable is bonded to the terminal, resulting in a continuous metal path for the current from every strand of the cable to the terminal.

Welding of these terminals to the cables may also be done with the tungsten electrode type of welding equipment with alternating-current power. The tungsten arc method is slower, but for small work gives somewhat better control.

The tongues or pads of the welding-type terminals, like the large compression type, are available with bolt holes to conform to the standards of the National Electrical Manufacturers Association (NEMA) for terminals to be used on equipment.

(7) *Procedure for Connecting Aluminum Conductors (Fig 124)*

(a) When cutting cable, avoid nicking the strands. Nicking makes the cable subject to easy breakage [Fig 124(a)].

(b) Contact surfaces must be cleaned. The abrasion of contact surfaces is helpful even with new surfaces, and is essential with weathered surfaces. Do not abrade plated surfaces [Fig 124(b)].

(c) Apply joint compound to conductor if the connector does not already have it [Fig 124(c)].

(d) Use only connectors specifically tested and approved for use on aluminum conductors.

(e) For mechanical-type connectors tighten the connector with screwdriver or wrench. Remove excess compound [Fig 124(d)].

NOTE: This type of aluminum connector should be used only where permitted by local codes.

(f) For compression-type connectors crimp the connector using proper tool and die. Remove excess compound [Fig 124(e)].

(g) Always use a joint compound compatible with the insulation and as recommended by the manufacturer. The oxide-film penetrating or removing properties of some compounds aid in obtaining high initial conductivity. The corrosion inhibiting and sealing properties of some compounds help ensure the maintenance of continued high conductivity and prevention of corrosion.

(h) In making an aluminum-to-copper connection which is exposed to moisture, place the aluminum conductor above the copper. This prevents soluble copper salts from reaching the aluminum conductor, which could result in corrosion. If there is no exposure to moisture, the relative position of the two metals is not important.

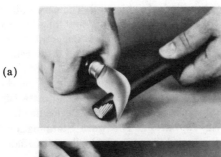

(a)

(b)

(c)

(d)

(e)

**Fig 124
Procedures for Connecting Aluminum
Conductors**

(i) When using insulated conductors outdoors, extend the conductor insulation or covering as close to the connector as possible to minimize weathering of the joint. Outdoors, whenever possible, joints should be completely protected by tape or other means. If outdoor joints are to be covered or protected, the protection should completely exclude moisture, as the retention of moisture could lead to severe corrosion.

10.6.3 *Connectors for Various Voltage Cables.* Standard mechanical or compression-type connectors are recommended for all primary voltages provided the bus is uninsulated. Welded connectors may also be used for conductors sized in circular mils. Up to 600 V, standard connector designs present no problem for insulated or uninsulated conductors. The standard compression-type connectors are recommended for use on insulated conductors up to 5 kV. Above 5 kV, stress considerations make it desirable to use tapered-end compression connectors or semiconducting tape construction to give the same effect.

10.6.4 *Performance Requirements.* Electric connectors for industrial plants are designed to meet the requirements of the NEC. They are evaluated on the basis of their ability to pass secureness, heating, heat-cycling, and pull-out tests as outlined in UL 486-1973, Electric-Wire Connectors and Soldering Lugs. This standard has been recently revised to incorporate more stringent requirements for aluminum terminating devices. The reader is cautioned to use only those lugs meeting the current UL standard.

Connectors must be able to meet electrical and mechanical operating requirements. Electrically, the connectors must carry the current without exceeding the temperature rise of the conductors being joined. Joint resistance not appreciably higher than that of an equal length of conductor being joined is recommended to assure continuous and satisfactory operation of the joint. In addition, the connector must be able to withstand momentary overloads or short-circuit currents to the same degree as the conductor itself. Mechanically, a connector must be able to withstand the effects of the environment within which it is operating. If outdoors, it must stand up against temperature extremes, wind, vibration, rain, ice, sleet, chemical attack, etc. If used indoors, any vibration from rotating machinery, corrosion caused by plating or manufacturing processes, high temperatures from furnaces, etc, must not materially affect the performance of the joint.

10.7 Terminations

10.7.1 *Purpose.* A termination for an insulated power cable must provide certain basic electrical and mechanical functions. These essential requirements include the following.

(1) Electrically connect the insulated cable conductor to electric equipment, bus, or uninsulated conductor.

(2) Physically protect and support the end of the cable conductor, insulation, shielding system, and overall jacket, sheath, or armor of the cable.

(3) Effectively control electrical stresses to provide both internal and external dielectric strength to meet desired insulation levels for the cable system.

The current-carrying requirements are the controlling factors in the selection of the proper type and size of connector or lug to be used. Variations in these components are related, in turn, to the base material used to make up the conductor within the cable, the type of termination used, and the requirements of the electric system.

The physical protection offered by the termination will vary considerably, depending on the requirements of the cable system, the environment, and the type of termination used. The termination must provide an insulating cover at the cable end to protect the cable components (conductor, insulation, and shielding system) from damage by any contaminants which may be present, including gases, moisture, and weathering.

Shielded high-voltage insulated cables are subject to unusual electrical stresses where the cable shield system is ended just short of the point of termination. The creepage distance that must be provided between the end of the cable shield, which is at ground potential, and the cable conductor, which is at line potential, will vary with the magnitude of the voltage, the type of terminating device used, and, to some

degree, the kind of cable used. The net result is the introduction of both radial and longitudinal voltage gradients which impose dielectric stress of varying magnitude at the end of the cable. The termination provides a means of reducing and controlling these stresses within the working limits of the cable insulation and materials used to make up the terminating device itself.

10.7.2 *Definition.* The word "termination" covers a number of individually identifiable types of units, such as

(1) Taped terminations

(2) Armor terminators

(3) Potheads

(4) Preassembled terminators

This is only a partial listing of the many types of terminations in use, and variations within each group listed are such that it would be impractical to attempt to give a specific definition. Requirements for each vary with voltage, cable construction, and nature of the application.

10.7.3 *Voltage Categories*

(1) *600 V and Below.* Cables usually have a copper or aluminum conductor with a rubber or plastic type insulating system with no shield. Terminations for these cables generally consist of a lug and may be taped. The lug is fastened to the cable by one of the several methods described in Section 10.6 and tape is applied over the lower portion of the barrel of the lug and down onto the cable insulation. Tapes used for this purpose are selected on the basis of compatibility with the cable insulation and suitability for application in the environmental exposure anticipated.

(2) *Over 600 V.* Cables above 600 V may have a copper or aluminum conductor with either an extruded solid-type insulation such as rubber, polyethylene, etc, or a laminated insulating system such as oil-impregnated paper tapes, varnished-cloth tapes, etc. A shielding system must be used on cables rated 8 kV or more. Jackets for these cables may be nonmetallic, such as neoprene, polyethylene, or polyvinyl chloride, or metallic, such as lead, aluminum, or galvanized steel. These latter two metallic jackets are generally furnished in a helical or corrugated form with sections joined either by an interlocking arrangement or by continuous weld. The terminations available for

use with these cables and techniques for applying them vary with the type of cable, its construction, its voltage rating, and the requirements for the installation.

The requirements of the installation dictate that the termination must be designed for a specific end use. The least imposing is an indoor installation such as within a building or inside a protective housing. Here the termination is subjected to a minimum exposure to the elements.

Outdoor installations expose the termination to the elements and require that features be included in its makeup to withstand this exposure. In some areas the air can be expected to carry a high percentage of gaseous contaminants, liquid or solid particles which may be conducting, either alone or in the presence of moisture. These environments impose an even greater demand on the termination for protection of the cable end from the damaging contaminants and for the termination itself to withstand exposure to these contaminants. The termination may be required to perform its intended function while partially or fully immersed in a liquid or gaseous dielectric. Liquid dielectrics include oils, such as transformer oil and askarel. These exposures impose upon the termination the necessity of complete compatibility between the liquids and exposed parts of the termination, including any gasket sealing material. Generally speaking, cork gaskets are used where there is a direct contact with askarel, but the more recent materials such as tetrafluoroethylene[15] and silicone will provide superior gasketing characteristics. The gaseous dielectrics may be nitrogen or any of the electronegative gases, such as sulfur hexafluoride, that are used for the dielectric property of the gas to fill electric equipment.

10.7.4 *Shielded Cables.* At the point of termination, the cutback of the cable shielding to provide necessary creepage distances between the conductor and shielding introduces a longitudinal stress over the surface of the exposed cable insulation. The resultant combination of radial and longitudinal electrical stress at the termination of the cable end results in

---

[15]For example, Teflon.

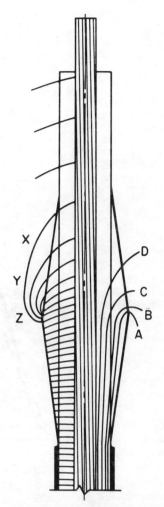

| X, Y, Z | Electric stress lines |
| A, B, C, D | Equipotential lines |

Fig 125
Stress-Relief Cone

maximum stress occurring at the point. However, these stresses can be controlled and reduced to values within safe working limits of the materials used to make up the termination. The most common method of reducing these stresses is to gradually increase the total thickness of insulation at the termination by adding insulating tapes to form a cone. The cable shielding is carried up the cone surface and terminated at a point approximately 1/8 in behind the largest diameter of the cone. This

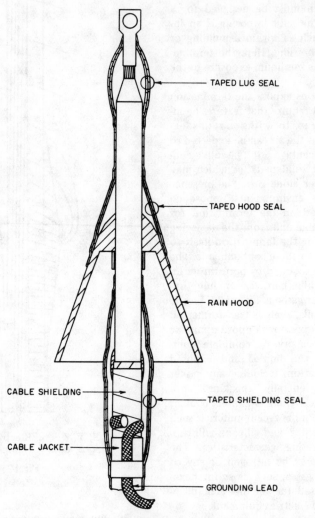

Fig 126
15 kV Termination

NOTE: Consult individual cable supplier for recommended
installation procedures and materials.

construction, illustrated in Fig 125, is commonly referred to as a stress relief cone.

Certain cable insulating systems are inherently more resistant to high levels of electrical stress than others, therefore obviating the need for, and cost of, stress relief cones. It is advisable to consult individual cable manufacturers for their recommendations in terminating and splicing shielded cables.

10.7.5 *Taped Terminations.* Taped terminations (Fig 126) may be used either indoors or outdoors and on shielded or unshielded cables. Generally, taped terminations are used at 15 kV and below; however, there are instances where taped terminations are used on some cable types to 69 kV. On unshielded cables the termination is made up with only a lug and tape, either indoors or outdoors. Terminations of shielded cables generally require the use of a stress relief cone and cover tapes in addition to the lug. The size and location of the stress cone is controlled primarily by the operating voltage and location of the termination, that is, indoors or outdoors.

Creepage of 1 inch per kilovolt of nominal system voltage is commonly used for indoor application, and from 1 to 2 inches or more per kilovolt of system voltage is allowed for outdoor installations. Additional creepage may be gained by using a rain hood of neoprene, rubber, plastic, or porcelain for outdoor installations. Insulating tapes for the stress relief cone are selected to be compatible with the cable insulation, and tinned copper braid or semiconducting tape is used as a conducting material for the cone. Cover tapes are applied over the stress cone, cable insulation, and up into the connecting lug. These cover tapes may be of an anhydrous material painted with a weatherproofing paint, vinyl, silicone rubber, tetrafluoroethylene,[16] or other taping materials, such as a laminated buildup of coarse-mesh tape saturated with epoxy.

The rain hoods which are generally used on the outdoor taped terminations are positioned in the creepage path between the lug and stress cone. On upright terminations the rain hood is usually placed directly over the stress

relief cone, and its primary function is to keep some portion of the cable insulation along the creepage path dry at all times. Some installations have been made with two or more hoods in the creepage path. Where unusual surface contamination is expected, potheads should be used.

10.7.6 *Preformed Stress Cones.* Indoor applications or equipment-type installations where the apparatus housing provides weather protection for the cable termination may be made up using a preformed stress relief cone. The most common preformed stress cone is a two-part elastomeric assembly consisting of a semiconducting lower section formed in the shape of a stress relief cone and an insulating upper section. In addition, some types are housed in a hard plastic-like insulating protective housing. Both types are applicable for use on shielded cables with extruded solid-type insulation, such as, rubber, polyethylene, etc.

10.7.7 *Armor Terminators.* Cables with a steel, aluminum, or copper metallic jacket of helical continuous weld or interlocking covering require, in addition to a taped termination (or other terminating device), an arrangement to serve and ground the armor. Fittings available for this purpose are generally called armor terminators. These armor terminators perform one or more of the following important functions:

(1) Provide mechanical termination and electrical ground of armor

(2) Provide a water-tight seal for the cable entrance to a box, compartment, pothead, or other piece of electric equipment

(3) Provide a housing to permit a compound seal over the cable core to exclude moisture and protect cable insulation from contamination

These armor terminators are sized to fit the cable armor and are designed for use on the cable alone, with brackets, or with locking nuts or adaptors for application to other pieces of equipment.

10.7.8 *Potheads (Fig 127).* A pothead is a hermetically sealed device used to enclose and protect a cable end. It consists of a metallic body with one or more porcelain insulators. The body is arranged to accept a variety of optional cable entrance sealing fittings, while the porcelain bushings, in turn, are designed to ac-

---

[16]For example, Teflon.

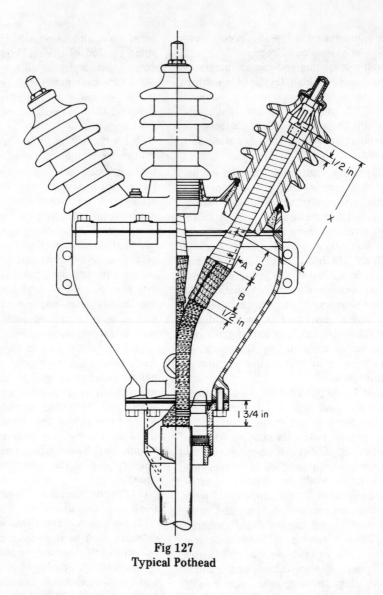

Fig 127
Typical Pothead

commodate a number of optional cable conductor and aerial connections. These parts would be field assembled to the prepared cable ends (taped stress relief cones required for shielded cables), and the assembled unit is filled with an insulating compound.

Potheads are available in ratings of 5 kV and above for either single-conductor or three-conductor installation and for indoor, outdoor, or liquid-immersed application. Mounting variations include bracket, plate, and flanged types.

Potheads offer a greater degree of protection to the cable elements (conductor, insulation, and shielding system) than the taped terminations and are generally preferred for applications involving exposure to a high degree of environmental contamination. Equipment applications where there is an exposure to liquid dielectrics usually employ potheads since these units may be directly attached to the equipment.

Both cable construction and type of application must be considered in the selection of a pothead. Voltage rating, desired basic impulse insulation level, conductor size, and current requirements are also basic considerations in the selection of a pothead. Cable construction is the controlling factor in the selection of proper entrance sealing fittings, stress relief cone materials, and filling compound. Application, in turn, is the prime consideration for selecting the type of pothead, method of mounting, and desired aerial connectors.

Potheads can be subdivided into a number of groups or types related to cable system requirements and construction of the devices. Cable system requirements may be arranged in two general groups, nonpressurized and pressurized. Most industrial power cable systems are of the nonpressurized type using solid dielectric insulated cables. The two most commonly used pothead types for these solid-type cables are the capnut and solder seal. The capnut pothead is made up of cast metal parts (iron, aluminum, or bronze) with gasketed joints between metal parts and the porcelains. The metal parts for the solder seal potheads are copper spinnings which are solder-bonded directly to the porcelain insulator, thus eliminating several or all gasketed joints. Arrangements other than terminal up, cable down may be expected to have higher failure rates.

Potheads are filled with a high-strength dielectric and designed to be electrically stronger internally than externally. However, they require greater care and skill in installation.

Dielectrics in common use for filling potheads include asphaltic based materials, resins, and oils. The asphaltic and resin materials must be heated to liquify them, then poured into the pothead, and allowed to cool. Special techniques are employed to control the rate of cooling so that voids which may result from the shrinkage of the compound with cooling can be forced to occur in areas of little or no electrical stress.

10.7.9 *Preassembled Terminators.* Advances in the art of terminating single-conductor cables include several types of units designed to reduce the required cable end preparation and eliminate the "hot fill with compound step" associated with potheads. One applies elastomeric materials directly to the cable end. This type is offered both with and without a metal—porcelain housing and is applicable only on solid dielectric cables. The other type consists of a metal—porcelain housing filled with a gelatin-like substance designed to be partially displaced as the terminator is installed on the cable. This latter unit may be used on any nonpressure-type cable.

Advantages of the preassembled terminators include simplified installation procedures and reduced installation time. Accordingly, they can be installed by less skilled workers, yet they offer a high degree of consistency to the overall quality and integrity of the installed system.

Preassembled terminators are available in ratings of 15 kV and above for most types of application. The porcelain-housed units include flanged mounting arrangements for equipment mounting and liquid-immersed applications. Although ratings below 15 kV are not offered, it is often desirable to use the 15 kV units on 5 kV systems when terminating shielded cables.

Selection of preassembled terminators is essentially the same as for potheads, with the exception that those units using solid elastomeric materials must be sized, with close tolerance, to the cable diameters to provide proper fit.

10.7.10 *Separable Insulated Connectors.* These are two-part devices used in conjunction with high-voltage electric apparatus. A bushing assembly is attached to the high-voltage apparatus (transformer, switch, or fusing device, etc), and a molded plug-in connector is used to terminate the insulated cable and connect the cable system to the bushing. The dead-front feature is obtained by fully shielding the plug-in connector assembly.

Two types of separable insulated connectors for application at 15 kV and 25 kV are available, one load break and the other non-load break. Both are essentially of a molded construction design for use on solid dielectric insulated cables (rubber, polyethylene, etc) and are suitable for submersible application. The connector section of the device assumes an elbow-type (90°) configuration to facilitate installation, improve separation, and save space. See IEEE Std 386-1974, Separable Insulated Connectors for Power Distribution Systems Above 600 V (ANSI C119.2-1974).

Electric apparatus may be furnished with a "universal bushing wall" only for future installation of bushings for either the load break or non-load break dead-front assemblies. Shielded elbow connectors may be furnished with a "voltage detection tap" to provide a means of determining whether or not the circuit is energized.

10.7.11 *Performance Requirements.* Design test criteria have been established for potheads and are listed in IEEE Std 48-1975, High-Voltage AC Cable Terminations, which outlines short-time alternating-current 60 Hz and impulse withstand requirements. Also listed in this design standard are maximum direct-current field proof test voltages. Individual types may safely withstand higher test voltages, and the manufacturer should be contacted for such information. All devices employed to terminate insulated power cables should meet these basic requirements. Additional performance requirements may include thermal load cycle capabilities of the current-carrying components, environmental performance of completed units, and long-time overvoltage withstand capabilities of the device.

10.8 Splicing Devices and Techniques. Splicing devices are subjected to a somewhat different set of voltage gradients and dielectric stress from that of a cable termination. In a splice, as in the cable itself, the highest stresses are around the conductor and connector area. Splicing design must recognize this fundamental consideration and provide the means to control these stresses to values within the working limits of the materials used to make up the splice.

In addition, on shielded cables the splice is in the direct line of the cable system and must be capable of handling any ground currents or fault currents that may pass through the cable shielding.

The connectors used to join the cable conductors together must be electrically capable of carrying full rated load, emergency overload, and fault currents without overheating as well as being mechanically strong enough to prevent accidental conductor pullout or separation.

Finally, the splice housing or protective cover must provide adequate protection to the splice, giving full consideration to the nature of the application and its environmental exposure.

(1) *600 V and Below.* An insulating tape is applied over the conductor connection to electrically and physically seal the joint. The same taping technique is employed in the higher voltages, but with more refinement to cable end preparation and tape applications.

Insulated connectors are used where several relatively large cables must be joined together. These terminators, called "moles" or "crabs," are, fundamentally, insulated buses with provision for making a number of tap connections which can be very easily taped or covered with an insulating sleeve. Connectors of this type enable a completely insulated multiple connection to be made without the skilled labor normally required for careful "crotch" taping or the expense of special junction boxes. One widely used type is a preinsulated multiple outlet joint in which the cable connections are made mechanically by compression cones and clamping nuts. Another type is a more compact preinsulated multiple joint in which the cable connections are made by standard compression tooling which indents the conductor to the tubular cable sockets. Also available are range-taking tap connectors having an independent insulating cover. After the connection is made, the cover is snapped closed to insulate the joint.

Insulated connectors lend themselves particularly well to underground services and industrial wiring where a large number of multiple-connection joints must be made.

(2) *Over 600 V.* Splicing of unshielded cables up to 8 kV consists of assembling a connector, usually soldered or pressed onto the cable conductors, and applying insulating tapes to build up an insulation wall to a thickness of 1½ to 2 times that of the factory-applied insulation on the cable. Some care must be exercised in applying the connector and insulating tapes to the cables, but it is not as critical with unshielded cables as with shielded cables since the unshielded cables operate at relatively low voltages.

Aluminum conductor cables require a waterproof joint to prevent moisture entry into the stranding of the aluminum conductors. Splices on solid dielectric cables are made with uncured tapes which will fuse together after application and provide a waterproof assembly. It is necessary, however, to use a moistureproof adhesive

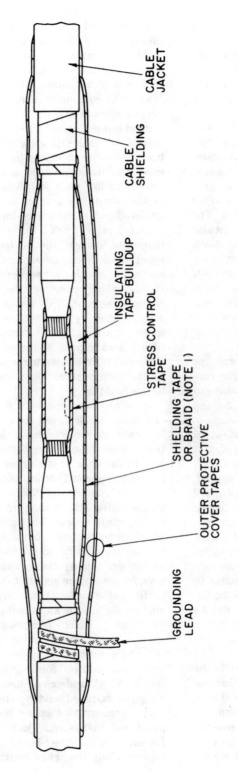

Fig 128
Typical Taped Splice in Shielded Cable

NOTES: (1) Heavy braid jumper should be used across splice to carry possible ground-fault current. Stress control tape should cover strands completely, lapping slightly onto insulation taper.
(2) Consult individual cable supplier for recommended installation procedures and materials.

between the cable insulation and the first layer of insulating tapes. Additional protection may be obtained through the use of a moisture-proof cover over the insulated splice. This cover may consist of additional moistureproof tapes and paint or a sealed weatherproof housing of some form.

10.8.1 *Taped Splices (Fig 128).* Taped splices for shielded cables have been used quite successfully for many years. Basic considerations are essentially the same as for unshielded cables. Insulating tapes are selected not only on the basis of dielectric properties but also for compatability with the cable insulation. The characteristics of the insulating tapes must also be suitable for the application of the splice. This latter consideration gives attention to details such as providing a moisture seal for splices subjected to water immersion or direct burial, thermal stability of tapes for splices subjected to high ambient and operating temperatures, and ease of handling for applications of tapes on wye- or tee-type splices.

Connector surfaces must be smooth and free from any sharp protrusions or edges. The connector ends are tapered, and indentations or distortion caused by pressing tools are filled and shaped to provide a round smooth surface. Semiconducting tapes are recommended for covering the connector and exposed conductor stranding to provide a uniform voidfree surface over which insulating tapes can be applied. Cables with a solid-type insulation are tapered and those with tape-type insulation are stepped to provide a gradual transition between conductor—connector diameter and cable insulation diameter prior to the application of insulating tapes. This is done to control the voltage gradients and resultant voltage stress to values within the working limits of the insulating materials. The splice should not be overinsulated to provide additional protection since this could restrict heat dissipation at the splice area and risk splice failure.

A tinned copper braid is used to provide the shielding function over the splice area. Grounding straps or heavy braid are applied to at least one end of the splice for grounding purposes, and a heavy braid jumper is applied across the splice to carry available ground-fault current.

Final cover tapes or weather barriers are ap-

lied over the built up splice to seal it against moisture entry. A splice on a cable with lead sheath is generally housed in a lead sleeve which is solder wiped to the cable sheath at each end of the splice. These lead sleeves are filled with compound in much the same manner as potheads.

Hand-taped splices may be made between lengths of dissimilar cables if proper precautions are taken to ensure the integrity of each cable's insulating system and the tapes used are compatible with both cables. One example of this would be a splice between a rubber-insulated cable and an oil-impregnated paper insulated cable. Such a splice must have an oil barrier to prevent the oil impregnate in the paper cable from coming in contact with the insulation on the rubber cable. In addition, the assembled splice must be made completely moistureproof. This requirement is usually accomplished by housing the splice in a lead sleeve with wiped joints at both ends. A close-fitting lead nipple is placed on the rubber cable and is tape or epoxy sealed to the jacket of the rubber cable. The solder wipe is made to this lead tube.

Three-way wye and tee splices and the several other special hand-taped splices that can be made all require special design considerations. In addition, a high degree of skill on the part of the installer is a prime requirement for proper makeup and service reliability.

10.8.2 *Preassembled Splices.* Similar to the preassembled terminators are several types of factory-made splices. The most elementary is an elastomeric unit consisting of a molded housing sized to fit the cables involved, a connector for joining the conductors, and tape seals for sealing the ends of the molded housing to the cable jacket. Other versions of elastomeric units include an overall protective metallic housing which completely encloses the splice. These preassembled elastomeric-type splices are available in two-way, three-way tee type, and multiple configurations for application to 35 kV and can be used on most cables having an extruded solid-type insulation.

The preassembled splice provides a waterproof seal to the cable jacket and is suitable for submersible, direct-burial, and other applications where the splice housing must provide protection for the splice to the same degree that the cable jacket provides protection to the cable insulation and shielding system. An advantage of these preassembled splices is the reduced time to complete the splice after cable end preparation. However, the solid elastomeric materials used for the splice must be sized, with close tolerance, to the cable diameters to provide proper fit.

10.9 Grounding of Cable Systems. For safety and reliable operation, the shields and metallic sheaths of power cables must be grounded. Without such grounding, shields would operate at a potential considerably above ground. Thus they would be hazardous to touch, and would incur rapid degradation of the jacket or other material intervening between shield and ground. This is caused by the capacitive charging current of the cable insulation which is of the order of 1 mA per foot of conductor length. This current normally flows, at power frequency, between the conductor and the earth electrode of the cable, normally the shield. In addition, the shield or metallic sheath provides the fault return path in the event of insulation failure, permitting rapid operation of the protection devices.

The grounding conductor, and its attachment to the shield or metallic sheath, normally at a termination or splice, needs to have an ampacity no lower than that of the shield. In the case of a lead sheath, the ampacity must be ample to carry the available fault current and duration without overheating. Attachment to shield or sheath is frequently by means of solder, which has a low melting point; thus an adequate area of attachment is required.

There is much disagreement as to whether the cable shield lengths should be grounded at both ends or at only one end. If grounded at only one end, any possible fault current must traverse the length from the fault to the grounded end, imposing high current on the usually very light shield conductor. Such a current could readily damage or destroy the shield, and require replacement of the entire cable rather than only the faulted section. With both ends grounded, the fault current would divide and flow to both ends, reducing the duty on the shield, with consequently less chance of

damage. There are modifications of both systems. In one, single-ended grounding may be attained by insulating the shields at each splice or sectionalizing point, and grounding only the source end of each section. This limits possible shield damage to only the faulted section. Multiple grounding, rather than just two-end grounding, is simply the grounding of the cable shield or sheath at all access points, such as manholes or pull boxes. This also limits possible shield damage to only the faulted section.

10.9.1 *Sheath Losses.* Currents are induced in the multigrounded shields and sheaths of cables by the current flow in the power conductor. These currents increase with the separation of the power conductors, and increase with decreasing shield or sheath resistance. With three-conductor cables this sheath current is negligible, but with single-conductor cables separated in direct-burial or separate ducts it can be appreciable. For example, with three single-conductor 500 kcmil cables, flat, on 8 in centers, with twenty spiral no. 16 copper shield wires, the ampacity is reduced by about 20 percent by this shield current. With single-conductor lead-sheathed cables in separate ducts, this current is important enough that single-end grounding is obligatory. As an alternate, the shields are insulated at each splice (at approximately 500 ft intervals) and crossbonded to provide sheath transposition. This neutralizes the sheath currents, but still provides double-ended grounding. Of course, these sheaths and bonding jumpers must be insulated; their voltage differential from ground may be in the 30—50 V range. For details on calculating sheath losses in cable systems, consult [1].

Difficulties may arise from current attempting to flow via the cable shield, unrelated to cable insulation failures. To prevent this, all points served by a multiple-grounded shield cable need to be interconnected with an ample grounding system. (Insulation between shield sections at splices of single-end grounded shield systems needs enough dielectric strength to withstand possible abnormal voltages as well.) This system requires interconnecting grounding conductors of suitably low impedance that fault or lightning currents will follow this path rather than the cable shield. Cable-shield ground connections must be made to this system, which must also connect to the grounded element of the source supplying the energy to the cable. Duct runs, or direct-burial routes, generally include a heavy grounding conductor to ensure such interconnection.

For further details, the reader is referred to Chapter 6, as well as to IEEE Std 142-1972, Grounding of Industrial and Commercial Power Systems (ANSI C114.1-1973), and the National Electrical Safety Code, ANSI C2.1-1971, Rules for the Installation and Maintenance of Electric Supply Stations and Equipment, Part 3.

10.10 Protection from Transient Overvoltage. Cables up through 35 kV used in industrial service have insulation strengths well above that of essentially all other types of electric equipment of similar voltage ratings. This is to compensate for installation handling and possibly a higher deterioration rate than insulation which is exposed to less severe ambient conditions. This high insulation strength may or may not exist in splices or terminations, depending on their design and construction. Except for deteriorated points in the cable itself, the splices or terminations are most affected by overvoltages of lightning and switching transients. The terminations of cable systems not provided with surge protection may flash over due to switching transients. In this event, the cable proper would be subjected to possible wave reflections of even higher levels, possibly damaging the cable insulation; however, this is unlikely in this medium-voltage class.

Like other electric equipment, the means employed for protection from these overvoltages is usually surge arresters. Distribution or intermediate type arresters are used, applied at the junctions of open-wire lines and cables, and at terminals where switches may be open. Surge arresters are not required at intermediate positions along the cable run as they are frequently required on open-wire lines.

It is recommended that surge arresters be connected between the conductor and the cable shielding system with short leads to maximize the effectiveness of the arrester. Similarly recommended is the direct connection of the shields and arrester ground wires to a substantial grounding system to prevent surge current propagation through the shield.

Aerial messenger-supported fully insulated cables and spacer type cables are subject to direct lightning strokes, and a number of such cases are on record. The incidence rate is, however, rather low, and in most cases no protection is provided. Where, for reliability, such incidents must be guarded against, a grounded shield wire similar to that used for bare aerial circuits should be installed on the poles a few feet above the cable. Down-pole grounding conductors need to be carried past the cable messenger with a lateral offset of about 18 inches to guard against side flashes consequent to direct strokes. Metal bayonets, where used to support the grounded shielding wire, should also be kept no less than 18 inches clear of the cables or messengers.

10.11 Testing

10.11.1 *Application and Utility.* Testing, particularly of elastomeric and plastic (solid) insulations, is a useful method of checking the ability of a cable to withstand service conditions for a reasonable future period. Failure to pass test will either cause in-test breakdown of the cable or otherwise indicate the need for its immediate replacement.

Whether or not to routinely test cables is a decision each user has to make. The following factors must be taken into consideration.

(1) If there is no alternate source for the load supplied, testing must be done when the load equipment is not in operation.

(2) The costs of possible service outages due to cable failures must be weighed against the cost of testing. With solid-type insulation, in-service cable failures may be reduced by about 90 percent by direct-current maintenance testing.

(3) Personnel with adequate technical capability must be available to do the testing and evaluate observations and results.

The procedures outlined here are intended as a guide, and many variations are possible. At the same time, variations made without sound technical basis can negate the usefulness of the test, or even damage equipment.

With solid dielectric cable types (elastomeric and plastic), the principal failure mechanism results from "alternating-current corona cutting" during service at locations of either manufacturing defects, installation damage, or accessory workmanship shortcomings. Initial tests reveal only gross damage, improper splicing or terminating, or cable imperfections. Subsequent use on alternating current usually causes progressive enlargement of such defects proportional to their severity.

Oil—paper (laminated) cable with lead sheath fails usually from water entrance at a perforation in the sheath, generally within three to six months after the perforation occurs. Periodic testing, unless very frequent, is therefore likely to miss many of these cases, making the method less effective with this type of cable.

Testing is not useful in detecting possible failure from moisture-induced tracking across termination surfaces, since this develops principally during periods of precipitation, condensation, or leakage failure of the enclosure or housing. However, terminals should be examined regularly for signs of tracking and the condition corrected whenever detected.

10.11.2 *Alternating Current Versus Direct Current.* Cable insulation can, without damage, sustain application of direct-current potential equal to the system basic impulse insulation level for very long periods. In contrast, most cable insulations will sustain degradation from alternating-current overpotential, proportional to a high power of overvoltage to time (and frequency) of the application. Hence it is desirable to utilize direct current for any testing that will be repetitive. While the manufacturers use alternating current for the original "factory" test, it is almost univeral practice to employ direct current for any subsequent testing. All discussion of field testing hereafter applies to direct-current high-voltage testing.

10.11.3 *Factory Tests.* All cable is tested by the manufacturer before shipment, normally with alternating voltage for a 5 min period. Unshielded cable is immersed in water (ground) for this test, shielded cable is tested using the shield as the ground return. Test voltages are specified by the manufacturer, by the applicable specification of the IPCEA, or by other specifications such as that of the Association of Edison Illuminating Companies[17] (AEIC). In

---

[17] 51 East 42nd Street, New York, NY 10017.

addition, a test may be made using direct voltage of two to three times the rms value used in the alternating-current test. On cable rated 3000 V and above, corona tests also may be made.

10.11.4 *Field Tests.* As well as having no deteriorating effect on good insulation, direct-current high voltage is most convenient for field testing since the test power sources or test sets are relatively light and portable.

Voltages for such testing should fulfill both of the following requirements:

(1) Not be high enough to damage sound cable or component insulation

(2) Be high enough to indicate incipient failure of unsound insulation which may fail in service before the next scheduled test

Test voltages and intervals require coordination to attain suitable performance. One large industrial company with cable testing experience of over twenty years has reached over 90 percent reduction of cable system service failures through use of IPCEA specified voltages. These are applied at installation, after about three years of service, and every five to six years thereafter. The majority of test failures occur at the first two tests; test (or service) failures after eight years of satisfactory service are less frequent. The importance of uninterrupted service should also influence the test frequency for specific cables. Tables 84 and 85 specify cable field test voltages.

The AEIC has specified test values for 1968 and later cables of approximately 20 percent higher than the IPCEA values.

Cables to be tested must have their ends free of equipment and clear from ground. All conductors not under test must be grounded. Since equipment to which cable is customarily connected may not withstand the test voltages allowable for cable, either the cable must be disconnected from this equipment, or the test voltage must be limited to levels which the equipment can tolerate. The latter constitutes a relatively mild test on the cable condition, and the predominant leakage current measured is likely to be that of the attached equipment. In essence, this tests the equipment, not the cable.

In field testing, in contrast to the "go—no go" nature of factory testing, the leakage cur-

Table 84
IPCEA Specified Direct-Current Cable Test Voltages
Pre-1968 Cable

Insulation Type	Grounding	Maintenance Test Rated Cable Voltage			
		5 kV	15 kV	25 kV	35 kV
Elastomeric:	Grounded	27	47	—	—
butyl, oil base, EPR	Ungrounded	—	67	—	—
Polyethylene,	Grounded	22	40	67	88
including cross-linked polyethylene	Ungrounded	—	52	—	—

Table 85
IPCEA Specified Direct-Current Cable Test Voltages
1968 and Later Cable

Insulation Type	Insulation Level (%)	Rated Cable Voltage							
		5 kV		15 kV		25 kV		35 kV	
		1	2	1	2	1	2	1	2
Elastomeric:	100	37*	27*	65	49	108	81	—	—
butyl and oil base	133	—	—	91	68	—	—	—	—
Elastomeric:	100	25	19	55	41	80	60	100	75
EPR†	133	25	19	65	49	100	75	—	—
Polyethylene,	100	25	19	55	41	80	60	100	75
including cross-linked polyethylene	133	25	19	65	49	100	75	—	—

NOTE: Columns 1 — Installation tests, made after installation, before service; columns 2 — maintenance tests, made after cable has been in service.

*225—1000 kcmil only. Smaller sizes take 10 percent lower test voltage; larger 10 percent higher.

†These test values are lower than for pre-1968 elastomeric cables, because the insulation is thinner. Hence the alternating-current test voltage is lower. The direct-current test voltage is specified as three times the alternating-current test voltage, so it is also lower than for older cables.

rent of the cable system must be closely watched and recorded for signs of approaching failure. The test voltage may be raised continuously and slowly from zero to the maximum value, or it may be raised in steps, pausing for 1 min or more at each step. Potential differences between steps are of the order of the alternating-current rms rated voltage of the cable. As the voltage is raised, current will flow at a relatively high rate to charge the capacitance, and to a much lesser extent to supply the dielectric absorption characteristics of the cable, as well as to supply the leakage current. The capacitance charging current subsides within a second or so, the absorption current subsides much more slowly and would continue to decrease for 10 min or more, ultimately leaving only the leakage current flowing.

At each step, and for the 5—15 min duration of the maximum voltage, the current meter (normally a microammeter) is closely watched. If, except when the voltage is being increased, the current starts to increase, slowly at first, then more rapidly, the last remnants of insulation at a weak point are failing, and total failure will occur shortly thereafter unless the voltage is reduced. This is characteristic of about 80 percent of all elastomeric-type test failures.

In contrast to this "avalanche" current increase to failure, sudden failure (flashover) can occur if the insulation is already completely (or nearly) punctured. In the latter case, voltage increases until it reaches the sparkover potential of the air gap length, then flashover occurs. Polyethylene cables exhibit the latter characteristic for all failure modes. Conducting leakage paths, such as at terminations or through the body of the insulation, exhibit a constant leakage resistance independent of time or voltage.

One advantage of step testing is that a 1 min "absorption-stabilized" current may be read at the end of each voltage step. The calculated resistance of these steps may be compared as the test progresses to higher voltage. At any step where the calculated leakage resistance decreases markedly (say to 50 percent of that of the next lower voltage level), the cable could be near failure and the test should be discontinued

short of failure as it may be desirable to retain the cable in serviceable condition until a replacement can be made ready. On any test in which the cable will not withstand the prescribed test voltage for the full test period (usually 5 min) without current increase, the cable is considered to have failed the test and is subject to replacement as soon as possible.

The polarization index is the ratio of the current after 1 min to the current after 5 min of maximum voltage test, and on good cable it will be between 1.25 and 2. Anything less than 1.0 should be considered a failure, and between 1.0 and 1.25 only a marginal pass.

After completion of the 5 min maximum test voltage step, the supply voltage control dial should be returned to zero and the charge in the cable allowed to drain off through the leakage of the test set and voltmeter circuits. If this requires too long a time, a bleeder resistor of 1 MΩ per 10 kV of test potential can be added to the drainage path, discharging the circuit in a few seconds. After the remaining potential drops below 10 percent of the original value, the cable conductor may be solidly grounded. All conductors should be left grounded when not on test during the testing of other conductors and for at least 30 min after the removal of a direct-current test potential. They may be touched only while the ground is connected to them; otherwise the release of absorption current by the dielectric may again raise their potential to a dangerous level.

10.11.5 *Procedure.* Load is removed from the cables either by diverting the load to an alternate supply, or by shutdown of the load served. The cables are de-energized by switching; they are tested to ensure voltage removal, then grounded and disconnected from their attached switching equipment. (In case they are left connected, lower test potentials are required.) Surge arresters, potential transformers, and capacitors must be disconnected.

All conductors and shields must be grounded. The test set is checked for operation, and after its power is turned off, the test lead is attached to the conductor to be tested. At this time (and not before) the ground should be removed from that conductor, and the bag or jar (see Section 10.11.6) applied over all of its terminals, covering all uninsulated parts at both ends of the run. The test voltage is then slowly applied, either continuously or in steps as outlined in Section 10.11.4. At completion of the maximum test voltage duration, the charge is drained off, the conductor grounded, and the test lead removed for connection to the next conductor. This procedure is repeated for each conductor to be tested. Grounds should be left on each tested conductor for no less than 30 min.

10.11.6 *Direct-Current Corona and Its Suppression.* Starting at about 10—15 kV and increasing at a high power of the incremental voltage, the air surrounding all bare conductor portions of the cable circuit becomes ionized from the test potential on the conductor and draws current from the conductor. This ionizing current indication is not separable from that of the normal leakage current, and degrades the apparent leakage resistance value of the cable. Wind and other air currents tend to blow the ionized air away from the terminals, dissipating the space charge, and allowing ionization of the new air, thus increasing the "direct corona current" as this may be called.

Enclosing the bare portions of both end terminations in plastic bags, or plastic or glass jars, prevents the escape of this ionized air; thus it becomes a captive space charge. Once formed, it requires no further current, so the "direct corona current" disappears. With this treatment, testing up to about 100 kV is possible. Above 100 kV, larger bags or a small bag inside a large one are required.

An alternate method to minimize corona is to completely tape all bare conductor surfaces with standard electrical insulating tape. This method is superior to the bag method for corona suppression, but it requires more time to adequately tape all exposed ends.

10.11.7 *Line-Voltage Fluctuations.* The very large capacitance of the cable circuit makes the microammeter extremely sensitive to even minor variations in 120 V 60 Hz supply to the test set. Normally it is possible only to read average current values or the near-steady current values. A low-harmonic-content constant-voltage transformer improves this condition moderately. Complete isolation and stability are attainable only by use of a storage battery and 120 V 60 Hz inverter to supply the test set.

10.11.8 *Resistance Evaluation.* High-voltage cable exhibits extremely high insulation resistance, frequently many thousands of megohms. While insulation resistance alone is not a primary indication of the condition of the cable insulation, the comparison of the insulation resistances of the three-phase conductors is useful. On circuits under 1000 ft long, a ratio in excess of 5:1 between any two conductors is indicative of some questionable condition. On longer circuits, a ratio of 3:1 should be regarded as maximum. Comparison of insulation resistance values with previous tests may be informative; but insulation resistance varies inversely with temperature, with winter insulation resistance measurements being much higher than those obtained under summer conditions. An abnormally low insulation resistance is frequently indicative of a faulty splice, termination, or a weak spot in the insulation. (Higher than standard test voltages have been found practical to locate these by causing a test failure where the standard voltage will not cause breakdown. Fault location methods may be used to locate the failure.)

10.11.9 *Megohmmeter Test.* Since the insulation resistance of a sound high-voltage cable circuit is generally in the order of thousands to hundreds of thousands of megohms, a megohmmeter test will reveal only grossly deteriorated insulation conditions of high-voltage cable. For low-voltage cable, however, the megohmmeter tester is quite useful, and is probably the only practicable test. Sound 600 V cable insulation will normally withstand 20 000 V or higher direct current. Thus a 1000 V or 2500 V megohmmeter is preferable to the lower 500 V testers for such cable testing.

For this low-voltage class, temperature-corrected comparisons of insulation resistances with other phases of the same circuit, with previous readings on the same conductor, and with other similar circuits are useful criteria for adequacy. Continued reduction in a cable's insulation resistance over a period of several tests is indicative of degrading insulation; however, a megohmmeter will rarely initiate final breakdown of such insulation.

10.12 Locating Cable Faults. In an industrial plant a wide variety of cable faults can occur.

The problem may be in a communication circuit or in a power circuit, either in the low- or high-voltage class. Circuit interruption may have resulted, or operation may continue with some objectionable characteristic. Regardless of the class of equipment involved or the type of fault, the one common problem is to determine the location of the fault so that repairs can be made.

The vast majority of cable faults encountered in an industrial power system occur between conductor and ground. Most fault-locating techniques are made with the circuit de-energized. In ungrounded or high-resistance-grounded low-voltage systems, however, the occurrence of a single line-to-ground fault will not result in automatic circuit interruption, and therefore the process of locating the fault may be carried out by special procedures with the circuit energized.

10.12.1 *Influence of Ground-Fault Resistance.* Once a line-to-ground fault has occurred, the resistance of the fault path can range from almost zero up to millions of ohms. The fault resistance has a bearing on the method used to locate the failure. In general a low-resistance fault can be located more readily than one of high resistance. In some cases the fault resistance can be reduced by the application of voltage sufficiently high to cause the fault to break down with sufficient current to cause the insulation to carbonize. The equipment required to do this is quite large and expensive, and its success is dependent to a large degree on the type of insulation involved. Large users indicate that this method is useful with paper and elastomeric cables, but generally of little use with plastic types.

The fault resistance which exists after the occurrence of the original fault depends on the type of cable insulation and construction, the location of the fault, and the cause of the failure. A fault which is immersed in water will generally exhibit a variable fault resistance and will not consistently arc over at a constant voltage. Damp faults behave in a similar manner until the moisture has been vaporized. In contrast, a dry fault will normally be much more stable and consequently can more readily be located.

For failures which have occurred in service, the type of system grounding and available fault current, as well as the speed of relay protection, will be influencing factors. Because of the greater carbonization and conductor vaporization, a fault resulting from an in-service failure can generally be expected to be of a lower resistance than one resulting from over-potential testing.

10.12.2 *Equipment and Methods.* A wide variety of commercially available equipment and a number of different approaches can be used to locate cable faults. The safety considerations outlined in Section 10.11 should be observed.

The method used to locate a cable fault depends on
(1) Nature of fault
(2) Type and voltage rating of cable
(3) Value of rapid location of faults
(4) Frequency of faults
(5) Experience and capability of personnel

(1) *Physical Evidence of the Fault.* Observation of a flash, sound, or smoke accompanying the discharge of current through the faulted insulation will usually locate a fault. This is more probable with an overhead circuit than with underground construction. The discharge may be from the original fault or may be intentionally caused by the application of test voltages. The burned or disrupted appearance of the cable will also serve to indicate the faulted section.

(2) *Megohmmeter Instrument Test.* When the fault resistance is sufficiently low that it can be detected with a megohmmeter, the cable can be sectionalized and each section tested to determine which contains the fault. This procedure may require that the cable be opened in a number of locations before the fault is isolated to one replaceable section. This could, therefore, involve considerable time and expense, and might result in additional splices. Since splices are often the weakest part of a cable circuit, this method of fault locating may introduce additional failures at a subsequent time.

(3) *Conductor Resistance Measurement.* This method consists of measuring the resistance of the conductor from the test location to the point of fault by using either the Varley loop or

the Murray loop test. Once the resistance of the conductor to the point of fault has been measured, it can be translated into distance by using handbook values of resistance per unit length of the size and type of conductor involved, correcting for temperature as required. Both of these methods give good results which are independent of fault resistance, provided the fault resistance is low enough that sufficient current for readable galvanometer deflection can be produced with the available test voltage. Normally a low-voltage bridge is used for this resistance measurement. For distribution systems using cables insulated with organic materials, relatively low resistance faults are normally encountered. The conductor resistance measurement method has its major application on such systems. Loop tests on large conductor sizes may not be sensitive enough to narrow down the location of the fault.

High-voltage bridges are available for higher resistance faults but have the disadvantage of increased cost and size as well as requiring a high-voltage direct-current power supply. High-voltage bridges are generally capable of locating faults with a resistance to ground of up to 1 or 2 MΩ, while a low-voltage bridge is limited to the application where this resistance is several kilohms or less.

(4) *Capacitor Discharge.* This method consists of applying a high-voltage high-current impulse to the faulted cable. A high-voltage capacitor is charged by a relatively low current capacity source such as that used for high-potential testing. The capacitor is then discharged across an air gap into the cable. The repeated discharging of the capacitor provides a periodic pulsing of the faulted cable. Where the cable is accessible, or the fault is located at an accessible position, the fault may be located simply by sound. Where the cable is not accessible, such as in duct or directly buried, the discharge at the fault may not be audible. In such cases, detectors are available to trace the signal to the point of fault. The detector generally consists of a magnetic pickup coil, an amplifier, and a meter to display the relative magnitude and direction of the signal. The direction indication changes as the detector passes beyond the fault. Acoustic detectors are also employed, particularly in situations where no appreciable magnetic field

external to the cable is generated by the tracing signal.

In applications where relatively high resistance faults are anticipated, such as with solid dielectric cables or through compound in splices and terminations, the impulse method is the most practical method presently available and is the one most commonly used.

(5) *Tone Signal* (May be used on energized circuits.) A fixed-frequency signal, generally in the audio frequency range, is imposed on the faulted cable. The cable route is then traced by means of a detector which consists of a pickup coil, receiver, and head set or visual display, to the point where the signal leaves the conductor and enters the ground return path. This class of equipment has its primary application in the low-voltage field and is frequently used for fault location on energized ungrounded type circuits. On systems over 600 V the use of a tone signal for fault location is generally unsatisfactory because of the relatively large capacitance of the cable circuit.

(6) *Radar System.* A short-duration low-energy pulse is imposed on the faulted cable and the time required for propagation to and return from the point of fault is monitored on an oscilloscope. The time is then translated into distance in order to locate the point of fault. Although this type of equipment has been available for a number of years, its major application in the power field has been on long-distance high-voltage lines. In older equipment the propagation time is such that it cannot be displayed with good resolution for relatively short cables encountered in industrial systems. However, recent equipment advances have largely overcome this problem. The major limitation to this method is the inability to adequately determine the difference between faults and splices on multitapped circuits. An important feature of this method is that it will locate an "open" in an otherwise unfaulted circuit.

10.12.3 *Selection.* The methods listed represent some of the means available to industrial plant power system operators to locate cable faults. They range from very simple to relatively complex. Some require no equipment, others require equipment which is inexpensive and can be used for other purposes, while still others require special equipment. As the complexity of the means used to locate a fault increases, so does the cost of the equipment, and also the training and experience required of those who are to use it.

In determining which approach is most practical for any particular plant, the size of the plant and the amount of circuit redundancy which it contains must be considered. The importance of minimizing the outage time of any particular circuit must be evaluated. The cable installation and maintenance practices and number and time of anticipated faults will determine the expenditure for test equipment which can be justified. Equipment which requires considerable experience and operator interpretation for accurate results may be satisfactory for a plant with frequent cable faults but ineffective where the number of faults is so small that adequate experience cannot be obtained. Because of these factors, many companies employ firms which offer the service of cable-fault locating. Such firms are established in large cities and cover a large area with mobile test equipment.

No single method of cable-fault locating can be considered to be most suitable for all applications. The final decision on which method or methods to use must depend upon evaluation of the merits and disadvantages of each in relation to the particular circumstances of the plant in question.

10.13 Cable Specification. Once the correct cable has been determined, it can be described in a cable specification. Cable specifications generally start with the conductor and progress radially through the insulation and coverings. The following is a check list which can be used in preparing a cable requirement:

(1) Number of conductors in cable, and phase identification required

(2) Conductor size (AWG, kcmil) and material

(3) Insulation type (rubber, polyvinyl chloride, polyethylene, EPR, etc)

(4) Voltage rating

(5) Shielding system; applicable to cable used on systems 8 kV and above, and may be required on 2.1 kV to 8 kV

(6) Outer finishes

(7) Installation (underground, tray, etc)

An alternate method of specifying cable is to furnish the ampacity of the circuit (amperes), the voltage (phase to phase, phase to ground, grounded, or ungrounded), and the frequency along with any other pertinent system data. Also required is the method of installation anticipated and the installation conditions (ambient temperature, load factor, etc). For either method, the total number of lineal feet of conductors required, the quantity desired shipped in one length, the pulling eyes, and whether it is desired to have several single-conductor cables paralleled on a reel should also be given.

10.14 Standards References. The following standards publications were used as references in preparing this chapter.

ANSI C2 (1976 ed), National Electrical Safety Code

IEEE Std 48-1975, High Voltage AC Cable Terminations

IEEE Std 142-1972, Grounding of Industrial and Commercial Power Systems (ANSI C114.1-1973)

IEEE Std 386-1974, Separable Insulated Connectors for Power Distribution Systems Above 600 V (ANSI C119.2-1974)

NFPA No 70, National Electrical Code (1975), (ANSI C1-1975)

UL 486-1973, Electric-Wire Connectors and Soldering Lugs

10.15 References

[1] *Underground Systems Reference Book.* New York: Association of Illuminating Companies, 1957, chap 10.

[2] IEEE S-135-1-1962, Power Cable Ampacities, Copper Conductors

[3] IEEE S-135-2-1962, Power Cable Ampacities, Aluminum Conductors

11. Busways

11.1 Origin. Busways originated as a result of a request of the automotive industry in Detroit in the late 1920s for an overhead wiring system that would simplify electrical connections for electric-motor-driven machines and permit a convenient rearrangement of these machines in production lines. From this beginning, busways have grown to become an integral part of the low-voltage distribution system for industrial plants at 600 V and below.

Busways are particularly advantageous when numerous current taps are required. Plugs with circuit breakers or fusible switches may be installed and wired without de-energizing the busway.

Power circuits over 1000 A are usually more economical and require less space with busways than with conduit and wire. Busways may be dismantled and reinstalled in whole or part to accommodate changes in the electric distribution system layout.

11.2 Busway Construction. Originally a busway consisted of bare copper conductors supported on inorganic insulators such as porcelain, mounted within a nonventilated steel housing. This type of construction was adequate for the current ratings of 225—600 A then used. As the use of busway expanded, and increased loads demanded higher current ratings, the housing was ventilated to provide better cooling at higher capacities. The bus bars were covered with insulation for safety and to permit closer spacing of bars of opposite polarity to achieve lower reactance and voltage drop.

In the late 1950s busways were introduced utilizing conduction for heat transfer by placing the insulated conductor in thermal contact with the enclosure. By utilizing conduction, current densities are achieved for totally enclosed busways that are comparable to those previously attained with ventilated busways. Totally enclosed busways of this type have the same current rating regardless of mounting position. A stack of one bus bar per phase is used where each bus bar is up to approximately 7 in wide (1600 A). Higher ratings will use two (3000 A) or three stacks (5000 A). Each stack will contain all three phases and neutral to minimize circuit reactance.

Early busway designs required multiple nuts, bolts, and washers to electrically join adjacent sections. The most recent designs use a single bolt for each stack (with bars up to 7 in wide). All hardware is captive to the busway section when shipped from the factory. Installation labor is greatly reduced with corresponding savings in installation costs.

Busways are available with either copper or aluminum conductors. Compared to copper, aluminum has lower electrical conductivity, less mechanical strength, and upon exposure to the

FEEDER   LIGHTING   PLUG-IN   TROLLEY

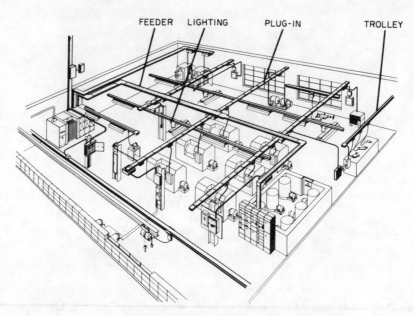

Fig 129
Illustration of Versatility of Busways, Showing Use of Feeder, Plug-In,
Lighting, and Trolley Types

atmosphere quickly forms an insulating film on the surface. For equal current-carrying ability aluminum is lighter in weight and less costly.

For these reasons aluminum conductors will have electroplated contact surfaces (tin or silver) and at electrical joints use Belleville springs and bolting practices which accommodate aluminum mechanical properties. Copper busway will be physically smaller (cross section) while aluminum busway is lighter in weight and lower in cost. Copper plug-in busway is more tolerant of cycling loads such as welding.

Busway is usually made in 10 ft sections. Since the busway must comform to the building, all possible combinations of elbows, tees, and crosses are available. Feed and tap fittings to other electric equipment such as switchboards, transformers, motor-control centers, etc, is provided. Plugs for plug-in busway use fusible switches and molded-case circuit breakers. Standard busway current ratings are 20—5000 A for single-phase and three-phase service. Neutral conductors may be supplied if required. Newer designs of busway including plug-in devices can incorporate a ground bar if specified.

Four types of busways are available, complete with fittings and accessories, providing a unified and continuous system of enclosed conductors (Fig 129):

(1) Feeder busway for low-impedance transmission of power

(2) Plug-in busway for easy connection or rearrangement of loads

(3) Lighting busway to provide electric power and mechanical support to fluorescent, high-intensity discharge, and incandescent fixtures

(4) Trolley busway for mobile power "tap-offs" to electric hoists, cranes, portable tools, etc

11.3 Feeder Busway. Feeder busway is used to transmit large blocks of power. It has a very low and balanced circuit reactance for control of voltage at the utilization equipment (Fig 130).

Feeder busway is frequently used between the source of power, such as a distribution transformer or service drop, and the service entrance equipment. Industrial plants use feeder busway from the service equipment to supply

Fig 130
Feeder Busway

large loads directly and to supply smaller current ratings of feeder and plug-in busway, which in turn supply loads through power takeoffs or plug-in units.

Available current ratings range from 600 to 5000 A, 600 V, alternating current. The manufacturer should be consulted for direct-current ratings. Feeder busway is available in single-phase and three-phase service with 50 and 100 percent neutral conductor. A ground bus is available with all ratings and types. Available short-circuit-current ratings are 50 000–200 000 A, symmetrical rms (see Section 11.8.2). The voltage drop of low-impedance feeder busway with the entire load at the end of the run ranges from 1 to 3 V per 100 ft, line to line, depending upon the type of construction and the current rating (see Section 11.8.3).

Feeder busway is available in indoor and weatherproof (outdoor) construction. Weatherproof construction is designed to shed water (rain). It should be used indoors where the busway may be subjected to water or other liquids. If National Electrical Manufacturers Association (NEMA) "3R" equipment is suitable, weatherproof busway should be used. Busway of any type is not suitable for immersion in water.

11.4 Plug-In Busway. Plug-in busway is used in industrial plants as an overhead system to supply power to utilization equipment. It serves as an elongated switchboard or panelboard running through the area with covered plug-in openings provided at closely spaced intervals to accommodate the plug-in devices placed on the busway near the loads which they supply.

Plug-in tapoffs and rearrangement are greatly facilitated by the use of flexible bus drop cable. The plus may be removed from the busway together with the bus drop cable and reinstalled with the machine in a minimum of time (Fig. 131).

Fig 131
Installation View of Small Plug-In Busway Showing
Individual Circuit Breaker Power Tapoff
and Flexible Bus-Drop Cable

Plug-in devices available include fusible switches, circuit breakers, static voltage protectors, ground indicators, combination starters, lighting contactors, and capacitor plugs.

Most plug-in busway is totally enclosed with current ratings from 100 to 4000 A. Usually plug-in and feeder busway sections of the same manufacturer above 600 A have compatible joints, so that they are interchangeable in a run. Plug-in busway may be inserted in a feeder run when a tapoff is desired. Plug-in tapoffs are generally limited to maximum ratings of 800 A.

Short-circuit-current ratings vary from 15 000 to 150 000 A, symmetrical rms (see Section 11.8.2). The voltage drop ranges from 1 to 3 V per 100 ft, line to line, for evenly distributed loading. If the entire load is concentrated at the end of the run, these values double (see Section 11.8.3).

A ground bar is often added to provide a lower resistance ground path than furnished by the busway housing. A neutral bar may be provided for single-phase loads such as lighting. Neutral bars vary from 25 to 100 percent of the capacity of the phase bars.

11.5 Lighting Busway. Lighting busway is rated a maximum of 60 A, 300 V to ground, with two, three or four conductors. It may be used on 480Y/277 or 208Y/120 V systems and is specifically designed for use with fluorescent and high-intensity discharge lighting (Fig 132).

Lighting busways provide power to the lighting fixture and also serve as the mechanical support for the fixture. Auxiliary supporting means called "strength beams" are available. Strength beams may be supported at maximum intervals of 16 ft. This permits the strength beam supports to conform to building column spacing. The strength beams provide supports for the lighting busway as required by NFPA No 70, National Electrical Code (1975), (ANSI C1-1975), (NEC).

Fluorescent lighting fixtures may be suspended from the busway or they may be ordered with plugs and hangers attached for close coupling of the fixture to the busway. The busway may also be recessed in or surface mounted to dropped ceilings.

Lighting busway is also used to provide power for light industrial applications.

Fig 132
Lighting Busway Supporting and Supplying Power to
High-Intensity Discharge Fixture

11.6 Trolley Busway. Trolley busway is constructed to receive stationary or movable take-off devices. It is used on a moving production line to supply electric power to a motor or a portable tool moving with a production line, or where operators move back and forth over a range of 10—20 ft to perform their specific operations.

11.7 Standards. Busways are designed to conform to

(1) The NEC, Article 364

(2) Underwriters Laboratories Inc, standard UL 857-1974, Electric Busways and Associated Fittings

(3) NEMA BU 1-1972, Busways

UL and NEMA are primarily manufacturing and testing standards. The NEMA standard is generally an extension of the UL standard to areas that UL does not cover. The most important areas are busway parameters resistance R, inductance X, and impedance Z, and short-circuit testing and rating.

The NEC is the most important standard for busway installation. Some of its most important areas are as follows.

(1) Busway may be installed only where located in the open and visible. Installation behind panels is permitted if access is provided and the following conditions are met:

(a) No overcurrent devices are installed on the busway other than for an individual fixture

(b) The space behind the panels is not used for air-handling purposes

(c) The busway is the totally enclosed non-ventilating type

(d) The busway is so installed that the joints between sections and fittings are accessible for maintenance purposes

(2) Busway may not be installed where subject to severe physical damage, corrosive vapors, or in hoistways.

(3) When specifically approved for the purpose, busway may be installed in a hazardous location or outdoors or in wet or damp locations.

(4) Busway must be supported at intervals not to exceed 5 ft unless otherwise approved. Where specifically approved for the purpose, horizontal busway may be supported at intervals up to 10 ft, and vertical busway may be supported at intervals up to 16 ft.

(5) Busway must be totally enclosed where passing through floors and for a minimum distance of 6 ft above the floor to provide adequate protection from physical damage. It may extend through walls if joints are outside walls.

State and local electrical codes may have specific requirements over and above UL and the NEC. Appropriate code authorities and manufacturers should be contacted to ensure that requirements are met.

11.8 Selection and Application of Busways.
To properly apply busways in an electric power distribution system, some of the more important items to consider are the following.

11.8.1 Current-Carrying Capacity.
Busways should be rated on a temperature rise basis to provide safe operation, long life, and reliable service.

Conductor size (cross-sectional area) must not be used as the sole criterion for specifying busway. Busway may have seemingly adequate cross-sectional area and yet have a dangerously high temperature rise. The UL requirement for temperature rise ($55°C$) should be used to specify the maximum temperature rise permitted. Larger cross-sectional areas can be used to provide lower voltage drop and temperature rise.

Although the temperature rise will not vary significantly with changes in ambient temperature, it may be a significant factor in the life of the busway. The limiting factor in most busway designs is the insulation life, and there is a wide range of types of insulating materials used by various manufacturers. If the ambient temperature exceeds $40°C$ or a total temperature in excess of $95°C$ is expected, then the manufacturer should be consulted.

11.8.2 Short-Circuit-Current Rating.
The bus bars in busways may be subject to electromagnetic forces of considerable magnitude by a short-circuit current. The generated force per unit length of bus bar is directly proportional to the square of the short-circuit current and is inversely proportional to the spacing between bus bars. Short-circuit-current ratings are generally assigned and tested in accordance with NEMA BU 1-1972. The ratings are based on (1) the use of an adequately rated protective device ahead of the busway that will clear the short circuit in 3 cycles and (2) application in a system with power factor not less than that given in Table 86.

If the system on which the busway is to be applied has a lower power factor (larger X/R ratio) or a protective device with a longer clearing time, the short-circuit-current rating may have to be reduced. The manufacturer should then be consulted.

The required short-circuit-current rating should be determined by calculating the avail-

**Table 86
Busway Ratings as a Function of Power Factor**

Busway Rating (symmetrical rms amperes)	Power Factor	X/R Ratio*
10 000 or less	0.50	1.7
10 001—20 000	0.30	3.2
Above 20 000	0.20	4.9

*X/R is load reactance X divided by load resistance R.

able short-circuit current and X/R ratio at the point where the input end of the busway is to be connected. The short-circuit-current rating of the busway must equal or exceed the available short-circuit current.

The short-circuit current may be reduced by using a current-limiting fuse at the supply end of the busway to cut it off before it reaches maximum value (see Chapter 4).

Short-circuit-current ratings are dependent on many factors such as bus bar center line spacing, size, and strength of bus bars and mechanical supports.

Since the ratings are different for each design of bus bar, the manufacturer should be consulted for specific ratings. Short-circuit-current ratings should include the ability of the ground return path (housing and ground bar if provided) to carry the rated short-circuit current. Failure of the ground return path to adequately carry this current can result in arcing and spitting at joints with attendant fire hazard. The ground-fault current can also be reduced to the point that the overcurrent protective device does not operate.

11.8.3 Voltage Drop.
Line-to-neutral voltage drop V_D in busways may be calculated by the following formulas. The exact formulas for concentrated loads at the end of the line are, with V_R known,

$$V_D = \sqrt{(V_R \cos \phi + IR)^2 + (V_R \sin \phi + IX)^2} - V_R \quad \text{(Eq 22)}$$

and with V_S known,

$$V_D = V_S + IR \cos \phi + IX \sin \phi - \sqrt{V_S^2 - (IX \cos \phi - IR \sin \phi)^2} \quad \text{(Eq 23)}$$

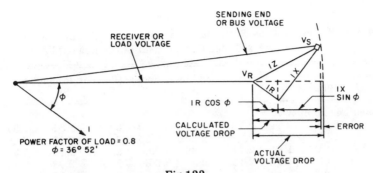

Fig 133
Diagram Illustrating Voltage Drop and Indicating Error
When Approximate Voltage-Drop Formulas Are Used

where

$$V_R = V_S \frac{Z_L}{Z_S}, \qquad V_D = V_S - V_R \qquad \text{(Eq 24)}$$

NOTE: Multiply the line-to-neutral voltage drop by $\sqrt{3}$ to obtain the line-to-line voltage drop in three-phase systems. Multiply the line-to-neutral voltage drop by 2 to obtain the line-to-line voltage drop in single-phase systems.

The approximate formulas for concentrated loads at the end of the line are

$$V_D = I(R \cos \phi + X \sin \phi) \qquad \text{(Eq 25)}$$

$$V_{pr} = \frac{S(R \cos \phi + X \sin \phi)}{10 \, V_k^2} \qquad \text{(Eq 26)}$$

The approximate formula for distributed load on a line is

$$V_{pr} = \frac{S(R \cos \phi + X \sin \phi) \, L}{10 \, V_k^2} \left(1 - \frac{L_1}{2L} \right) \quad \text{(Eq 27)}$$

where

V_D = Voltage drop, in volts

V_{pr} = Voltage drop, in percent of voltage at sending end

V_S = Line-to-neutral voltage at sending end, in volts

V_R = Line-to-neutral voltage at receiving end, in volts

ϕ = Angle whose cosine is the load power factor

R = Resistance of circuit, in ohms per phase

X = Reactance of circuit, in ohms per phase

I = Load current, in amperes

Z_L = Load impedance, in ohms

Z_S = Circuit impedance, in ohms, plus load impedance, in ohms, added vectorially

S = Three-phase apparent power for three-phase circuits or single-phase apparent power for single-phase circuits, in kilovolt-amperes

V_k = Line-to-line voltage, in kilovolts

L_1 = Distance from source to desired point, in feet

L = Total length of line, in feet

The foregoing formulas for concentrated loads may be verified by a trigonometric analysis of Fig 133. From this figure it can be seen that the approximate formulas are sufficiently accurate for practical purposes. In practical cases the angle between V_R and V_S will be small (much smaller than in Fig 133 which has been exaggerated for illustrative purposes). The error in the approximate formulas diminishes as the angle between V_R and V_S decreases and is zero if that angle is zero. This latter condition will exist when the X/R ratio (or power factor) of the load is equal to the X/R ratio (or power factor) of the circuit through which the load current is flowing.

In actual practice, loads may be concentrated at various locations along the feeders, uniformly distributed along the feeder, or any combination of the same. A comparison of the approximate formulas for concentrated end loading and uniform loading will show that a uniformly loaded line will exhibit one half the voltage drop as that due to the same total load concentrated at the end of the line. This aspect of

Table 87
Voltage-Drop Values for Three-Phase Busways with Copper Bus Bars, in Volts per 100 ft,
Line to Line, at Rated Current with Entire Load at End*

Current Rating (amperes)	Load Power Factor (Percent, lagging)									
	20	30	40	50	60	70	80	90	95	100
Totally Enclosed Feeder Busway										
600	2.28	2.51	2.73	2.93	3.09	3.23	3.31	3.31	3.23	2.83
800	1.75	1.93	2.08	2.23	2.35	2.44	2.49	2.48	2.42	2.10
1000	1.51	1.81	2.11	2.39	2.66	2.92	3.15	3.33	3.39	3.29
1350	1.60	1.87	2.13	2.37	2.60	2.80	2.98	3.11	3.13	2.96
1600	1.90	2.10	2.27	2.43	2.56	2.67	2.73	2.72	2.66	2.31
2000	1.82	2.00	2.16	2.30	2.43	2.52	2.57	2.55	2.49	2.15
2500	1.75	1.91	2.06	2.18	2.29	2.36	2.40	2.37	2.30	1.96
3000	1.96	2.14	2.30	2.43	2.55	2.63	2.67	2.63	2.55	2.17
4000	1.84	2.01	2.16	2.29	2.40	2.49	2.53	2.49	2.42	2.07
5000	1.67	1.83	1.98	2.11	2.22	2.30	2.35	2.33	2.27	1.96
Totally Enclosed Plug-In Busway										
225	1.92	2.08	2.22	2.36	2.46	2.54	2.56	2.52	2.42	2.04
400	2.26	2.40	2.52	2.60	2.66	2.70	2.66	2.54	2.40	1.90
600	4.91	5.03	5.10	5.11	5.04	4.89	4.62	4.11	3.67	2.38
800	5.75	5.91	6.00	6.02	5.96	5.80	5.50	4.92	4.42	2.92
1000	4.77	4.91	4.98	5.02	4.98	4.84	4.60	4.12	3.70	2.46
1350	3.72	3.84	3.92	3.94	3.94	3.84	3.68	3.32	3.01	2.06
1600	3.58	3.70	3.78	3.82	3.80	3.72	3.54	3.22	2.92	2.00
2000	4.67	4.79	4.86	4.86	4.82	4.68	4.42	3.94	3.52	2.30
2500	4.08	4.20	4.26	4.30	4.26	4.14	3.94	3.54	3.18	2.12
3000	3.76	3.87	3.92	3.94	3.90	3.80	3.60	3.24	2.90	1.92
4000	4.64	4.74	4.80	4.79	4.73	4.57	4.30	3.81	3.38	2.15
5000	3.66	3.75	3.78	3.78	3.78	3.62	3.40	3.02	2.70	1.76
Lighting, Single Phase, Distributed Loading										
30	0.84	1.11	1.38	1.65	1.89	2.13	2.40	2.51	2.20	2.75
60	1.08	1.38	1.62	1.98	2.22	2.46	2.70	2.88	3.00	3.00
Trolley										
100	1.16	1.38	1.56	1.74	1.90	2.06	2.20	2.28	2.30	2.18

NOTE: Voltage-drop values are based on bus bar resistance at 75°C (room ambient temperature 25°C plus average conductor temperature at full load of 50°C rise).

*Divide values by 2 for distributed loading.

the approximate formula is mathematically exact and entails no approximation. Therefore, in calculations of composite loading involving approximately uniformly loaded sections and concentrated loads, the uniformly loaded sections may be treated as end-loaded sections having one half normal voltage drop of the same total load. Thus the load can be divided into a number of concentrated loads distributed at various distances along the line. The voltage drop in each section may then be calculated for the load which it carries.

Three-phase voltage drops may be determined with reasonable accuracy by the use of Tables 87 and 88. These are typical values for the particular types of busway shown. The voltage drops will be different for other types of busway and will vary slightly by manufacturer within each type. The voltage drop shown is three phase, line to line, per 100 ft, at rated load on a concentrated loading basis for feeder, plug-in, and trolley busway. Lighting busway values are single phase, distributed loading. For other loading and distances use the formula

Table 88
Voltage-Drop Values for Three-Phase Busways with Aluminum Bus Bars, in Volts per 100 ft, Line to Line, at Rated Current with Entire Load at End*

Current Rating (amperes)	Load Power Factor (Percent, lagging)									
	20	30	40	50	60	70	80	90	95	100
Totally Enclosed Feeder Busway										
600	1.64	1.93	2.21	2.48	2.73	2.96	3.16	3.30	3.34	3.17
800	1.69	1.95	2.21	2.44	2.66	2.86	3.03	3.14	3.15	2.94
1000	1.51	1.81	2.11	2.39	2.66	2.92	3.15	3.33	3.39	3.29
1350	1.60	1.87	2.13	2.37	2.60	2.80	2.98	3.11	3.13	2.96
1600	1.70	1.97	2.22	2.45	2.67	2.87	3.04	3.14	3.15	2.94
2000	1.57	1.81	2.03	2.23	2.42	2.59	2.73	2.81	2.81	2.60
2500	1.56	1.78	1.98	2.18	2.35	2.51	2.63	2.70	2.69	2:48
3000	1.64	1.94	2.14	2.37	2.58	2.78	2.94	3.04	3.05	2.85
4000	1.60	1.83	2.04	2.24	2.42	2.59	2.71	2.79	2.78	2.56
Totally Enclosed Plug-In Busway										
100	2.05	2.63	3.20	3.76	4.30	4.83	5.33	5.79	5.98	6.01
225	1.94	2.22	2.49	2.73	2.96	3.15	3.31	3.41	3.40	3.13
400	3.47	3.66	3.81	3.92	3.99	3.99	3.92	3.69	3.45	2.64
600	4.62	4.89	5.12	5.30	5.41	5.45	5.37	5.10	4.80	3.76
800	4.09	4.34	4.54	4.70	4.81	4.84	4.78	4.54	4.28	3.36
1000	3.22	3.43	3.61	3.75	3.85	3.89	3.86	3.70	3.50	2.79
1350	2.92	3.10	3.12	3.36	3.44	3.48	3.44	3.28	3.08	2.44
1600	3.98	4.20	4.38	4.51	4.59	4.61	4.52	4.27	3.99	3.07
2000	3.48	3.68	3.85	3.99	4.07	4.09	4.04	3.83	3.60	2.81
2500	2.83	3.00	3.13	3.24	3.30	3.32	3.27	3.10	2.92	2.27
3000	3.68	3.85	3.99	4.09	4.14	4.12	4.01	3.74	3.47	2.60
4000	3.11	3.27	3.40	3.50	3.55	3.55	3.47	3.26	3.04	2.31

NOTE: Voltage-drop values are based on bus bar resistance at 75°C (room ambient temperature 25°C plus average conductor temperature at full load of 50°C rise).

*Divide values by 2 for distributed loading.

voltage drop V_D

$$= \text{table } V_D \left(\frac{\text{actual load}}{\text{rated load}}\right)\left(\frac{\text{actual distance (feet)}}{100 \text{ ft}}\right)$$

The voltage drop for a single-phase load connected to a three-phase busway is 15.5 percent higher than the values shown in the table. Typical values of resistance and reactance are shown in Table 89. Resistance is shown at normal room temperature (25°C). This value should be used in calculating the short-circuit current available in systems since short circuits can occur when busway is lightly loaded or initially energized. To calculate the voltage drop when fully loaded (75°C), the resistance of copper and aluminum should be multiplied by 1.19.

11.8.4 *Thermal Expansion.* As load is increased, the bus bar temperature will increase and the bus bars will expand. The lengthwise expansion between no load and full load will range from 1/2 to 1 in per 100 ft. The amount of expansion will depend on the total load, the size and location of the tapoffs, and the size and duration of varying loads. To accommodate the expansion, the busway should be mounted using hangers which permit it to move. It may be necessary to insert expansion lengths in the busway run. To locate expansion lengths, the method of support, the location of power takeoffs, the degree of movement at each end of the run which is permissible, and the orientation of the busway must be known. The manufacturer can then make recommendations as to the location and number of expansion lengths.

Table 89
Typical Busway Parameters, Line to Neutral, in Milliohms per 100 ft, 25°C

| Current Rating (amperes) | Feeder Busway | | | | Plug-In Busway | | | |
| | Aluminum | | Copper | | Aluminum | | Copper | |
	R	X	R	X	R	X	R	X
100	—	—	—	—	29.1	5.0	—	—
225	—	—	—	—	6.74	3.45	4.44	3.94
400	—	—	—	—	3.20	4.33	2.31	2.76
600	2.56	0.99	2.28	1.68	3.03	3.80	1.92	4.35
800	1.78	0.81	1.27	0.98	2.03	2.52	1.78	3.80
1000	1.59	0.50	1.05	0.82	1.35	1.57	1.20	2.52
1350	1.06	0.44	0.76	0.65	0.88	1.06	0.75	1.44
1600	0.89	0.41	0.70	0.53	0.93	1.24	0.61	1.17
2000	0.63	0.31	0.52	0.41	0.68	0.86	0.56	1.24
2500	0.48	0.25	0.38	0.32	0.44	0.56	0.42	0.86
3000	0.46	0.21	0.35	0.30	0.43	0.62	0.32	0.66
4000	0.31	0.16	0.25	0.21	0.28	0.39	0.26	0.62
5000	—	—	0.19	0.15	—	—	0.17	0.39
Lighting								
30	—	—	—	—	—	—	79.0	3.0
60	—	—	—	—	—	—	51.0	3.0
Trolley								
100	—	—	12.6	4.3	—	—	—	—

NOTE: Resistance values increase as temperature increases. Reactance values are not affected by temperature. The above values are based on conductor temperature of 25°C (normal room temperature) since short circuits may occur when busway is initially energized or lightly loaded. To calculate voltage drop when fully loaded (75°C), multiply resistance of copper and aluminum by 1.19.

11.8.5 *Building Expansion Joints.* Busway, when crossing a building expansion joint, must include provision for accommodating movement of the building structure. Fittings providing for 6 inches of movement are available.

11.8.6 *Welding Loads.* The busway and the plug-in device must be properly sized when plug-in busway is used to supply power to welding loads. The plug sliding contacts (stabs) and protective device (circuit breaker or fused switch) must have sufficient thermal rating to carry the welder current. This is normally done by determining the equivalent continuous current of the welder based on the maximum peak welder current, the duration of the welder current, and the duty cycle. Values may be obtained from the welder manufacturer. Loads 600 A and greater require special attention including consideration of bolted taps. As previously stated, copper busway is more tolerant of cycling loads than aluminum busway. When using aluminum busway, cycling loads above 200 A should be referred to the manufacturer.

11.9 Layout. Busway must be tailored to the building in which it is installed. Once the basic engineering work has been completed and the busway type, current rating, number of poles, etc, determined, a layout should be made for all but the simplest straight runs. The initial step in the layout is to identify and locate the building structure (walls, ceilings, columns, etc) and other equipment which is in the busway route. A layout of the busway to conform to this route is made. Although the preliminary layout (drawings for approval) can be made from architectural drawings, it is essential that field measurements be taken to verify building and busway dimensions prior to the release of the busway for manufacture. Where dimensions are critical, it is recommended that a section be held for field check of dimensions and manufactured after the remainder of the run has been installed. Manufacturers will provide quick delivery on limited numbers of these "field-check" sections.

Busway has great physical and electrical flexibility. It may be tailored to almost any layout requirement. However, some users find it a good practice to limit their busway installations to a minimum number of current ratings and maintain as many 10 ft lengths as possible. This enables them to reuse the busway components to maximum advantage where production line changes, etc, require relocation of the busway.

Another important consideration when laying out busway is coordination with other trades. Since there is a finite time lapse between job measurement and actual installation, other trades may use the busway "clear area" if coordination is lacking. Again, standard components can help since they are more readily available (sometimes from stock). By reducing the time between final measurement and installation, in addition to proper coordination, the chances of interference from other trades can be reduced to a minimum.

Finally, terminations are a significant part of busway layout considerations. For ratings 600 A and above, direct bused connections to the switchboard, motor-control center, etc, can reduce installation time and problems. For ratings up to 600 A, direct bused terminations are generally not practical or economical. These lower current-ratings of busway are usually fed by short cable runs.

11.10 Installation. Busway installs quickly and easily. When compared with other distribution methods, the reduced installation time for busway can result in direct savings on installation costs. In order to ensure maximum safety, reliability and long life from a busway system, proper installation is a must. The guide lines below can serve as an outline from which to develop a complete installation procedure and time table.

11.10.1 *Procedure Prior to Installation*

(1) Manufacturers supply installation drawings on all but the simplest of busway layouts. Study these drawings carefully. Where drawings are not supplied, make your own.

(2) Verify actual components on hand against those shown on installation drawing to be sure that there are no missing items. Drawings identify components by catalog number and location in the installation. Catalog numbers appear on section nameplate and carton label. Location on the installation (item number) will also be on each section.

(3) During storage (prior to installation) all components, even the weatherproof type, should be stored in a clean, dry area and protected from physical damage.

(4) Read manufacturer's instructions for installation of individual components. If you are still in doubt, ask for more information; never guess.

(5) Finally preposition hanger supports (drop rods, etc) and hangers if of the type which can be prepositioned. You are now ready to begin the actual installation of busway components.

(6) Electrical testing of individual components prior to installation should be done (see Section 11.11). Identification of defective pieces prior to installation will save considerable time and money.

11.10.2 *Procedure During Installation*

(1) Almost all busway components are built with two dissimilar ends which are commonly called "bolt" end and "slot" end. Refer to the installation drawing to properly orient the bolt and slot ends of each component. This is important because it is not possible to properly connect two slot ends or bolt ends.

(2) Lift individual components into position and attach to hangers. It is generally best to begin this process at the end of the busway run which is most rigidly fixed (for example, the switchboards).

(3) Pay particular attention to "TOP" labels and other orientation marks where applicable.

(4) As each new component is installed in position, tighten the joint bolt to proper torque per manufacturer's instructions. Also install any additional joint hardware which may be required.

(5) Finally, on plug-in busway installations, attach plug-in units in accordance with manufacturer's instructions and proceed with wiring.

(6) Outdoor busway may require removal of "weep hole" screws and addition of joint shields. Pay particular attention to installation instructions to ensure that all steps are followed.

Table 90
Voltage, Insulation, Continuous-Current, and Momentary-Current Ratings of
Nonsegregated-Phase Metal-Enclosed Bus

Voltage (kV, rms)		Continuous Current (amperes)	Insulation, Withstand Level (kV)			Momentary Current (kA, asymmetrical)
Nominal	Rated Maximum		Power Frequency (rms), 1 min	DC Withstand, 1 min	Impulse	
4.16	4.76	1200	19.0	27.0	60	19—78
13.8	15.00	2000	36.0	50.0	95	19—78
23.0	25.80	3000	60.0	—	125	58
34.5	38.00	—	80.0	—	150	58

11.10.3 *Procedure After Installation.* Be sure to recheck all steps to ensure that you have not forgotten anything. Be particularly sure that all joint bolts have been properly tightened.

At this point the busway installation should be almost complete. Before energizing, however, the complete installation should be properly tested.

11.11 **Field Testing.** The completely installed busway run should be electrically tested prior to being energized. The testing procedure should first verify that the proper phase relationships exist between the busway and associated equipment. This phasing and continuity test can be performed in the same manner as similar tests on other pieces of electric equipment on the job.

All busway installations should be tested with a megohmmeter or high-potential voltage to be sure that excessive leakage paths between phases and ground do not exist. Megohmmeter values depend on the busway construction, type of insulation, size and length of busway, and atmospheric conditions. Acceptable values for a particular busway should be obtained from the manufacturer.

If a megohmmeter is used, it should be rated 1000 V direct current. Normal high-potential test voltages are twice rated voltage plus 1000 V for 1 min. Since this may be above the corona starting voltage of some busway, frequent testing is undesirable. 120 percent of twice rated voltage plus 100 V for 1 s is a satisfactory alternative.

11.12 **Busways over 600 V (Metal-Enclosed Bus).** Busway over 600 V is referred to as metal-enclosed bus and consists of three types, isolated phase, segregated phase, and nonsegregated phase. Isolated phase and segregated phase are utility-type busways used in power generation stations. Industrial plants outside of power generation areas use nonsegregated phase for connection of transformers and switchgear and interconnection of switchgear lineups. The advantage of metal-enclosed bus over cable is a simpler connection to equipment (no potheads required). It is rarely used to feed individual loads.

11.12.1 *Standards.* Metal-enclosed bus is covered for the first time in the 1975 NEC. The NEC requires that the metal-enclosed bus nameplate specify its rated

(1) Voltage
(2) Continuous current
(3) Frequency
(4) 60 Hz withstand voltage
(5) Momentary current

The NEC further requires that metal-enclosed bus be constructed and tested in accordance with IEEE Std 27-1974, Switchgear Assemblies Including Metal-Enclosed Bus (ANSI C37.20-1969 and Supplements).

11.12.2 *Ratings.* IEEE Std 27-1974 specifies the voltage, insulation, and the continuous- and momentary-current levels for metal-enclosed bus (Table 90). The ratings are equal to the corresponding values for metal-enclosed switchgear.

11.12.3 *Construction.* Metal-enclosed (nonsegregated phase) bus consists of aluminum or copper conductors with bus supports usually of glass polyester or porcelain. Bus bars are insulated with sleeves or by fluid bed process.

After installation joints are covered with boots or tape. Metal-enclosed bus is totally enclosed. The enclosure is fabricated from steel in lower continuous-current ratings and aluminum or stainless steel in higher ratings. Normal lengths are 8—10 ft with a cross section of approximately 16 in by 26—36 in, depending on conductor size and spacing. Electrical connection points are electroplated with either silver or tin. Indoor and outdoor (weatherproof) constructions are available.

11.12.4 *Field Testing.* After installation the metal-enclosed bus should be electrically tested prior to being energized. Phasing and continuity tests can be performed with other associated electric equipment on the job. Megohmmeter tests can be made similar to those described for busway under 600 V. High-potential tests should be conducted at 75 percent of the values shown in Table 90.

Continuous-current ratings are based on a maximum temperature rise of 65°C of the bus (30°C if joints are not electroplated). Insulation temperature limits vary with the class of insulating material. Maximum total temperature limits for metal-enclosed bus are based on 40°C ambient. If the ambient temperature will exceed 40°C, the manufacturer should be consulted.

The momentary-current rating is the maximum rms total current (including direct-current component) which the metal-enclosed bus can carry for 10 cycles without electrical, thermal, or mechanical damage.

11.13 Standards References. The following standards publications were used in preparing this chapter.

IEEE Std 27-1974, Switchgear Assemblies Including Metal-Enclosed Bus (ANSI C37.20-1969 and Supplements)

NEMA BU 1-1972, Busways

NFPA No 70, National Electrical Code (1975), (ANSI C1-1975)

UL 857-1974, Electric Busways and Associated Fittings

Appendix
ANSI Protective Device Designations[18]

1. *Master Element* is the initiating device, such as a control switch, voltage relay, float switch, etc, which serves either directly or through such permissive devices as protective and time-delay relays to place an equipment in or out of operation.

2. *Time-Delay Starting or Closing Relay* is a device that functions to give a desired amount of time delay before or after any point of operation in a switching sequence or protective relay system, except as specifically provided by device functions 48, 62, and 79.

3. *Checking or Interlocking Relay* is a relay that operates in response to the position of a number of other devices (or to a number of predetermined conditions) in an equipment, to allow an operating sequence to proceed, or to stop, or to provide a check of the position of these devices or of these conditions for any purpose.

4. *Master Contactor* is a device, generally controlled by device function 1 or the equivalent and the required permissive and protective devices, that serves to make and break the necesssary control circuits to place an equipment into operation under the desired conditions and to take it out of operation under other or abnormal conditions.

5. *Stopping Device* is a control device used primarily to shut down an equipment and hold it out of operation. [This device may be manually or electrically actuated, but excludes the function of electrical lockout (see device function 86) on abnormal conditions.]

6. *Starting Circuit Breaker* is a device whose principal function is to connect a machine to its source of starting voltage.

7. *Anode Circuit Breaker* is a device used in the anode circuits of a power rectifier for the primary purpose of interrupting the rectifier circuit if an arc-back should occur.

8. *Control Power Disconnecting Device* is a disconnecting device, such as a knife switch, circuit breaker, or pull-out fuse block, used for the purpose of respectively connecting and disconnecting the source of control power to and from the control bus or equipment.

NOTE: Control power is considered to include auxiliary power which supplies such apparatus as small motors and heaters.

9. *Reversing Device* is a device that is used for the purpose of reversing a machine field or for performing any other reversing functions.

10. *Unit Sequence Switch* is a switch that is used to change the sequence in which units may be placed in and out of service in multiple-unit equipments.

11. Reserved for future application.

12. *Over-Speed Device* is usually a direct-connected speed switch which functions on machine overspeed.

13. *Synchronous-Speed Device* is a device such as a centrifugal-speed switch, a slip-frequency relay, a voltage relay, an under-current relay, or any type of device that operates at approximately the synchronous speed of a machine.

14. *Under-Speed Device* is a device that functions when the speed of a machine falls below a predetermined value.

15. *Speed or Frequency Matching Device* is a device that functions to match and hold the speed or the frequency of a machine or of a

---

[18] From ANSI C37.2-1970, Manual and Automatic Station Control, Supervisory, and Associated Telemetering Equipments

system equal to, or approximately equal to, that of another machine, source, or system.

16. Reserved for future application.

17. *Shunting or Discharge Switch* is a switch that serves to open or to close a shunting circuit around any piece of apparatus (except a resistor), such as a machine field, a machine armature, a capacitor, or a reactor.

NOTE: This excludes devices that perform such shunting operations as may be necessary in the process of starting a machine by devices 6 or 42, or their equivalent, and also excludes device function 73 that serves for the switching of resistors.

18. *Accelerating or Decelerating Device* is a device that is used to close or to cause the closing of circuits which are used to increase or decrease the speed of a machine.

19. *Starting-to-Running Transition Contactor* is a device that operates to initiate or cause the automatic transfer of a machine from the starting to the running power connection.

20. *Electrically Operated Valve* is an electrically operated, controlled or monitored valve used in a fluid line.

NOTE: The function of the valve may be indicated by the use of the suffixes in Section 2-9.4.

21. *Distance Relay* is a relay that functions when the circuit admittance, impedance, or reactance increases or decreases beyond predetermined limits.

22. *Equalizer Circuit Breaker* is a breaker that serves to control or to make and break the equalizer or the current-balancing connections for a machine field, or for regulating equipment, in a multiple-unit installation.

23. *Temperature Control Device* is a device that functions to raise or lower the temperature of a machine or other apparatus, or of any medium, when its temperature falls below, or rises above, a predetermined value.

NOTE: An example is a thermostat that switches on a space heater in a switchgear assembly when the temperature falls to a desired value as distinguished from a device that is used to provide automatic temperature regulation between close limits and would be designated as device function 90T.

24. Reserved for future application.

25. *Synchronizing or Synchronism-Check Device* is a device that operates when two ac circuits are within the desired limits of frequency, phase angle, or voltage, to permit or to cause the paralleling of these two circuits.

26. *Apparatus Thermal Device* is a device that functions when the temperature of the shunt field or the amortisseur winding of a machine, or that of a load limiting or load shifting resistor or of a liquid or other medium, exceeds a predetermined value; or if the temperature of the protected apparatus, such as a power rectifier, or of any medium decreases below a predetermined value.

27. *Undervoltage Relay* is a relay that functions on a given value of undervoltage.

28. *Flame Detector* is a device that monitors the presence of the pilot or main flame in such apparatus as a gas turbine or a steam boiler.

29. *Isolating Contactor* is a device that is used expressly for disconnecting one circuit from another for the purposes of emergency operation, maintenance, or test.

30. *Annunciator Relay* is a nonautomatically reset device that gives a number of separate visual indications upon the functioning of protective devices, and which may also be arranged to perform a lockout function.

31. *Separate Excitation Device* is a device that connects a circuit, such as the shunt field of a synchronous converter, to a source of separate excitation during the starting sequence; or one that energizes the excitation and ignition circuits of a power rectifier.

32. *Directional Power Relay* is a device that functions on a desired value of power flow in a given direction or upon reverse power resulting from arc-back in the anode or cathode circuits of a power rectifier.

33. *Position Switch* is a switch that makes or breaks contact when the main device or piece of apparatus which has no device function number reaches a given position.

34. *Master Sequence Device* is a device such as a motor-operated multi-contact switch, or the equivalent, or a programming device, such as a computer, that establishes or determines the operating sequence of the major devices in an equipment during starting and stopping or during other sequential switching operations.

35. *Brush-Operating or Slip-Ring Short-Circuiting Device* is a device for raising, lowering, or shifting the brushes of a machine, or for short-circuiting its slip rings, or for engaging or disengaging the contacts of a mechanical rectifier.

36. *Polarity or Polarizing Voltage Device* is a device that operates, or permits the operation of, another device on a predetermined polarity only, or verifies the presence of a polarizing voltage in an equipment.

37. *Undercurrent or Underpower Relay* is a relay that functions when the current or power flow decreases below a predetermined value.

38. *Bearing Protective Device* is a device that functions on excessive bearing temperature, or on other abnormal mechanical conditions associated with the bearing, such as undue wear, which may eventually result in excessive bearing temperature or failure.

39. *Mechanical Condition Monitor* is a device that functions upon the occurrence of an abnormal mechanical condition (except that associated with bearings as covered under device function 38), such as excessive vibration, eccentricity, expansion, shock, tilting, or seal failure.

40. *Field Relay* is a relay that functions on a given or abnormally low value or failure of machine field current, or on an excessive value of the reactive component of armature current in an ac machine indicating abnormally low field excitation.

41. *Field Circuit Breaker* is a device that functions to apply or remove the field excitation of a machine.

42. *Running Circuit Breaker* is a device whose principal function is to connect a machine to its source of running or operating voltage. This function may also be used for a device, such as a contactor, that is used in series with a circuit breaker or other fault protecting means, primarily for frequent opening and closing of the circuit.

43. *Manual Transfer or Selector Device* is a manually operated device that transfers the control circuits in order to modify the plan of operation of the switching equipment or of some of the devices.

44. *Unit Sequence Starting Relay* is a relay that functions to start the next available unit in a multiple-unit equipment upon the failure or nonavailability of the normally preceding unit.

45. *Atmospheric Condition Monitor* is a device that functions upon the occurrence of an abnormal atmospheric condition, such as damaging fumes, explosive mixtures, smoke, or fire.

46. *Reverse-Phase or Phase-Balance Current Relay* is a relay that functions when the polyphase currents are of reverse-phase sequence, or when the polyphase currents are unbalanced or contain negative phase-sequence components above a given amount.

47. *Phase-Sequence Voltage Relay* is a relay that functions upon a predetermined value of polyphase voltage in the desired phase sequence.

48. *Incomplete Sequence Relay* is a relay that generally returns the equipment to the normal, or off, position and locks it out if the normal starting, operating, or stopping sequence is not properly completed within a predetermined time. If the device is used for alarm purposes only, it should preferably be designated as 48A (alarm).

49. *Machine or Transformer Thermal Relay* is a relay that functions when the temperature of a machine armature or other load-carrying winding or element of a machine or the temperature of a power rectifier or power transformer (including a power rectifier transformer) exceeds a predetermined value.

50. *Instantaneous Overcurrent or Rate-of-Rise Relay* is a relay that functions instantaneously on an excessive value of current or on an excessive rate of current rise, thus indicating a fault in the apparatus or circuit being protected.

51. *AC Time Overcurrent Relay* is a relay with either a definite or inverse time characteristic that functions when the current in an ac circuit exceeds a predetermined value.

52. *AC Circuit Breaker* is a device that is used to close and interrupt an ac power circuit under normal conditions or to interrupt this circuit under fault or emergency conditions.

53. *Exciter or DC Generator Relay* is a relay that forces the dc machine field excitation to build up during starting or which functions when the machine voltage has built up to a given value.

54. Reserved for future application.

55. *Power Factor Relay* is a relay that operates when the power factor in an ac circuit rises above or falls below a predetermined value.

56. *Field Application Relay* is a relay that automatically controls the application of the field excitation to an ac motor at some predetermined point in the slip cycle.

57. *Short-Circuiting or Grounding Device* is a primary circuit switching device that functions to short-circuit or to ground a circuit in response to automatic or manual means.

58. *Rectification Failure Relay* is a device that functions if one or more anodes of a power rectifier fail to fire, or to detect an arcback, or on failure of a diode to conduct or block properly.

59. *Overvoltage Relay* is a relay that functions on a given value of overvoltage.

60. *Voltage or Current Balance Relay* is a relay that operates on a given difference in voltage, or current input or output, of two circuits.

61. Reserved for future application.

62. *Time-Delay Stopping or Opening Relay* is a time-delay relay that serves in conjunction with the device that initiates the shutdown, stopping, or opening operation in an automatic sequence or protective relay system.

63. *Pressure Switch* is a switch which operates on given values, or on a given rate of change, of pressure.

64. *Ground Protective Relay* is a relay that functions on failure of the insulation of a machine, transformer, or of other apparatus to ground, or on flashover of a dc machine to ground.

NOTE: This function is assigned only to a relay that detects the flow of current from the frame of a machine or enclosing case or structure of a piece of apparatus to ground, or detects a ground on a normally ungrounded winding or circuit. It is not applied to a device connected in the secondary circuit of a current transformer, or in the secondary neutral of current transformers, connected in the power circuit of a normally grounded system.

65. *Governor* is the assembly of fluid, electrical, or mechanical control equipment used for regulating the flow of water, steam, or other medium to the prime mover for such purposes as starting, holding speed or load, or stopping.

66. *Notching or Jogging Device* is a device that functions to allow only a specified number of operations of a given device, or equipment, or a specified number of successive operations within a given time of each other. It is also a device that functions to energize a circuit periodically or for fractions of specified time intervals, or that is used to permit intermittent acceleration or jogging of a machine at low speeds for mechanical positioning.

67. *AC Directional Overcurrent Relay* is a relay that functions on a desired value of ac overcurrent flowing in a predetermined direction.

68. *Blocking Relay* is a relay that initiates a pilot signal for blocking of tripping on external faults in a transmission line or in other apparatus under predetermined conditions, or cooperates with other devices to block tripping or to block reclosing on an out-of-step condition or on power swings.

69. *Permissive Control Device* is generally a two-position, manually-operated switch that, in one position, permits the closing of a circuit breaker, or the placing of an equipment into operation, and in the other position prevents the circuit breaker or the equipment from being operated.

70. *Rheostat* is a variable resistance device used in an electric circuit, which is electrically operated or has other electrical accessories, such as auxiliary, position, or limit switches.

71. *Level Switch* is a switch which operates on given values, or on a given rate of change, of level.

72. *DC Circuit Breaker* is a circuit breaker that is used to close and interrupt a dc power circuit under normal conditions or to interrupt this circuit under fault or emergency conditions.

73. *Load-Resistor Contactor* is a contactor that is used to shunt or insert a step of load limiting, shifting, or indicating resistance in a power circuit, or to switch a space heater in circuit, or to switch a light or regenerative load resistor of a power rectifier or other machine in and out of circuit.

74. *Alarm Relay* is a relay other than an annunciator, as covered under device function 30, that is used to operate, or to operate in connection with, a visual or audible alarm.

75. *Position Changing Mechanism* is a mechanism that is used for moving a main device from one position to another in an equipment; as for example, shifting a removable circuit breaker unit to and from the connected, disconnected, and test positions.

76. *DC Overcurrent Relay* is a relay that functions when the current in a dc circuit exceeds a given value.

77. *Pulse Transmitter* is used to generate and transmit pulses over a telemetering or pilot-wire circuit to the remote indicating or receiving device.

78. *Phase-Angle Measuring or Out-of-Step Protective Relay* is a relay that functions at a predetermined phase angle between two voltages or between two currents or between voltage and current.

79. *AC Reclosing Relay* is a relay that controls the automatic reclosing and locking out of an ac circuit interrupter.

80. *Flow Switch* is a switch which operates on given values, or on a given rate of change, of flow.

81. *Frequency Relay* is a relay that functions on a predetermined value of frequency (either under or over or on normal system frequency) or rate of change of frequency.

82. *DC Reclosing Relay* is a relay that controls the automatic closing and reclosing of a dc circuit interrupter, generally in response to load circuit conditions.

83. *Automatic Selective Control or Transfer Relay* is a relay that operates to select automatically between certain sources or conditions in an equipment, or performs a transfer operation automatically.

84. *Operating Mechanism* is the complete electrical mechanism or servomechanism, including the operating motor, solenoids, position switches, etc, for a tap changer, induction regulator, or any similar piece of apparatus which otherwise has no device function number.

85. *Carrier or Pilot-Wire Receiver Relay* is a relay that is operated or restrained by a signal used in connection with carrier-current or dc pilot-wire fault directional relaying.

86. *Locking-Out Relay* is an electrically operated hand, or electrically, reset relay or device that functions to shut down or hold an equipment out of service, or both, upon the occurrence of abnormal conditions.

87. *Differential Protective Relay* is a protective relay that functions on a percentage or phase angle or other quantitative difference of two currents or of some other electrical quantities.

88. *Auxiliary Motor or Motor Generator* is one used for operating auxiliary equipment, such as pumps, blowers, exciters, rotating magnetic amplifiers, etc.

89. *Line Switch* is a switch used as a disconnecting, load-interrupter, or isolating switch in an ac or dc power circuit, when this device is electrically operated or has electrical accessories, such as an auxiliary switch, magnetic lock, etc.

90. *Regulating Device* is a device that functions to regulate a quantity, or quantities, such as voltage, current, power, speed, frequency, temperature, and load, at a certain value or between certain (generally close) limits for machines, tie lines or other apparatus.

91. *Voltage Directional Relay* is a relay that operates when the voltage across an open circuit breaker or contactor exceeds a given value in a given direction.

92. *Voltage and Power Directional Relay* is a relay that permits or causes the connection of two circuits when the voltage difference between them exceeds a given value in a predetermined direction and causes these two circuits to be disconnected from each other when the power flowing between them exceeds a given value in the opposite direction.

93. *Field-Changing Contactor* is a contactor that functions to increase or decrease, in one step, the value of field excitation on a machine.

94. *Tripping or Trip-Free Relay* is a relay that functions to trip a circuit breaker, contactor, or equipment, or to permit immediate tripping by other devices; or to prevent immediate reclosure of a circuit interrupter if it should open automatically even though its closing circuit is maintained closed.

95.
96.
97.
98.
99.
Used only for specific applications in individual installations where none of the assigned numbered functions from 1 to 94 are suitable.

Numbers from 95 to 99 should be assigned only for those functions in specific cases where none of the assigned standard device function numbers are applicable. Numbers which are "reserved for future application" should not be used.

NOTE: When alternate names and descriptions are included under the function, only the name and description which applies to each specific case should be used. In general, only one name for each device, such as *relay, contactor, circuit breaker, switch, monitor, or other device* is included in each function designation. However, when the function is not inherently restricted to any specific type of device and where the type of device itself is thus merely incidental, any one of the above listed alternative names, as applicable, may be substituted. For example, if for device function 6 a contactor is used for the purpose in place of a circuit breaker, the function name should be specified as Starting Contactor.

Index